Process Operations

Second Edition

Technical Editor

Daniel J. Schmidt
Bismarck State College
Bismarck, ND

Director, Employability Solutions: Kelly Trakalo
Director, Digital Studio and Content Production: Brian Hyland
Media Producer: Jose Carchi
Content Producer: Alma Dabral
Development Editor: Rachel Bedard
Instructor and Student Supplement Development: Perci LLC dba/Publisher's Resource Center
Product Marketing Manager: Melissa Natali
Manufacturing Buyer: Maura Zaldivar-Garcia, LSC Communications
Cover Designer: Carie Keller
Cover Image Credit: Oil and Gas Photographer/Shutterstock
Editorial and Full-Service Production and Composition Services: SPi Global
Editorial Project Manager: Gowri Duraiswamy
Full-Service Project Manager: SPi Global
Printer/Bindery: LSC Communications Owensville
Cover Printer: Phoenix Color

Library of Congress Cataloging-in-Publication Data
Names: North American Process Technology Alliance.
Title: Process operations / North American Process Technology Alliance ;
technical editor, Daniel J. Schmidt, Bismarck State College, Bismarck, ND.
Other titles: Process operations (North American Process Technology
Alliance)
Description: Second edition. | Hoboken, New Jersey : Pearson Education, Inc., 2021. |
Series: NAPTA series for process technology | Includes index.
Identifiers: LCCN 2021003013 | ISBN 9780136419914 (paperback) | ISBN
9780136420026 (epub)
Subjects: LCSH: Process operations. | Process control. | Manufacturing processes—Operation control. |
Operations research. | Chemical plants--Equipment and supplies.
Classification: LCC TP155.75 .P763 2021 | DDC 660/.2815--dc23
LC record available at https://lccn.loc.gov/2021003013

178 2024

www.pearsonhighered.com

ISBN 10: 0-13-641991-7
ISBN 13: 978-0-13-641991-4

Preface

The Process Industries Challenge

In the early 1990s, the process industries recognized that they would face a major staffing shortage because of the large number of "baby boomer" employees who would be retiring. Industries partnered with community colleges, technical colleges, and universities to remedy this situation. These collaborators in education and industry recognized that pretraining for process technicians would benefit industry by reducing the costs associated with training and traditional hiring methods. They recognized that teachers needed consistent curriculum content and exit competencies in order to produce process technology graduates who would be knowledgeable, competent, and able to take over the demands of the field. This was how the NAPTA series for Process Technology was born.

To achieve consistency of exit competencies among graduates from different schools and regions, the North American Process Technology Alliance identified a core technical curriculum for the Associate Degree in Process Technology. This core consists of eight technical courses and is taught in institutions throughout North America.

Instructors who teach the process technology core curriculum, and who are recognized in industry for their years of experience and depth of subject matter expertise, requested that a textbook be developed to match the standardized curriculum. Reviewers from a broad range of process industries and educational institutions participated in the production of these materials so that the presentation of content would address the widest audience possible. This textbook is intended to provide a common national standard reference for the *Process Operations* course in the Process Technology degree program.

This textbook is intended for use in high schools, community colleges, technical colleges, universities, and corporate settings and by anyone desiring an understanding of basic operations concepts and practices. Current and future process technicians will use the information within this textbook as their foundation for work in the process industries. This knowledge will make them better prepared to meet the ever-changing roles and responsibilities within their specific process industry.

What's New!

The second edition has been thoroughly updated and revised.

- **New Content Organization** to align with apprenticeship and on-the-job training.
- **New Chapter Objectives** aligned with NAPTA core objectives, with links from chapter objective to text page.
- **All New Full Color Art Program** with more than 90 fully revised drawings and photos, including all new visuals.
- **New Key Terms definitions** at the beginning of each chapter and on text pages where content appears.
- **Updated Chapter Review questions and new Appendix** Checking Your Knowledge questions have been updated to meet new chapter objectives. Answers are included in the Appendix.
- **ALL NEW Instructor Resource Package** including lesson plans, test banks, review questions, PowerPoints, a correlation guide to NAPTA curriculum, and a Transition Guide from first edition to second edition.
- **ALL NEW Pearson eText** available to facilitate digital learning!

Organization of the Textbook

Each chapter has the same organization.

- **Objectives** for each chapter are aligned with the revised NAPTA curriculum and can cover one or more sessions in a course.
- **Key Terms** list important words or phrases and their definitions, which students should know and understand before proceeding to the next chapter.
- The **Introduction** provides a simple introductory paragraph or might introduce concepts necessary to the development of the chapter's content.
- **Key Topics** and subtopics address the objectives stated at the beginning of each chapter.
- The **Summary** is a restatement of important information in the chapter.
- **Checking Your Knowledge** questions are designed to help students do self-testing on essential information in the chapter.
- **Student Activities** provide opportunities for individual students or small groups to apply some of the knowledge they have gained from the chapter. These activities generally should be performed with instructor involvement.

Acknowledgments

A particular and special thank-you goes to Technical Editor Dan Schmidt for his key role in the revision of this title. His insight and his consistent, careful overview of materials have improved every chapter of this book.

The second edition of this series would not have been possible without the support of the entire NAPTA Board and, in particular, without the leadership and dedication of Executive Director Eric Newby and interim Curriculum Committee Co-Chair Martha McKinley. Thanks to all members of this team for your dedication and hard work.

Contributors

Sheri Bankston, Bankston Consulting Services
Baton Rouge, LA (Instructor supplements)

Gregg Curry, Learning Development, The Dow Chemical Company
Freeport, TX (Text revisions)

Ron Gamble, BASF Corporation
Freeport, TX (Text revisions)

Michael Kean, Los Medanos Process Technology Department Coordinator
Pittsburg, CA (Instructor supplements)

Karen Kupsa, College of the Mainland
Texas City, TX (Text revisions)

Eric Mack, BASF Corporation
Freeport, TX (Text revisions)

Ray Player, Texas Learning Services, Eastman Chemical Company
Longview, TX (Text revisions)

Reviewers

The following organizations and their dedicated personnel supported the development of the first and second editions of this textbook. Their contributions are greatly appreciated.

Industry Content Developers and Reviewers

Steve Ames, Dayton, Texas
Lisa Arnold, Marathon Ashland, Texas
BASF Corporation, New Jersey
Ted Borel, Equistar Chemical, Texas
Gerald Canady, The Dow Chemical Company, Louisiana
Tim Carroll, BASF Corporation, Louisiana
Lester Chin, Hovensa, LLC, U.S. Virgin Islands
Sheldon Cooley, Westlake Group, Louisiana
Regina Cooper, Marathon Petroleum, Texas
Ed Couvillion, Sterling Chemicals, Texas
Gregg Curry, The Dow Chemical Company, Texas
Lloyd Davis, Halliburton, Oklahoma
Greg Dearwater, Corys, Houston, Texas
Steve Erickson, GCPTA, Texas
Steve Ernest, ConocoPhillips, Oklahoma
Billy Fridelle, ExxonMobil, Texas
Thomas Germusa, Shell Polymers, Pennsylvania
Don Glaser, PetroSkills, New Jersey
Richard Honea, The Dow Chemical Company, Texas
Lee Hughes, BASF Corporation, Texas
INEOS, Texas
Glenn Johnson, BASF Corporation, Texas
Tim Judge, Simtronics, New Jersey
Nathan Leonard, Corys, Houston, Texas
Jennifer Martin, OG&E Horseshoe/Mustang Plant, Oklahoma
Mike McBride, BASF, Texas
Martha McKinley, Eastman Chemical (retired), Texas
Jason Oxley, OG&E Horseshoe/Mustang Plant, Oklahoma
Aracelli Palomo, The Dow Chemical Company, Texas
Allen Parker, Crompton Corporation, Louisiana
Shawn Parker, Marathon Ashland, Texas
Clemon Prevost, BP, Texas
Robert Rabalaise, The Dow Chemical Company, Louisiana
Raymond Robertson, PPG, Louisiana
Matt Scully, BP, Alaska
Roy St. Romain, Eastman Chemical, Texas
Shell International, Texas
Simtronics Corporation, New Jersey
Paul Summers, The Dow Chemical Company, Texas
Dennis Thibodeaux, Citgo, Louisiana
Robert Toups, Lyondell Chemical, Louisiana
John Zink Company LLC, Oklahoma

Education Content Developers and Reviewers

Sazar Ali, Houston Community College, Texas
Louis Babin, ITI Technical College, Louisiana
Allen Baragar, Brazosport College, Texas
Carrie Braud, Baton Rouge Community College, Louisiana
Tommie Ann Broome, Mississippi Gulf Coast Community College, Mississippi
Nathaniel "Nat" Byrom CT, HRDC, Training Manager (Ret.), Flint Hills Resources, Port Arthur, Texas

Larry Callaway, Bellingham Technical College, Washington

Chuck Carter, Lee College, Texas

Lou Caserta, Alvin Community College, Texas

Ken Clark, Perry Technical Institute, Washington

Michael Connella III, McNeese State University, Louisiana

Richard Cox, Baton Rouge Community College, Louisiana

Mark Demark, Alvin Community College, Texas

Ned Duffey, South Arkansas Community College, Arkansas

Jerry Duncan, College of the Mainland, Texas

Tommy Edgar, Texas State Technical College, Texas

Steve Ernst, Northern Oklahoma College, Oklahoma

Charles Gaffen, Elizabeth High School, New Jersey

Gary Hicks, Brazosport Community College, Texas

Kevin Holmstrom, Bismarck State College, North Dakota

Raymond Johns, Northern Oklahoma College, Oklahoma

Larry Johnson, Nova Scotia Community College, Nova Scotia

Mike Kukuk, College of the Mainland, Texas

Tony Kuphaldt, Bellingham Technical College, Washington

Linton LeCompte, Sowela Technical Community College, Louisiana

Michael Murray, Copiah-Lincoln Community College, Mississippi

Vicki F. Newby, Lamar Institute of Technology, Texas

Richard Ortloff, Bellingham Technical College, Washington

Anthony Pringle, Remington College, Texas

Denise Rector, Del Mar College, Texas

Robert Robertus, Montana State University - Billings, Montana

Paul Rodriguez, Lamar Institute of Technology, Texas

Dean Schwarz, Southwestern Illinois College, Illinois

Mike Speegle, San Jacinto College, Texas

Wayne Stephens, Wharton County Junior College, Texas

Mark Stoltenberg, Brazosport Community College, Texas

Robert Walls, Del Mar College, Texas

Scott Wells, Brazosport Community College, Texas

This material is based upon work supported, in part, by the National Science Foundation under Grant No. DUE 0532652. Any opinions, findings, and conclusions or recommendations expressed in this material are those of the authors and do not necessarily reflect the views of the National Science Foundation.

Contents

Chapter 1
Introduction to Operations

Objectives

After completing this chapter, you will be able to:

1.1 Identify key concepts from the process equipment course, process systems course, and process instrumentation courses. (NAPTA Introduction to Operations 1–3*) p. 2

1.2 Discuss the term *operations* and describe a general operations overview. (NAPTA Introduction to Operations 4) p. 6

1.3 List the various process technician roles and responsibilities within an operating unit:
- Operate and monitor the unit from the control room
- Operate and monitor the unit from the outside
- Take and analyze samples
- Perform housekeeping activities
- Conduct safety inspections
- Handle materials
- Prepare for, assist with, and/or perform maintenance as required
- Commit to a process of lifelong learning to keep up with changes in operations in process industry. (NAPTA Introduction to Operations 5–7) p. 7

Key Terms

Distributed control system (DCS)—a subsystem of a supervisory control system used to control a process unit: consists of field instruments and field controllers connected by wiring that carries a signal from the controller transmitter to a central control monitoring screen, **p. 7**

*North American Process Technology Alliance (NAPTA) developed curriculum to ensure that Process Technology courses will produce knowledgeable graduates to become entry-level employees in process technology. Objectives from that curriculum are named here in abbreviated form. For example, "(NAPTA Introduction to Operations 1–3)" means that this chapter's objective 1 relates to objectives 1, 2, and 3 of NAPTA's introductory curriculum about operations.

Hazard and operability (HAZOP)—formal and structured review and study method used to determine potential hazards associated with process systems, equipment, process materials, and work processes, **p. 8**

Instrumentation—any device used to measure or control flow, temperature, level pressure, analytical data, and so on, **p. 5**

Lockout/tagout (LOTO)—a safety term used to describe the isolation of equipment for maintenance; this is a federally mandated safety precaution; procedure of tagging valves, breakers, etc. in preparation of equipment for maintenance, **p. 3**

Pre-startup safety review (PSSR)—an element of the process safety management program; the PSSR helps to ensure that the new or modified process or facility is safe and operable before startup, **p. 8**

Procedures—series of actions that must be done in the specified manner and sequence to obtain the desired result under the same circumstances each time the work is performed, **p. 3**

Process hazard analysis (PHA)—systematic assessment of the potential hazards associated with an industrial process, taking into consideration specific hazards and locations of highest potential for exposure, **p. 8**

Process technician—worker in a process facility who monitors and controls mechanical, physical, and/or chemical changes throughout a process in order to create a product from raw materials; term reflects increased competence and skills required; also called *process operator* or *operator*, **p. 2**

System—set of interacting or interdependent equipment and process elements that work together to deliver a specific process function, **p. 4**

1.1 Introduction

Within the refining and petrochemical process industries, the term *operations* refers to the personnel group that makes up the facility operating team and includes process technicians, process engineers, and management. Other personnel such as maintenance, safety, human resources, and information technology are often supporting groups to the operations team.

In this chapter, we will focus on the process technician's role in operations. The **process technician** performs the tasks required to operate a process facility safely and to maintain product yield and unit parameters while protecting the environment and community where the facility is located. The process technician may also be referred to as an *operator*, *process operator*, or a *plant operator*.

Process technician worker in a process facility who monitors and controls mechanical, physical, and/or chemical changes throughout a process in order to create a product from raw materials; term reflects increased competence and skills required; also called *process operator* or *operator*.

Process technicians receive site and unit specific training for the area, or areas, of the facility to which they are assigned to work. Performance reviews, written exams, and other testing methods are used to evaluate the technician's understanding of process operations within his or her area of responsibility and to determine the qualifications to operate in specific areas within the facility. The qualified process technician is responsible for monitoring the process operation, making necessary process adjustments, and maintaining desired unit conditions throughout the designated shift period.

The role of the process technician is extremely important in the safe and efficient operation of process facilities. The technician needs to be familiar with all aspects of the assigned area and is responsible for the safety of any personnel in the area and for any of the work being done in that section of the facility, even if the work is actually being performed by someone else.

Equipment Review

A process unit is made up of various pieces of equipment, piping, instrumentation, and vessels. In order to safely operate and maintain the unit, process technicians are responsible for having a clear understanding of the equipment associated with their assigned process area—its components, operating limits, and how they function together to produce the desired end product(s).

Common equipment used in process industries includes: distributed control systems (DCSs), valves, pumps, compressors, turbines, motors, heat exchangers, cooling towers, furnaces, boilers, reactors, reformers, tanks, separators, distillation towers, absorbers, strippers, extraction vessels, adsorbers, rotary kilns, calciners, control systems, and filters.

The clear understanding of the equipment in process technicians' assigned area of responsibility includes its function, potential problems, safety, health, and environmental (SHE) concerns, potential quality issues, and related operating and emergency **procedures**. Their understanding must also include standard operating procedures (SOPs) and company requirements in the assigned process unit.

Procedures series of actions that must be done in the specified manner and sequence to obtain the desired result under the same circumstances each time the work is performed.

Knowledge of how each piece of equipment, piping systems, and associated instrumentation is integrated into the process is crucial to process technicians' overall understanding of their area of responsibility. To further this understanding, they must be proficient in reading process flow diagrams (PFDs) and piping and instrumentation diagrams (P&IDs), shown in Figure 1.1A. They must also be able to identify the symbols used in these types of drawings, which stand for various types of equipment in actual application (such as, Figure 1.1B).

Figure 1.1 **A.** Symbols for flow orifice and control valve. **B.** Photo of a flow orifice protruding from the flange.

CREDIT: B. Oil and Gas Photographer/Shutterstock.

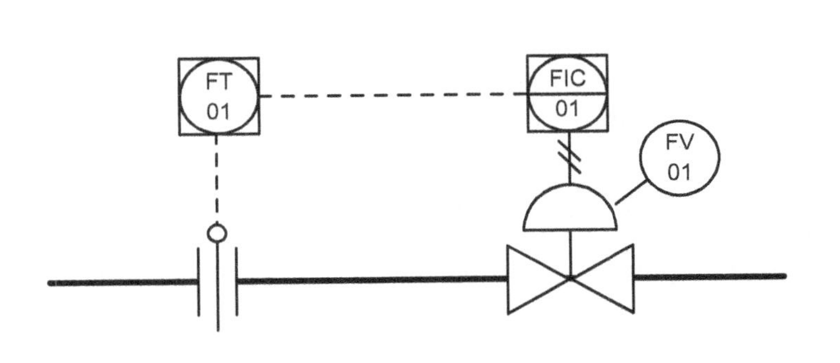

A.

B.

Process technicians receive training in equipment-specific procedures relative to startup, normal operation, and shutdown, as well as emergency operating procedures and responses to abnormal conditions. Part of the training process includes procedure rehearsals, or "walk-throughs," to develop understanding and proficiency in all phases of operation. Knowledge of the process, the equipment, and how each part affects upstream and downstream equipment and processes, along with understanding of procedures, is crucial to safe and efficient operation.

Additionally, process technicians should be trained to follow routine operating procedures by following the steps exactly as written, unless otherwise instructed by supervisors. However, in an emergency situation, the initial steps may have to be performed without a written procedure in hand so that the unit/equipment can be safely secured until a procedure can be obtained. Procedure walkthroughs and scenario based training exercises prepare process technicians to respond to emergency situations to prevent personnel injury or equipment damage. Much of the training and emergency planning is mandated by government agencies to prevent catastrophes and workplace fatalities.

Process technicians are also responsible for hazardous energy **lockout/tagout (LOTO)**. The technician must understand lockout/tagout procedures for the various pieces of equipment and piping in the assigned area to ensure personnel safety when maintenance is to be performed.

Lockout/tagout (LOTO) a safety term used to describe the isolation of equipment for maintenance; this is a federally mandated safety precaution; procedure of tagging valves, breakers, etc. in preparation of equipment for maintenance.

Systems Review

System set of interacting or interdependent equipment and process elements that work together to deliver a specific process function.

Understanding the equipment is important not only to operate and maintain the equipment but also to understand the larger **system** the equipment comprises. Depending on how a facility is laid out, a process technician may be assigned to an area that has multiple systems.

Envision a plant where Unit A feeds its byproduct stream to Unit B, which refines the Unit A byproduct into one or more products for marketing. In this example, Unit A and B are both interacting and interdependent systems that are part of the whole process facility. There may also be several systems within a single unit, such as a cooling or refrigeration system. The coolant is piped from a pump or compressor through various pieces of equipment to exchange heat and cool the process in that system. Both systems are interdependent and interact with each other to achieve the desired process temperature.

To review, the process technician learns about key systems within a process facility. Systems include the following:

- *Distillation system*—process that separates feed stream components by repeated vaporization and condensation with separate recovery of vapor and liquids. Distillation systems work well where the boiling points for the separated components are not too close.
- *Reactor system*—process that chemically alters materials by the application of heat and pressure, usually in the presence of a specific catalyst that initiates, speeds, or intensifies the chemical reaction.
- *Steam generation system*—process that converts high-purity water to high-pressure, high-temperature steam for heating process streams, used as a motive agent for electrical power generation systems and/or a motive agent for mechanical drives.
- *Refrigeration system*—system designed for the removal of heat. The system typically consists of a compressor that circulates a refrigerant through a condenser, an expansion valve or orifice, and an evaporator. The refrigerant may provide process cooling or cool a secondary system, such as water. A pump circulates the chilled water for process cooling.
- *Water systems*—system that includes fire water, process water, potable (drinkable) water, cooling water, demineralized water, and boiler feedwater systems, among others. All these water systems serve unique purposes within the process and are equally important to unit and process operation.
- *Utility systems*—systems that may include nitrogen, steam, plant air, instrument air, natural gas, compressed gas, and so on. The various utility systems within the facility are critical to the operating unit and facility. The utility systems also include the wastewater disposal, process sewer, and flare systems that safely dispose of liquid and gaseous wastes in an environmentally sound manner. The wastewater and process sewer systems transfer waste liquids to a treatment facility (either local or offsite) where water and hydrocarbons (or chemicals) are separated. The water is cleaned and distilled for reuse and the recovered process materials are either stored and processed or disposed of.
- *Relief valve system*—safety devices designed to open if the pressure in a closed space, such as a vessel or a pipe, exceeds a preset level. It is a system designed to protect personnel, equipment, and the environment by venting excess equipment pressure through relief valves. Personnel and the environment are protected from hazardous releases and equipment is protected from exceeding design pressure limits.
- *Flare system*—device to burn unwanted process gasses before they are released into the atmosphere. Relief valves vent to the flare system, which is designed to protect site personnel and the environment from exposure to harmful chemicals or hydrocarbons.

There are many systems within a process facility that are interdependent. Unit specific training gives the process technician a better understanding of the interdependency of various systems.

Instrumentation Review

Instrumentation is a system of pneumatics, electronic instruments, digital logic devices, and computer based process controls that make up the measurement and control system for equipment for the purpose of safe, efficient, and cost effective unit operation. Instruments include devices that measure, transmit, and indicate variables such as flow, temperature, level, or pressure. Figure 1.2 shows examples of temperature indicators (TIs).

Instrumentation any device used to measure or control flow, temperature, level pressure, analytical data, and so on.

Figure 1.2 Simple temperature indicator (TI). **A.** Fahrenheit (°F). **B.** Celsius (°C).

CREDIT: A. Eaum M/Shutterstock. **B.** engineer story/Shutterstock.

A. Fahrenheit

B. Celsius

Instrumentation also includes complex devices and configurations such as interconnected multiple controllers, analyzers, logic devices, and computers that automatically operate valves to establish and maintain desired conditions. Process control is one of the main branches of applied instrumentation.

Process technicians need to know about different types of instrumentation for the measurement of pressure, flow, level, and temperature. The control loop terminology, nomenclature, and symbolism as they relate to the process technician are also important to know.

Key instrumentation review information includes the following elements:

- *Control loop*—group of instruments working together to control a single process variable such as temperature, flow, pressure, or level. Typical components in a control loop include a sensor/indicator; a controller; an I/P transducer, which usually converts the signal from the controller to a pneumatic signal; and a final control element, such as a control valve, an electrical switch, or a motor.
- *Motor control center (MCC)*—enclosure that houses the feeder breakers, motor control units, variable frequency drives, programmable controllers, and metering devices needed to supply power safely to unit equipment. Typically, the MCC provides a safe, pressurized enclosure with one or more sections having a common power bus.
- *Programmable logic controllers (PLC)*—computer-based controller that uses multiple inputs to monitor processes and automated outputs to control processes at desired parameters. These controllers are relatively low in cost and typically control specific pieces of equipment or systems within a process unit. Also, PLCs may operate independently of a DCS, and most are local to the equipment being controlled.
- *Transmitter*—instrumentation device that transmits a specified measurement signal from the measuring element to the control device, indicator, or recorder, such as from a temperature sensing element to a DCS indicator.

- *Uninterruptible power supply (UPS)*—auxiliary power supply consisting of batteries that automatically provide temporary power, typically for control systems and lighting, when the normal power supply is interrupted. In some cases, a generator may augment the UPS.

The process technician is required to have a working knowledge of the specific instrumentation within his or her unit or facility.

1.2 Operations Organizational Structure

The process operations organizational structure (shown in Figure 1.3) is comparable across most production facilities, although titles may vary. Most facilities include the following:

- *Facility management team*—generally consists of the plant manager, operations manager, safety, health, and environment (SHE) manager, human resources manager, information technology manager, engineering manager, maintenance manager, security manager (if the facility is large enough), and perhaps a project manager, if applicable.

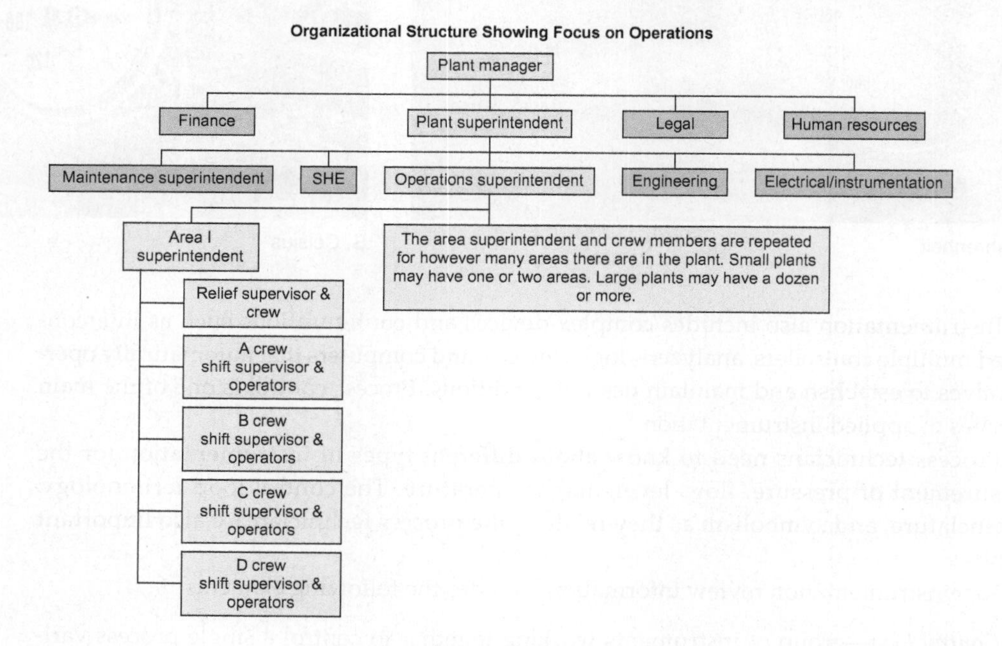

Figure 1.3 Sample operations organizational structure.

- *Operations superintendent*—reports directly to the plant superintendent. In larger process facilities, including refineries and petrochemical facilities, there may be multiple superintendents assigned by location within the facility (i.e., north, west, etc.) or by unit. In some facilities, this layer of supervision may be called the operations supervisor.
- *Area superintendent*—reports directly to the operations superintendent and is responsible for coordinating a specific area's operation overall. Coordinates activities from shift to shift and plans plant operation with superintendents from other areas in the plant.
- *Shift supervisor*—also called *process supervisor* or *team leader*; directly responsible for unit operation during the shift. The process technicians report directly to the shift supervisor, who, in turn, reports directly to the area superintendent.
- *Process technician*—operator directly responsible for running and maintaining the process unit. A process technician reports directly to the shift supervisor.

These positions make up the "operations" group. The remainder of the personnel inside an operating facility falls under different categories such as maintenance, information technology support, safety, human resources, security, engineering (mechanical, electrical, and civil), and administration.

1.3 Process Technician's Roles and Responsibilities

Process technicians (also commonly called *operators*) have different roles and responsibilities for each section of the operating unit. For example, a process technician assigned to operate the control board has different responsibilities than a process technician assigned to monitor field activities. In many facilities, process technicians cross-train to operate all areas of a unit.

Distributed control system (DCS) a subsystem of a supervisory control system used to control a process unit: consists of field instruments and field controllers connected by wiring that carries a signal from the controller transmitter to a central control monitoring screen.

Control Board Operator Responsibilities

The **distributed control system (DCS)** is the interface that allows the control board operator to monitor and control the process via a computer or a PC where process diagrams and variables are displayed and controlled (Figure 1.4).

Figure 1.4 Control board technician optimizing facility operations.

CREDIT: Sebastian Kahnert/dpa-Zentralbild/dpa/Alamy Live News/Dpa picture alliance/Alamy Stock Photo.

Automation systems such as the DCS allow greater control and optimization of one process or many processes simultaneously. They also offer ease of communication between the field and control room and easy transmission of large amounts of data to and from a central location. Generally, the duties of a control board operator include the following:

- Optimize facility operation to maximize production, minimize cost, minimize impact to the environment, maintain product specifications, and ensure personnel safety.
- Perform necessary corrective actions to keep operating parameters within control guidelines.
- Record performance data (readings) as required by the operating facility.
- Interpret laboratory analysis and adjust process parameters to maintain product quality specifications.
- Participate in a thorough exchange of information from one shift or work team to another (called shift change or turnover). The exchange should provide the oncoming shift with information regarding the following:
  - Safety and environmental issues that exist or were corrected
  - Process and equipment problems, including corrective actions taken
  - Material transfers in progress

- Special operating instructions
- Items being coordinated with other process areas
- Ongoing or upcoming unit maintenance or contract work
- Technical support personnel working on the unit

- Monitor alarm reports and take corrective action as required.
- Coordinate process activities with the field technician as required.
- Coordinate maintenance, contractor, and technical department activities with the field technician as needed.
- Record all laboratory analysis data as required.
- Record shift activities in the unit logbook (paper or electronic).
- Participate in **process hazard analysis (PHA)**. There is more than one type of PHA.
- Participate in **hazard and operability (HAZOP)** studies. HAZOP is only one type of process hazards analysis.
- Participate in **pre-startup safety review (PSSR)**. The pre-startup team is typically made up of representatives from various departments or crafts, and each member must sign off on the PSSR before startup can be performed.
- Detect and troubleshoot process operation problems.
- Maintain the qualifications and training requirements required by regulatory agencies and assigned by each facility.
- Perform other duties as directed by the facility management.

Process hazard analysis (PHA) systematic assessment of the potential hazards associated with an industrial process, taking into consideration specific hazards and locations of highest potential for exposure.

Hazard and operability (HAZOP) formal and structured review and study method used to determine potential hazards associated with process systems, equipment, process materials, and work processes.

Pre-startup safety review (PSSR) an element of the process safety management program; the PSSR helps to ensure that the new or modified process or facility is safe and operable before startup.

Outside Operator Responsibilities

The outside operator, also called the *field technician*, has a wide range of daily duties. Each operating facility develops guides and checklists that fit the facility operating requirements. The field technician's routine duties vary by company and the type of process, but may include the following:

- Participate in a thorough exchange of information from one shift to another (called *shift change* or *turnover*). The exchange should provide the oncoming shift with information regarding the following:
  - Safety and environmental issues that exist or were corrected
  - Process and equipment problems, including corrective actions taken
  - Material transfers in progress
  - Special operating instructions
  - Items being coordinated with other process areas
  - Maintenance or contractor work occurring on the unit
  - Technical support personnel working on the unit
- Make a thorough inspection of the technician's area of responsibility and equipment at the beginning of the shift and at regular intervals throughout the shift (referred to as *rounds*).
- Oversee and assist maintenance personnel, contractors, and technical personnel working in the field.
- Perform safety verification checks as required by the facility management.
- Perform equipment inspections/surveys as directed by the facility management.
- Check the technician's area of responsibility for leaks.
- Check rotating equipment for proper lubrication and operation.
- Check the cooling tower and other auxiliary systems.

- Prepare equipment for maintenance using accepted practices and guidelines.
- Collect routine samples and special samples as needed.
- Receive and store supplies and materials for the unit (lubricating oils, specialty chemicals, and other supplies as required).
- Alert the control board technician of process or equipment abnormalities and suggest corrective actions.
- Perform equipment preventive maintenance as directed by site policies.
- Perform housekeeping as required.
- Record normal duties performed in the unit logbook (paper or electronic).
- Participate in hazard and operability (HAZOP) studies.
- Participate in process hazard analysis (PHA).
- Participate in pre-startup safety review (PSSR).
- Wear appropriate personal protective equipment (PPE).
- Maintain qualifications and training requirements required by regulatory agencies and assigned by each facility.
- Prepare equipment for maintenance.

Process Technicians of the Future

The business environment of process related industries, including refining and petrochemical processing, is constantly changing. In order to compete in world markets, new technologies and imaginative applications for them must be implemented to maintain adequate profit margins. With these improvements, the role and responsibilities of the future process technician will continue to evolve.

Because technology has expanded so rapidly, the process technician will be required to work more intimately with unit process controls. Industrial automation systems will require tighter integration between devices in the plant and the field. Remote control of process units will continue to evolve over the next several years.

Whatever the future holds for the process industries, the process technician will remain an important position. Skills gained from a process technology (PTEC) degree, especially academic skills, will aid process technicians to improve their operating techniques quickly and to keep up with the industry's technological innovations.

Summary

Process technicians play an important role in maintaining safe, reliable, environmentally responsible, and profitable operations. They are integral members of the operations team and are primarily responsible for optimizing the process. These individuals are also the first line of defense in preventing unsafe conditions, environmental releases, and equipment malfunctions.

The major requirement of process technicians is that they must understand process systems, including equipment and instrumentation, in order to operate and monitor the process safely. Without this knowledge, process technicians would be unable to perform their basic duties.

The variety of roles and responsibilities process technicians have in the process industry include operating and monitoring the unit from the control room and from the outside, conducting safety inspections, preparing equipment for maintenance, and performing various housekeeping duties. Other duties may be assigned as needed.

Advanced technology will allow technicians, as individuals and in teams, to provide ever greater business and technical competencies to help the industry run more efficiently and effectively in the future.

Checking Your Knowledge

1. Define the following terms:
   a. Distributed control system (DCS)
   b. Hazard and operability (HAZOP)
   c. Instrumentation
   d. Lockout/tagout (LOTO)
   e. Pre-startup safety review (PSSR)
   f. Procedure
   g. Process hazard analysis (PHA)
   h. Process technician
   i. System

2. Process technicians must have a clear understanding of the equipment in their area of responsibility. Name three things they need to understand about this equipment.

3. Match each of these key systems with the correct description.

Term	Description
I. Distillation system	a. System designed for the removal of heat.
II. Flare system	b. Process that separates feed stream. components by repeated vaporization and condensation with separate recovery of vapor and liquids.
III. Reactor system	c. Device to burn unwanted process gasses before they are released into the atmosphere.
IV. Refrigeration system	d. System designed to open if the pressure in a closed space, such as a vessel or a pipe, exceeds a preset level.
V. Relief valve system	e. Process that converts high-purity water to high-pressure, high-temperature steam for heating process streams.
VI. Steam generation system	f. Process that chemically alters materials by the application of heat and pressure
VII. Utility systems	g. System that includes fire water, process water, potable (drinkable) water, cooling water, demineralized water, and boiler feedwater systems, among others.
VIII. Water systems	h. Systems critical to the operating unit and facility such as wastewater disposal, process sewers, and systems that safely dispose of liquid and gaseous wastes in an environmentally sound manner.

4. Which individuals, also called team leaders, are directly responsible for unit operation during their shift?
   a. Operations superintendents
   b. Area superintendents
   c. Shift supervisors
   d. Process technicians

5. Which of the following report directly to the plant superintendent? (Select all that apply.)
   a. SHE department head
   b. Electrical/instrumentation department head
   c. Relief supervisor
   d. Process technicians
   e. Maintenance superintendent

6. Which of the following is the duty of the *outside operator (field technician)*?
   a. Adjust control set points in the DCS.
   b. Monitor alarm reports and take corrective action as required.
   c. Perform safety verification checks as required by the facility management.
   d. Organize and lead HAZOP workshops.

7. Which of the following is a duty of the process technician who is a control room operator?
   a. Check rotating equipment for proper lubrication and operation.
   b. Interpret laboratory analysis and adjust process parameters to maintain product specifications.
   c. Check the cooling tower and other auxiliary systems.
   d. Collect routine samples and special samples as needed.

8. Which of the following are duties common to both control room operators and outside operators (field technicians)? (Select all that apply.)
   a. Participating in pre-startup safety review (PSSR).
   b. Performing necessary corrective actions to keep operating parameters within control guidelines.
   c. Performing safety verification checks as required by the facility management.
   d. Preparing equipment for maintenance.
   e. Participating in HAZOP studies.

9. Pre-startup safety reviews are needed to determine whether the unit __________.
   a. is ready to shut down
   b. has been started up successfully
   c. is ready or not ready for a safe startup
   d. has safely been shut down

10. List four topics that process technicians cover in shift change communication.

11. Having a thorough exchange of information from one shift to another is a key responsibility of the __________.
   a. laboratory staff
   b. chemical engineer
   c. process technician
   d. design team

12. The control room process technician may utilize a __________ to control the process.
   a. manual valve
   b. distributed control system (DCS)
   c. automatic valve
   d. instrument to pneumatic converter

13. Systems are defined as a set of interacting or interdependent equipment and process elements that work together to deliver a __________.
   a. specific process function
   b. workable solution
   c. final solution
   d. specific process parameter

14. The process technician will record all relevant operations activities in the operations__________.
   a. directory
   b. logbook
   c. file cabinet
   d. manual

NOTE: Answers to Checking Your Knowledge questions are in the Appendix.

Student Activities

1. Perform research on the roles and responsibilities of the process technician. Using the researched materials and the information from this chapter, write a one to two page paper detailing what you believe to be the most critical responsibilities of a process technician.
2. Together with a classmate, write a two page report on the possible future for the process technician.
3. Research the role of process technicians from several different companies with different types of process facilities. Identify the tasks they have most in common and make note of differences in their tasks and responsibilities. Share with the class what you think makes these roles similar and different.
4. Research key activities taken during a PSSR, and write a one to two page paper on them.

Chapter 2
On-the-Job Training

Objectives

After completing this chapter, you will be able to:

2.1 State the purpose and levels of skill development involved in on-the-job training (OJT). (NAPTA Introduction to Operations 7*) p. 13

2.2 Explain the proper method for demonstrating a task, observing a trainee, and providing feedback to a trainee. (NAPTA Introduction to Operations 5–7) p. 14

2.3 Describe the proper materials for organizing and preparing to conduct OJT. (NAPTA Introduction to Operations 5, 6) p. 16

Key Terms

Computer-based training (CBT)—delivers training material through a facility computing system, **p. 17**

Mentor—influential senior sponsor or trainer, usually in the form of a training coordinator or a chief or lead operator, who delivers the training material, tracks material completion, provides feedback, and conducts written and performance evaluations to verify knowledge, **p. 14**

On-the-job training (OJT) programs—objective-oriented training and qualification programs for process technicians to master process equipment, control systems, safety, and hazard management, **p. 13**

Process simulator—stand-alone, computer-generated simulation of a process unit or process system that emulates process equipment, piping systems, control mechanisms, and behaviors that control the process, **p. 16**

Safety data sheet (SDS)—a document that contains information related to the safety, hazards, and handling of a specific material, **p. 18**

Standards—guidelines established by authority as a rule to measure quantity, weight, extent, value, or quality; OSHA standards are rules that describe the methods employers must use to protect their employees from hazards, **p. 13**

*North American Process Technology Alliance (NAPTA) developed curriculum to ensure that Process Technology courses will produce knowledgeable graduates to become entry-level employees in process technology. Objectives from that curriculum are named here in abbreviated form. For example, "(NAPTA Introduction to Operations 7)" means that this chapter's objective 1 relates to objective 7 of NAPTA's introductory curriculum about operations.

2.1 Introduction

New process technicians will typically spend the first one to three months of their job in training. This chapter provides an overview of on-the-job training (OJT) programs that process technicians will complete upon entering the work force. Process facilities are required by federal law to ensure that employees are trained in their operations and have developed extensive training programs that reflect their responsibility toward people, the environment, and the communities in which they operate. An associate's degree in process technology or equivalent experience in the field of process technology may be required of an individual seeking a position in the process industry. Once hired, the trainee will complete intensive training to learn the site's specific processes, procedures, and equipment.

Personnel in the process industry must be fully capable of performing their assigned tasks, without error, based on the hazards found throughout the industry and in their particular facility. Tasks and hazards that are not managed responsibly can quickly result in severe harm to personnel, the environment, process facilities, and surrounding communities. Industry training programs include content and instruction required to comply with federal regulations and to meet site or production facility expectations in the presence of these hazards.

The Occupational Safety and Health Administration (OSHA) is a U.S. government agency created to establish and enforce workplace safety and health standards, conduct workplace inspections and propose penalties for noncompliance, and investigate serious workplace incidents. OSHA has developed **standards** that require employers to train employees in the safety and health aspects of their jobs.

Standards guidelines established by authority as a rule to measure quantity, weight, extent, value, or quality; OSHA standards are rules that describe the methods employers must use to protect their employees from hazards.

Additional OSHA standards make it the employer's responsibility to limit certain job assignments to employees who are certified, competent, or qualified. These terms mean that employees have had special training in or outside of the workplace. These standards reflect the OSHA belief that training is an essential part of every employer's safety and health program for protecting workers from injury and illness.

An example of OSHA safety and health training requirements is the Process Safety Management of Highly Hazardous Chemicals standard (Title 29 Code of Federal Regulations Part 1910.119). This standard was issued under the requirements of the Clean Air Act Amendments of 1990. It contains the requirements for management of hazards associated with processes using highly hazardous materials. It requires the employer to evaluate and verify that employees comprehend the training given to them. This means that training must have established goals and objectives. Subsequent to training, an evaluation must verify that the employees understand the material presented and have acquired the desired skills and knowledge.

Purpose and Importance of On-the-Job Training

Throughout the process industry, **on-the-job training (OJT) programs** are the primary training method utilized to ensure that operating personnel become demonstrably qualified. Trainees (Figure 2.1) learn how process equipment and systems are designed, integrated, and controlled for the manufacture of one or more products or byproducts.

On-the-job training (OJT) programs objective-oriented training and qualification programs for process technicians to master process equipment, control systems, safety, and hazard management.

Training programs are designed to teach the trainee how to operate safely while managing the hazards present in the unit or facility. Every training program has several goals, the first being to ensure that personnel are certified or qualified to safely and consistently complete the job tasks for which they are responsible.

Trainees in process operations must complete a training program consisting of predefined learning objectives within a curriculum that prepares the trainee for specific job assignments. Qualification for a specific job assignment requires the trainee to demonstrate knowledge and skills of training objectives.

Figure 2.1 An on-the-job training program helps process technicians learn about specific processes, procedures, and equipment.

CREDIT: Shutterstock.

Sound knowledge of the unit equipment, piping, and controls is an absolute necessity. The trainee must understand the inner workings and technological details of the process and equipment, as well as the scientific and physical attributes. In other words, process technicians are as much students of math, science, and physics as they are students of the mechanical requirements that are necessary throughout the process industry.

Communication is a vital factor in the day-to-day operation of a process unit. Operators, process and technical engineers, and managers all have the responsibility of developing communication skills to share information and knowledge. At the same time, they need to develop a philosophy of continuous self-improvement by constantly paying attention and staying engaged in daily operation activities. Staying engaged, communicating and sharing information, as well as paying attention to detail will enable process personnel to safely manage the hazards associated with the process industry.

Once initial process certification or qualification is achieved, refresher training or a system of continuous education begins. Continuous education training requirements for process technicians and those employees who are directly responsible for process operations are defined at the site or facility level and by the federal government.

2.2 Training Methods, Skill Development, and Observing the Trainee

Process technology (PTEC) trainees should enter a facility with an appreciation for the training program. Trainees with an associate degree in process technology have already learned much about the process industry. They know the purpose of many types of equipment. They have a fundamental understanding of the chemistry, physics, and science involved. They also are aware of many of the hazards present throughout the industry. These are the building blocks. Once hired, trainees will learn the site's specific processes, procedures, and equipment.

Mentor influential senior sponsor or trainer, usually in the form of a training coordinator or a chief or lead operator, who delivers the training material, tracks material completion, provides feedback, and conducts written and performance evaluations to verify knowledge.

A **mentor** delivers training material, tracks material completion, provides feedback, and conducts written performance evaluations to verify knowledge. In some facilities, depending on staff availability, a shift operator may be assigned as a mentor to work closely with the trainee throughout the training process. A mentor is a qualified person and a subject matter expert (SME) the trainee can depend on to answer questions, demonstrate tasks, and provide guidance and instruction for daily unit activities.

Mentorship is one of the most important pieces of a successful training program. Working closely with trained personnel enables the trainee to become directly involved in the operation and with the staff. Through task demonstration, a mentor can engage the trainee in the daily activities required to operate the unit safely. Under the supervision of an experienced mentor, a trainee can begin to execute simple tasks. Participation in more difficult tasks takes place as the trainee progresses through the training program and demonstrates enhanced abilities.

Gaining knowledge of the process through this type of supervised hands-on training supplements the trainee's ability to understand and learn the new material. The sights, sounds, and smells that are experienced through direct contact with the equipment in the operating unit provide invaluable learning experiences for trainees. Allowing a trainee to become actively engaged in the operation, with the guidance of a mentor, is the true meaning of on-the-job training. Figure 2.2 shows a mentoring situation.

Figure 2.2 Supervised on-the-job training.

CREDIT: Christian Lagerek/Shutterstock.

While coordinating the trainee's progression through the training modules, written evaluations, and job task (performance) evaluations, the training coordinator or mentor will also be carefully observing and evaluating the trainee in the area of people skills, sometimes referred to as *soft skills.*

Trainees' ability to establish working relationships and communicate effectively will be evaluated, along with their ability to learn and retain knowledge. Difficulties in any of these areas can be addressed and corrected early in the training program for the benefit of trainees as well as for the current staff.

While trainees progressively learn about the inner workings of the process from the training modules, they are also expected to become familiar with equipment, piping, control valve location, and applicable job tasks. Progress through every training program is based on successful completion of each assigned module, passing written evaluations, and demonstrating assigned tasks (performance evaluations). Completion of the module and knowledge of the content are tested in several ways:

- Written evaluations on a given subject are administered either manually or by computer, and trainees are required to achieve at least a minimum score.
- Performance evaluations require trainees to perform assigned tasks to demonstrate proof of knowledge and competence. These evaluation tasks are administered by the training coordinator, chief, lead operator, or mentor.

Trainee observation through a mentor program provides the opportunity to evaluate a trainee's:

- Work ethic, the ability to establish working relationships, and the skill to communicate effectively.
- Ability to learn, retain, and demonstrate knowledge and skills.

Providing feedback to trainees is one of the roles for the training coordinator or mentor. Some trainees may find it difficult to meet required training expectations or to accept feedback in these situations. Corrective feedback is given with the goal of enabling a trainee to understand a problem better, create a solution to the problem, and improve performance. Some elements of corrective feedback follow:

- Be detailed and honest.
- Focus on step-by-step actions and not simply a summary.
- Zero in on the issue and do not place blame.
- Communicate and be sure to listen.

Providing positive reinforcement and feedback is also critical to the success of every trainee. A few examples of positive reinforcement are given below:

- Recognizing a trainee for having completed modules and written evaluations on schedule
- Congratulating a trainee for the ability to learn quickly and effectively
- Recognizing a trainee for attention to detail and safety
- Recognizing a trainee for having completed job tasks proficiently
- Assigning additional job tasks that are progressively more complicated
- Recognizing when a trainee has qualified for on-the-job tasks for which he or she is responsible
- Welcoming questions from the trainee.

2.3 Training Materials

Most unit-specific training programs are divided into training modules. These modules are subdivided into parts that typically include written training material, reference material, written evaluations, and performance evaluations.

Unit-specific training modules are developed to teach the trainee about the systems and equipment found within the process unit. The subject matter for these modules typically consists of:

- Description of the process
- Control of the process
- Auxiliary equipment
- Specialty equipment
- Flare and vent systems
- Safety systems.

Process simulator stand-alone, computer-generated simulation of a process unit or process system that emulates process equipment, piping systems, control mechanisms, and behaviors that control the process.

Another training tool that is available in some facilities is the **process simulator**. A process simulator gives the trainee the ability to demonstrate tasks associated with control board or critical system operation without affecting the process. Simulator display graphics are designed to represent the process unit or system with the same level of detail that the trainee will experience using a typical distributed control system (DCS). Computer software within the simulator enables the trainee to manipulate process control variables and experience real-time results that demonstrate the correct methods of controlling a process or system. Figure 2.3 shows a process simulator.

Practicing on the actual equipment can jeopardize production and safety. Using a process simulator allows the trainee to practice controlling the equipment and unit during normal and emergency situations but to do so in a safe environment. A process simulator

Figure 2.3 A process simulator.

CREDIT: Andrei Kholmov/Shutterstock.

provides valuable experience not otherwise available due to associated risks. The trainee can practice such tasks as:

- Basic control methods for flow, temperature, pressure, and level
- Basic and advanced control methods that are process specific
- Real-time results that reflect how the process is affected by manipulated variables (i.e., flow, temperature, pressure, and level)
- How to troubleshoot equipment and control a malfunction
- How to adjust process variables to achieve a desired outcome
- How to respond to emergency situations.

Although training materials and delivery methods vary from site to site, **computer-based training (CBT)** is a common method of delivering training. An example of a computer-based training slide can be found in Figure 2.4.

Computer-based training (CBT) delivers training material through a facility computing system.

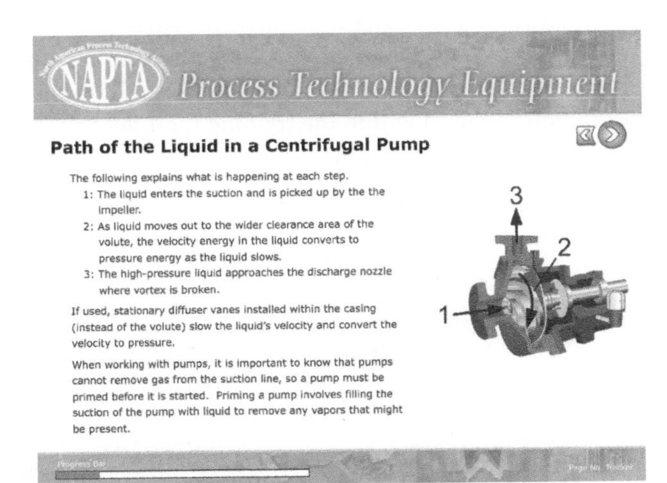

Figure 2.4 Example of a computer-based training slide.

CREDIT: Courtesy of NAPTA.

Skill development of the trainee within a predetermined window of time is the goal. Using a systematic approach, the goal of the training program is training, certification, and finally release of the trainee to operate or perform specific job tasks. Training programs usually begin with a process description, which typically contains chemistry, process equipment descriptions, operating parameters, and stream compositions. Then, more difficult subjects like process control and advanced controls are addressed. This approach is designed to increase the level of difficulty gradually so that the trainee can establish basic knowledge and then build upon it. This approach helps the trainee retain what is learned.

Training modules to teach the trainee about the hazards found within utility systems are delivered early in most training programs. The subject matter for these modules typically includes hazards related to the following:

- Steam
- Water
- Nitrogen
- Electricity

In addition to initial process training, examples of government-required training modules, developed relative to the hazards found throughout the industry, include the following examples:

Safety data sheet (SDS) a document that contains information related to the safety, hazards, and handling of a specific material.

- Hazardous Waste Operations and Emergency Response (HAZWOPER)
- Hazard communication
- **Safety data sheet (SDS)**
- Accident prevention signs and tags
- Respiratory protection
- Control of hazardous energy (lockout/tagout)
- Confined space permitting
- Portable fire extinguishers
- Bloodborne pathogens
- Toxic and hazardous substances
  - Benzene awareness
  - Asbestos awareness
  - Vinyl chloride awareness
  - Lead awareness

Additional reference materials are also made available to the trainee, as well as experienced technicians, in either hard copy or electronic form. These reference materials contain valuable information for new and experienced process technicians, and they supplement the information found within the training modules.

Reference materials that will be readily available to all unit personnel include:

- Operating manual (Figure 2.5)—unit specific material providing detailed process information such as:
  - The technical description of the process

Figure 2.5 Reference materials reinforce learning.

CREDIT: RAGMA IMAGES/Shutterstock.

- Process control and advanced control descriptions
- Instrument and alarm summary sheets complete with manufacturer data
- Technical description and manufacturer data for rotating equipment (i.e., pumps, compressors, mixers, and fin fan exchangers)
- Technical description and manufacturer data for fixed equipment (i.e., heat exchangers, towers, drums, and storage tanks)
- Technical description and manufacturer data for specialty equipment (i.e., centrifuges, reactors, furnaces, driers, and API separators).

- Unit standard operating procedures (SOPs)—unit-specific procedures for equipment and system startup, shutdown, and normal and emergency operations.
- Safety, health, and environmental (SHE) policies—policies implemented by process facilities in order to minimize or prevent the risks and/or hazards associated with the process industry, and to ensure that the facility is in compliance with applicable regulatory agencies.
- Piping and instrument diagrams (P&IDs)—detailed drawings that graphically represent the equipment, piping, and instrumentation contained within a process facility. They are a family of functional one-line diagrams showing process equipment, mechanical and electrical systems like piping, and cable block diagrams. Abbreviated as P&IDs, they show the interconnection of process equipment and the instrumentation used to control the process. They are the primary schematic drawings used for laying out a process control installation in a process facility.

  In addition to showing the interconnection of process equipment and the instrumentation used to control the process, P&IDs are used to teach the trainee:

  - The unit layout
  - Equipment identification tag numbers
  - Instrument identification tag numbers
  - Piping and equipment design criteria
  - Piping and equipment material of construction
  - Piping and equipment design temperature
  - Piping and equipment design pressure

Finally, the training program should include practice with operational technique. Practice may be provided for the following duties and expectations:

- Effective and efficient methods for performing safe and reliable process lineups
- Following safe work permit procedures
- Consistent and reliable radio communication
- Communicating with contractor work crews
- Application of safety regulations
- Application of mindfulness and situational awareness while working through a job procedure
- Operational contingency planning
- Unit emergency response
- Process optimization and troubleshooting
- Equipment inspection and plant rounds
- Managing time and prioritizing work based on plant needs

Summary

Continual learning is an expectation for process technicians. This chapter has covered the new process technician's expectations and responsibilities for learning once hired in the process industry. The required skills are needed because of the significant amount of science and technology involved in processing and also because of the hazards that are found throughout the industry. Training requirements are driven by the process facilities as well as responsible government agencies.

Content and delivery of the training material, written and performance evaluations to provide proof of knowledge, and continuous education such as refresher training and mentorship are all important to the success of every training program. Training modules are usually subdivided into parts, which include written training material, reference material, written evaluations, and performance evaluations. Key attributes of skilled workers in the process industry include personal motivation, communication skills, attention to detail, understanding the inner workings of the process, and understanding how to safely manage process hazards.

The goal of on-the-job training is to get new hires trained, certified, and finally released to operate or perform specific job tasks.

The ability of training staff and mentors to implement training as effectively as possible is also important to the success of every training program. Senior operators, training coordinators, and mentors with the knowledge, communication skills, desire, and ability to coach trainees can almost always motivate trainees to succeed.

Checking Your Knowledge

1. Define the following terms:
   a. Computer-based training (CBT)
   b. Mentor
   c. On-the-job training programs
   d. Process simulator
   e. Safety data sheet (SDS)
   f. Standards

2. ___ ___ ___ ______________ programs are the primary method used throughout the process industry to ensure that the personnel who operate process facilities are trained to perform the job tasks for which they are responsible.

3. Operators, process and technical engineers, and managers all have the responsibility of developing ______________ skills that support the sharing of information and knowledge.

4. Trainee observation through a ______________ program provides the opportunity to evaluate a trainee's work ethic and abilities and is one of the most important pieces of successful training programs.

5. Reference material that should be readily available to all unit personnel includes: (Select all that apply.)
   a. The operating manual
   b. P&IDs
   c. Consistent, reliable radio communication
   d. Personnel records
   e. Safety, health, and environmental (SHE) policies

6. Piping and instrument diagrams (P&IDs) provide information on which one of the following?
   a. Application of mindfulness and situational awareness
   b. Application of safety regulations
   c. Steps for following safe work permit procedures
   d. Piping and equipment design pressure.

7. In addition to showing the interconnection of process equipment and the instrumentation used to control the process, P&IDs teach the trainee: (Select all that apply.)
   a. Consistent and reliable radio communication
   b. The unit layout
   c. Piping and equipment material of construction
   d. Time management based on plant needs
   e. Piping and equipment lubrication techniques

8. Which training tools allow the trainee to demonstrate control board or critical system operation tasks without affecting the process?
   a. Process and instrument diagrams
   b. Process simulators
   c. Government-required training modules
   d. Safety data sheets

9. Which of the following are considered toxic substances? (Select all that apply.)
   a. Benzene
   b. Nitrogen
   c. Asbestos
   d. Steam

10. Process technicians' skill requirements are based on an understanding of the science and technology involved in processing and on knowing the risks and _______________ that are found throughout the facility

NOTE: Answers to Checking Your Knowledge questions are in the Appendix.

Student Activities

1. Work with a classmate to teach each other a skill that you currently do at work, home, or school.
2. Write a two-page paper describing what you think is the best delivery method for on-the-job training. Include examples.
3. Have students read and interpret sample BFDs, PFDs, and P&IDs.

Chapter 3
Reading Process Drawings

Objectives

After completing this chapter, you will be able to:

3.1 Identify the common types of drawings used in the process industry and their purpose. (NAPTA Operations, Diagrams for the Operating Unit 1–11*) p. 23

3.2 Explain the symbology used in process drawings to communicate information. (NAPTA Process Technology Operations, Diagrams for the Operating Unit 9) p. 33

Key Terms

American National Standards Institute (ANSI)—organization that oversees and coordinates the voluntary standards in the United States. ANSI develops and approves norms and guidelines that impact many business sectors. The coordination of U.S. standards with international standards allows American products to be used worldwide, **p. 41**

American Petroleum Institute (API)—trade association that represents the oil and natural gas industry in the areas of advocacy, research, standards, certification, and education, **p. 41**

American Society of Mechanical Engineers (ASME)—organization that specifies requirements and standards for pressure vessels, piping, and their fabrication, **p. 42**

Application block—main part of a drawing that contains symbols and defines elements such as relative position, types of materials, descriptions, and functions, **p. 32**

Block flow diagrams (BFDs)—simple drawings that show a general overview of a process, indicating the parts of a process and their relationships, **p. 24**

*North American Process Technology Alliance (NAPTA) developed curriculum to ensure that Process Technology courses will produce knowledgeable graduates to become entry-level employees in process technology. Objectives from that curriculum are named here in abbreviated form. For example, "(NAPTA Operations, Diagrams for the Operating Unit 1-11)" means that this chapter's objective 1 briefly reviews objectives 1 through 11 of NAPTA's curriculum about diagrams used by operating units.

Electrical diagrams—diagrams that help process technicians understand power distribution, and how it relates to the process, **p. 29**

Equipment symbols—set of symbols located on one sheet of a set of process flow diagrams (PFDs) for every piece of equipment found in industry for the user to review, **p. 25**

ISA—a global, nonprofit technical society that develops standards for automation, instrumentation, control, and measurement, **p. 41**

Isometric drawings (Isoms)—perspective drawings that depict objects, such as equipment and piping, as a 3D image, as they would appear to the viewer, **p. 30**

Legend—section of a drawing that defines the information or symbols contained within the drawing, **p. 32**

National Electric Code (NEC)—a standard that specifies electrical cable sizing requirements and installation practices, **p. 42**

National Fire Protection Association (NFPA)—international organization that specifies fire codes including building construction codes, fire suppression systems, and firefighting capabilities required at facilities, **p. 42**

One-line diagram—a single page document that represents a facility's electrical distribution infrastructure; also known as the single-line diagram, **p. 29**

Piping and instrumentation diagrams (P&IDs)—detailed drawings that graphically represent the equipment, piping, and instrumentation contained within a process facility, **p. 26**

Plot plans—diagrams that show the layout and dimensions of equipment, units, and buildings, drawn to scale, so that everything is of the correct relative size, **p. 31**

Process drawings—diagrams that provide a visual description and explanation of the processes, equipment, and other important items in a facility, **p. 23**

Process flow diagrams (PFDs)—basic drawings that use symbols and directional arrows to show primary product flow through a process, including such information as operating conditions, the location of main instruments, and major pieces of equipment, **p. 25**

Safety, health, environmental diagrams—a visual layout of the emergency access, personnel safety equipment, fire protection systems, and environmental systems, **p. 32**

Symbology—various graphical representations used to identify equipment, lines, instrumentation, or process configurations, **p. 33**

Symbols—letters used to designate chemical elements or equipment classes; figures used to designate types of equipment, **p. 24**

Title block—section of a drawing that contains information such as drawing title, drawing number, revision number, sheet number, originator signature, and approval signatures, **p. 33**

Utility flow diagrams (UFD)—drawings that provide process technicians a P&ID-type view of the utilities used for a process, **p. 28**

3.1 Introduction

Process drawings is a general term used to describe diagrams that show how a process is set up. They are also used for review by process technicians and other employees before additions or revisions to the facility are incorporated.

Process drawings diagrams that provide a visual description and explanation of the processes, equipment, and other important items in a facility.

There are several different types of drawings. Each drawing type represents different aspects of the process and different levels of detail. The combination of these drawings provides a more complete picture of the process and the facility than does any single drawing. Without process drawings, it would be difficult for process technicians to understand the process, and how it operates. It's important to remember that all process drawings have three common functions:

Symbols letters used to designate chemical elements or equipment classes; figures used to designate types of equipment.

- *Simplifying*—using **symbols** to make processes easier to understand.
- *Explaining*—describing how all of the parts or components of a system work together (drawings can quickly and clearly show the details of a system that might otherwise take many written pages to explain).
- *Standardizing*—using a common set of lines and symbols to represent components (while efforts are constantly made to standardize drawings and symbols across various industries, there is still a wide variance between the many industries).

Diagrams are used extensively for process technicians learning to troubleshoot and during startups and shutdowns. They are used when preparing equipment, piping, and/or valves for maintenance. They are also used before and after initial commissioning.

Each process drawing must meet several requirements to be considered a proper industrial drawing. These requirements include specific, universal rules governing:

- How lines are drawn
- How proportions are used
- What measurements are used
- What components are included
- What industrial application is targeted.

Common Process Drawings

Process technicians must recognize a variety of drawings and understand how to use them. The most commonly encountered drawings include:

- Block flow diagrams (BFDs)
- Process flow diagrams (PFDs)
- Piping and instrumentation diagrams (P&IDs)
- Utility flow diagrams (UFDs)
- Electrical diagrams (schematics)
- Isometric drawings (Isoms)
- Plot plans.

Block flow diagrams (BFDs) simple drawings that show a general overview of a process, indicating the parts of a process and their relationships.

BLOCK FLOW DIAGRAM (BFD) A **block flow diagram (BFD)** provides a simplistic set of sequences that move from left to right and show the primary flow path of a process. BFDs are used to represent unit operations (shown in Figure 3.1). BFDs consist of blocks connected by straight lines. The blocks represent processes or subsections of a process, and the lines represent process flow streams between the blocks. These streams may be liquids or gases, which flow through pipes and ducts, or solids that are moved on conveyors. Block flow diagrams show a "big picture" view of a process operation from the introduction of a raw material to the output of the final product. They have few specifics, and they do not describe how steps are to be accomplished. Instead, they show what is done in different sections of the facility. These drawings are helpful to individuals who are not expected to operate the unit but do need to understand the flow paths and the equipment/units involved.

Block flow diagrams follow a set of rules. Unit operations may be represented as a single block. The block may represent specific pieces of equipment such as a reactor, a distillation column, boiler, and so on, or a larger system, such as a production unit's reaction system, distillation operation, or packaging operation. Process flow streams flowing into and out of the blocks are represented by straight lines. They may be either horizontal or vertical with directional arrows indicating the direction of the process flow. They are often numbered in order to represent the process sequence from start to finish. Block flow diagrams should be arranged so that the process flow is basically from left to right.

PROCESS FLOW DIAGRAMS (PFDS) Process technicians make use of different types of industrial drawings on the job. The two most useful types of drawings to the process technician are process flow diagrams (PFDs) and piping and instrumentation diagrams (P&IDs).

Figure 3.1 Block flow diagram.

CREDIT: Courtesy of Corys.

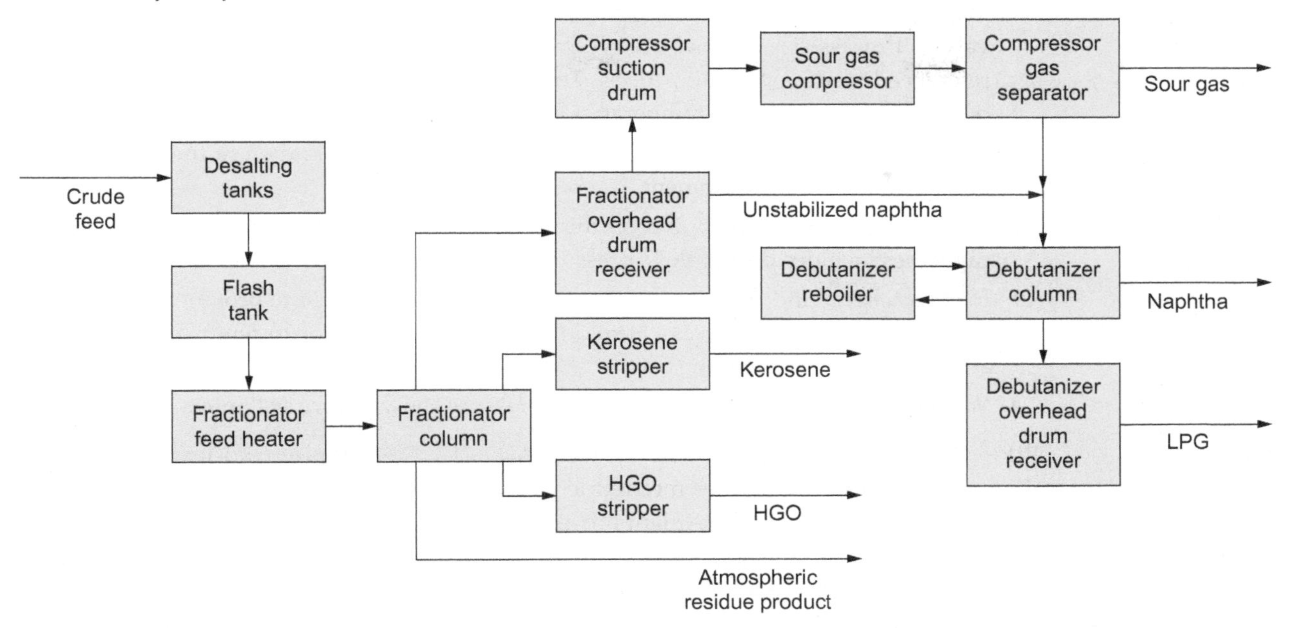

Process flow diagrams (PFDs) (shown in Figure 3.2) give the process technician much more detail than the BFD, including specific conditions and equipment. New technicians find PFDs helpful when learning equipment and flows on their assigned operating unit. The PFDs contain symbols that represent the major pieces of equipment and piping systems used in the process. Directional arrows show the path of the process from the beginning to the end. Process operating parameters may be included.

Process flow diagrams (PFDs) basic drawings that use symbols and directional arrows to show primary product flow through a process, including such information as operating conditions, the location of main instruments, and major pieces of equipment.

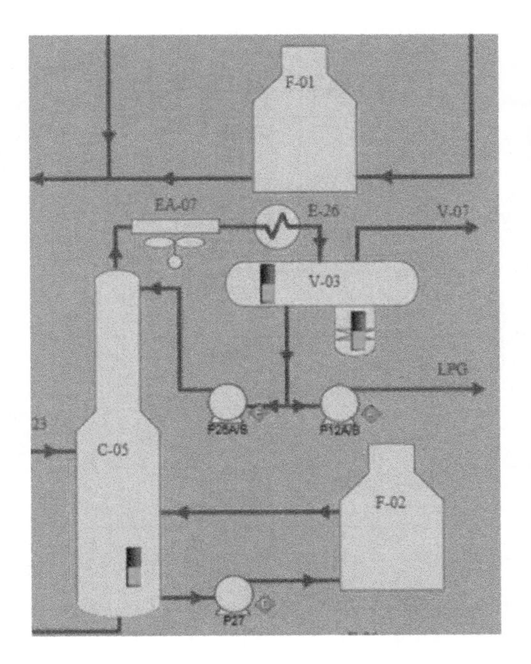

Figure 3.2 Process flow diagram.

CREDIT: Courtesy of Corys.

The process flow is typically drawn from left to right, with feed products or raw materials entering the process on the left, and ending with finished products on the right. Other information found on a PFD includes:

- **Equipment symbols**, to identify every piece of equipment on the diagram.
- *Equipment designations*, including all major vessels, pumps, compressors, and other equipment with some sort of descriptive designation. For example, with a two-stage reactor, the

Equipment symbols set of symbols located on one sheet of a set of process flow diagrams (PFDs) for every piece of equipment found in industry for the user to review.

designation might be "First Stage Reactor" and "Second Stage Reactor." At some point in the design of the facility, equipment numbers are designated and added to the drawings.

- *Major process piping*, which is indicated as lines on the PFD. It contains major control valves. Process stream lines are typically numbered and the numbers may be cross-referenced to stream compositions. The material included in the stream composition data also includes design temperatures, pressures, and stream flow rates.
- *Control instruments*, especially if they are essential to the operation of the process or if they contain special equipment.
- *Pump capacities* often included on the process flow diagrams. Information for pump capacities includes design flow, pressure, temperature, and density.
- *Heat exchangers* (input or output) and/or furnaces (input) used to provide heat input to or remove heat from a process. Heat duties, or the heat addition or removal requirement at design rates, are included on the PFD.
- *Variables* such as flow, temperature, and pressure shown at critical points.

Piping and instrumentation diagrams (P&IDs) detailed drawings that graphically represent the equipment, piping, and instrumentation contained within a process facility.

PIPING AND INSTRUMENTATION DIAGRAMS (P&IDS) **Piping and instrumentation diagrams (P&IDs)**, sometimes referred to as *process and instrument drawings* are the most important representation of the plant process—so much so, that OSHA requires a company to maintain an accurate, up-to-date, copy (set) of the plant's P&IDs in the control room, or easily accessible plant location (e.g., electronic format). P&IDs show more detailed information about the equipment, piping, and control systems than do PFDs. P&IDs are slightly different from company to company; however, they use similar symbol conventions such that the process technician should be able to understand any P&ID.

A vital part of a P&ID, as shown in Figure 3.3, is the instrumentation information. This information gives the technician a firm understanding of how the process is controlled, how

Figure 3.3 Sample P&ID of a debutanizer system.

CREDIT: Courtesy of Corys.

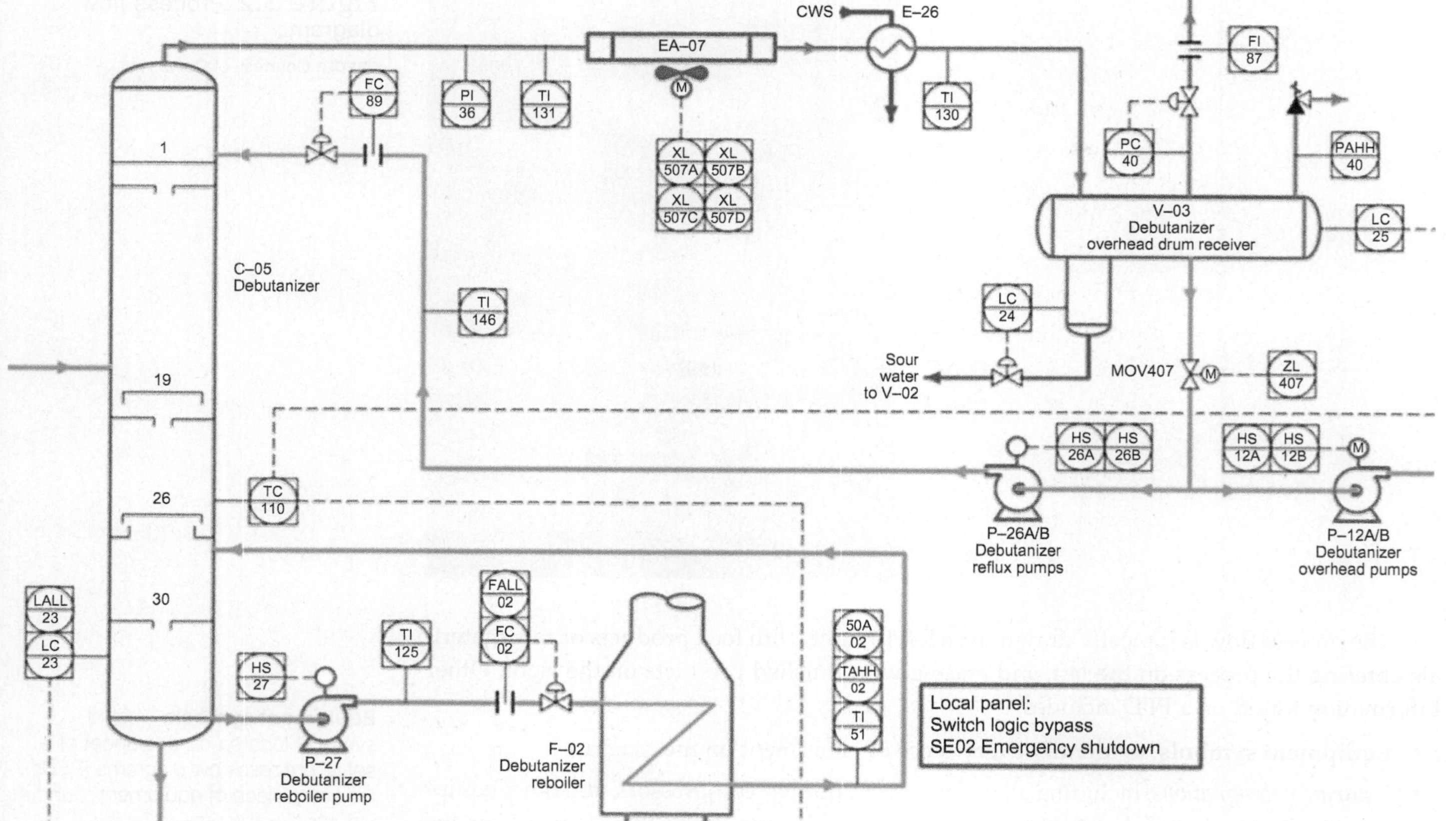

a product flows through the process, and how it can be monitored and controlled. Engineers and maintenance may also use P&IDs for troubleshooting, plant modifications, and upgrades; although maintenance more often uses mechanical, isometric, and electrical drawings.

Information found on P&IDs includes:

- *Equipment* symbols and numbers
- *Equipment designations*, including all major vessels, pumps, compressors, and other equipment. P&IDs should show all piping and control systems related in space—that is, a pump should be shown below the vessel from which it takes suction. Some, but not all, P&IDs have a brief description of the equipment to include its name, equipment design specifications (vessel maximum allowable working pressure [MAWP], pump motor horsepower [hp], pump net positive suction head [NPSH], etc.), and material of construction, and they may have a unique equipment number. They will often show internal components such as reactor baffles, agitators, mixers, and distillation trays. Examples of P&ID equipment information include:
  - *Pump information* may include material of construction (e.g., 304 stainless steel [304SS]), flow rate at the design pressure (e.g., 200 gpm @ 150 ft head), motor horsepower (e.g., 25 hp), and usually an equipment number unique to the unit or plant.
  - *Vessel information* may include the vessel's maximum allowable working pressure (MAWP), material of construction (CS—carbon steel; SS—stainless steel), the size of the vessel (height and diameter), and the vessel's unique number based on the plant naming conventions.
  - *Heat exchanger information* may include all the information described above for a vessel; however, the shell and tube side are typically treated separately for construction material and MAWP.
- *Major process piping*, indicated by lines on the P&ID and designated with a line number or descriptor. Line numbers or descriptors include the line service designation (such as CWS for cooling water supply); a line number, which follows company convention; and pipe size and a piping specification number, which is also unique to the company. The piping specification is determined by the process flowing through the line or its service and includes the range of pipe diameters covered by the specification, the material of construction, and the acceptable valve types. The line description will also indicate the presence of electric or steam heat tracing, and, if required, the insulation specification.
- *All instrumentation* is shown on P&IDs, using symbols consistent with the ISA-5.1 standard (discussed later in the chapter). Common control system details shown on P&IDs include the following:
  - Control valve type: Can be characterized in terms of the number of plugs present, as single-seated valve and double-seated valve
  - Control and emergency isolation valves along with the failure position, which is indicated as fails closed (FC) or fails open (FO). Failure is the valve position taken when the valve actuator loses pneumatic or electrical power.
    - Flow control valve (FCV)—an automated valve used to regulate and throttle flow; typically provides the final control element of a control loop
    - Temperature control valve (TCV)—an automated valve used to regulate process gas or fluid temperature; typically provides the final control element of a temperature control loop
    - Pressure control valve (PCV)—an automated valve used to regulate process gas pressure; typically provides the final control element of a pressure control loop
    - Level control valve (LCV)—an automated valve used to measure process fluid level; typically provides the final control element of a level control loop
    - Emergency block valve (EBV)—automatic valve, typically controlled by an operating parameter and/or hand switches for process isolation when the parameter

approaches unsafe conditions or equipment limitations; also known as emergency isolation valve (EIV)

- XV—a valve that automatically fully opens or fully closes
  - Controller type (pneumatic or solenoid) and type of signal (electronic or pneumatic)
  - Control valve interlocks and emergency shutdown systems
- Pressure indicator (PI); pressure indicating controller (PIC); pressure indicating alarm (PIA)
- Level indicator (LI); level indicating controller (LIC); level indicating alarm (LIA)
- Temperature indicator (TI); temperature indicating controller (TIC); temperature indicating alarm (TIA); temperature element (TE)
- Flow indicator (FI); flow indicating controller (FIC); flow indicating alarm (FIA); flow element (FE)

- *General Notes* that are included on the left side of the P&ID to add clarification to the drawing. They may include curio notations (a circle with a number enclosed) that more clearly describes a symbol or an instruction—for example, "Locate close to the pump" or "Locate at low point."

Utility flow diagrams (UFD) drawings that provide process technicians a P&ID-type view of the utilities used for a process.

UTILITY FLOW DIAGRAMS (UFDS) **Utility flow diagrams (UFDs)** focus on the utilities used in the process (Figure 3.4). UFDs indicate utility supply and connections to process equipment, along with the piping and main instrumentation for operating the utilities.

Figure 3.4 Utility flow diagram (UFD).

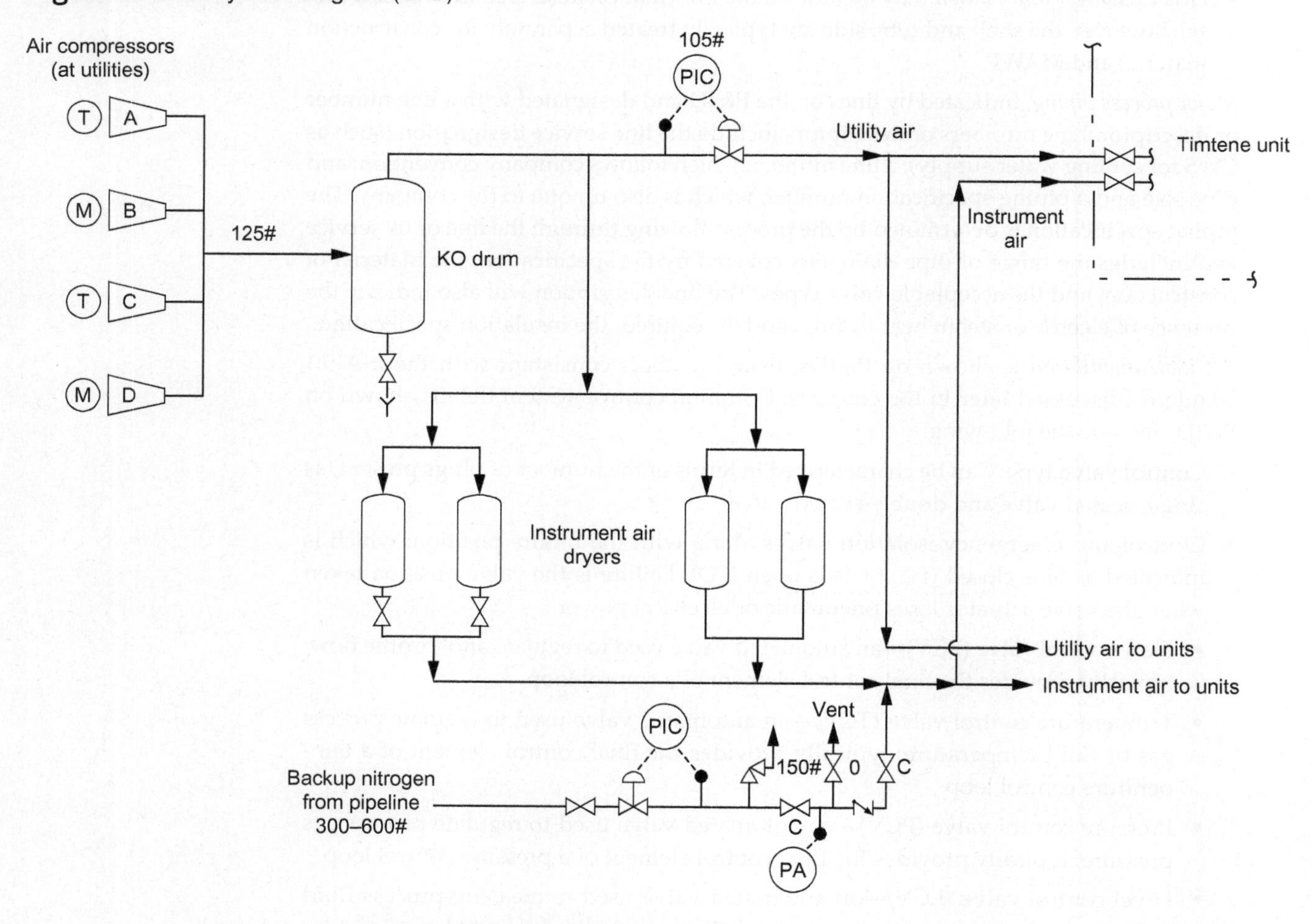

Typical utilities shown on a UFD include:

- Steam
- Condensate

- Cooling water
- Instrument air
- Plant air
- Nitrogen
- Fuel gas
- Boiler feedwater
- Potable water
- Fire water.

ELECTRICAL DIAGRAMS Many processes rely on electricity, so it is important for process technicians to understand electrical systems, and how they work. This is the purpose of **electrical diagrams**. A firm understanding of the relationship between power distribution and its use in the process is critical when performing lockout/tagout (LOTO) procedures (i.e., isolating hazardous energy sources) and monitoring various electrical measurements.

Electrical diagrams diagrams that help process technicians understand power distribution, and how it relates to the process.

Electrical diagrams (a type of *schematic*) show the various electrical components and their relationships. For example:

- Switches used to stop, start, or change the flow of electricity in a circuit
- Power sources provided by transmission lines, transformers, busses, and MCCs
- Loads (the components that use the power)
- Coils or wire used to increase the voltage of a current
- Inductors (coils of wire that generate a magnetic field and are used to create a brief current in the opposite direction of the original current) that can be used for surge protection
- Transformers (used to make changes in electrical power by means of electromagnetism)
- Resistors (coils of wire used to provide resistance in a circuit)
- Contacts used to join two or more electrical components

Process technicians are frequently required to perform or assist with LOTO of electrical power supply switchgear for process equipment. This requires the technician to *open the circuit* supplying power to the equipment—for example, an electrical motor. The switchgear room, sometimes called the motor control center (MCC), houses the electrical switchgear and/or breakers. The key electrical drawing used by the process technician is the **one-line diagram**, also known as a *single-line diagram* (shown in Figure 3.5).

One-line diagram a single page document that represents a facility's electrical distribution infrastructure; also known as the single-line diagram.

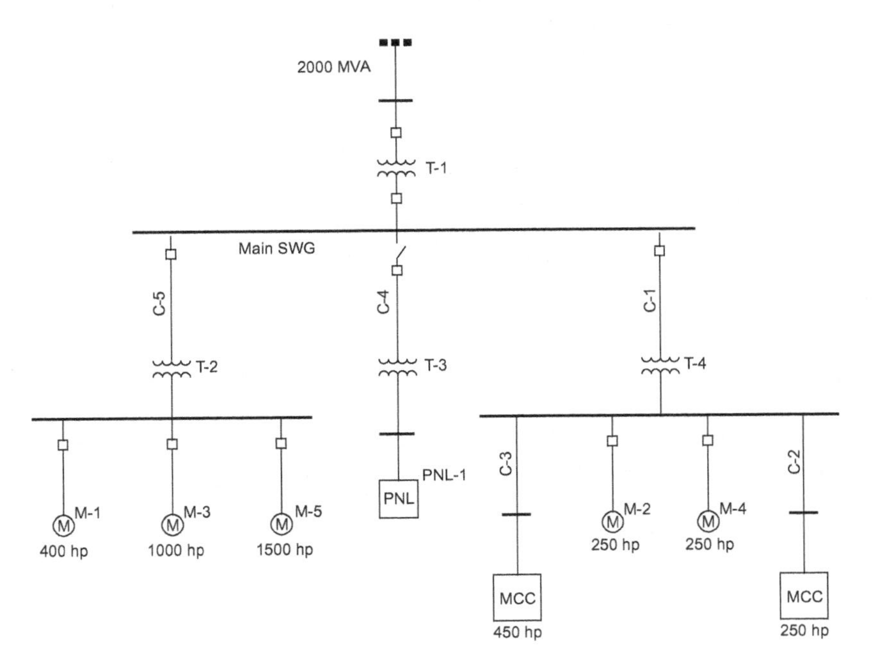

Figure 3.5 One-line diagram (single-line diagram).

The one-line diagram is the process technician's most important electrical drawing because it shows the entire electrical system of interconnecting generators, transformers, transmission and distribution lines, loads, circuit breakers, and so on. In this drawing, essentially a block diagram, a single line is used to represent a three-phase power system, from the incoming power source to each load (switchgear or breaker). It includes the ratings and sizes of each piece of electrical equipment supplied by the incoming power source. As with P&IDs, there is no single universally accepted set of symbols for one-line drawings. However, some often used symbols are shown in Figure 3.6.

Figure 3.6 Commonly used electrical symbols.

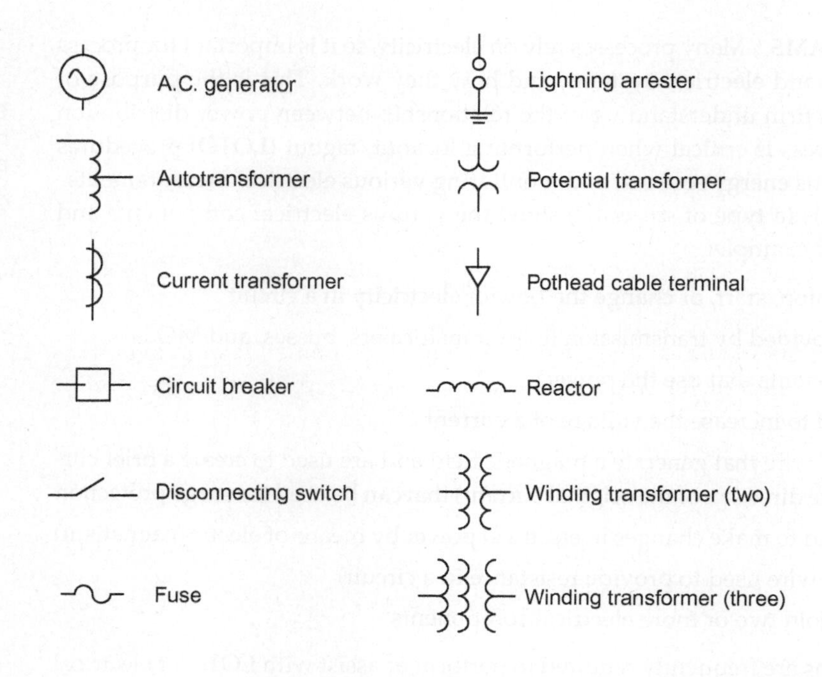

Isometric drawings (Isoms) perspective drawings that depict objects, such as equipment and piping, as a 3D image, as they would appear to the viewer.

ISOMETRICS (PIPING) **Isometric drawings (Isoms)** present object images drawn at a 30-degree angle to show three sides of the object to the viewer. Isometric drawings may also contain cutaway views to show an object's inner workings. Figure 3.7 shows an example of an isometric drawing. Typically, isometrics are provided during new unit construction and prove useful to new process technicians learning both to identify equipment and piping and to understand the inner workings.

Figure 3.7 Isometric drawing.

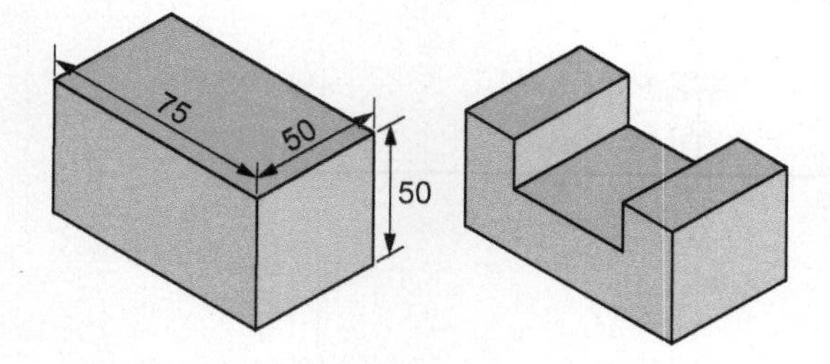

A common isom that a process technician may come to view is the piping isometric. Most isometric drawings generated in construction packages are piping isometrics that process technicians may be asked to review for comment on the line routing, the location of valves, and sample points for operator access. Figure 3.8 is an example of a piping isometric drawing.

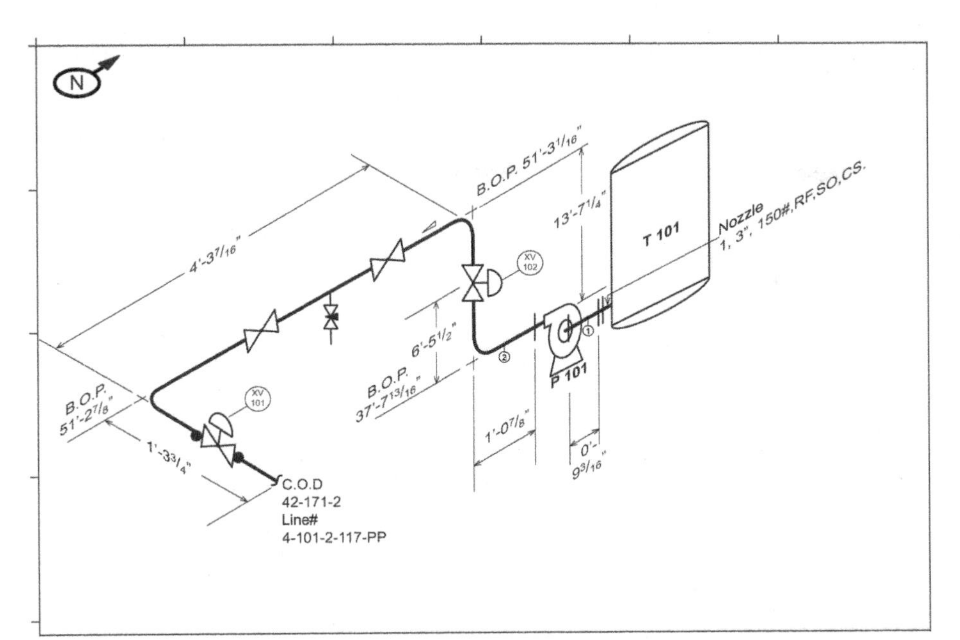

Figure 3.8 Piping isometric drawing.

PLOT PLANS **Plot plans** are drawn to scale so that equipment, units, and buildings are of the correct relative size, as shown in Figure 3.9. For example, plot plans show the location of machinery (e.g., pumps and heat exchangers) in an equipment room. On a larger scale, a plot plan shows the location and dimensions of process units, buildings, roads, and other site constructions such as fences. A site plot plan also shows elevations and grades of the ground surface.

Plot plans diagrams that show the layout and dimensions of equipment, units, and buildings, drawn to scale, so that everything is of the correct relative size.

Figure 3.9 Plot plan.

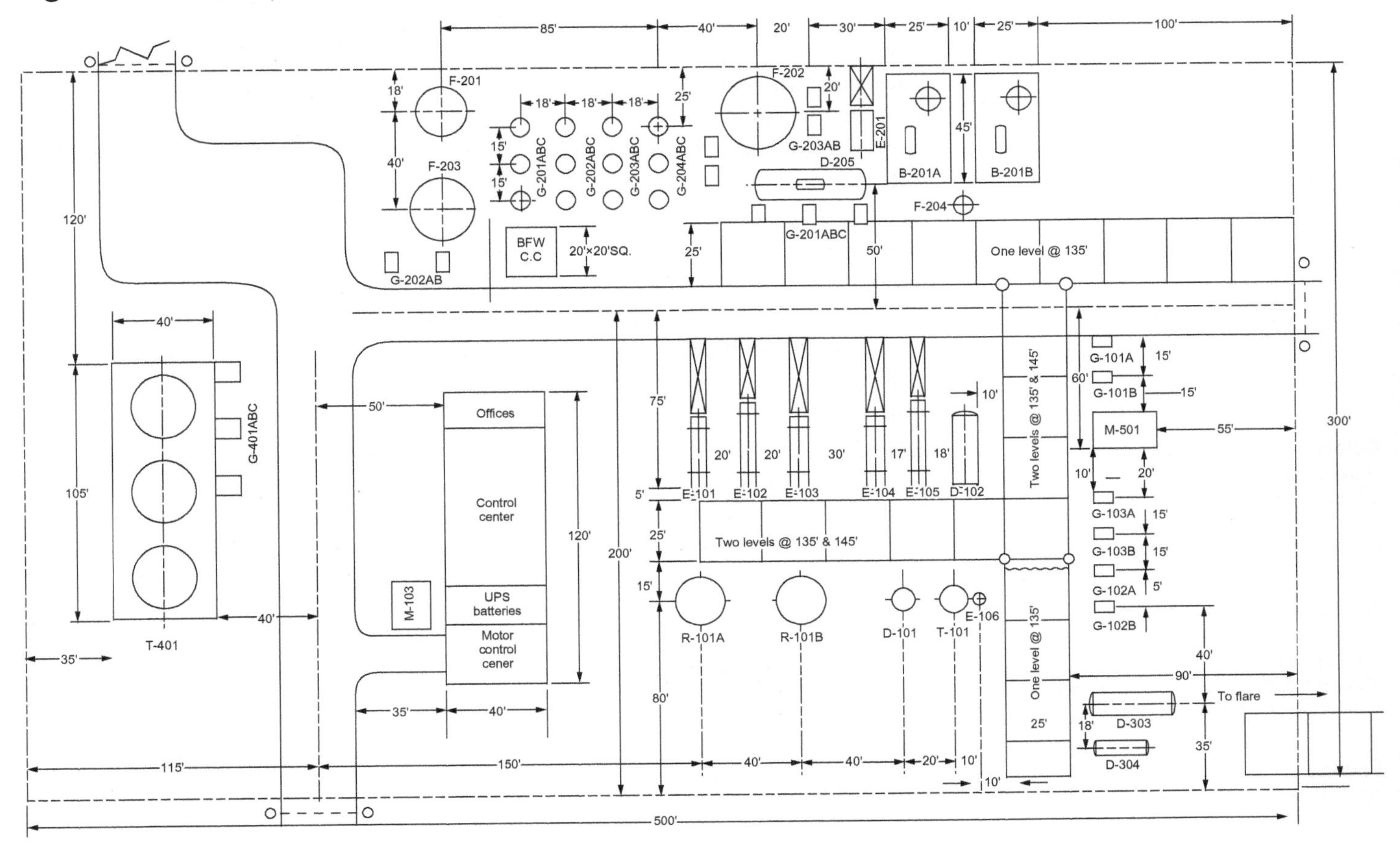

OTHER DRAWINGS Along with the drawings mentioned previously in this chapter, process technicians might encounter other types of drawings such as:

- *Elevation diagrams*—represent the relationship of equipment to ground level and other structures (Figure 3.10).

Figure 3.10 Elevation diagram.

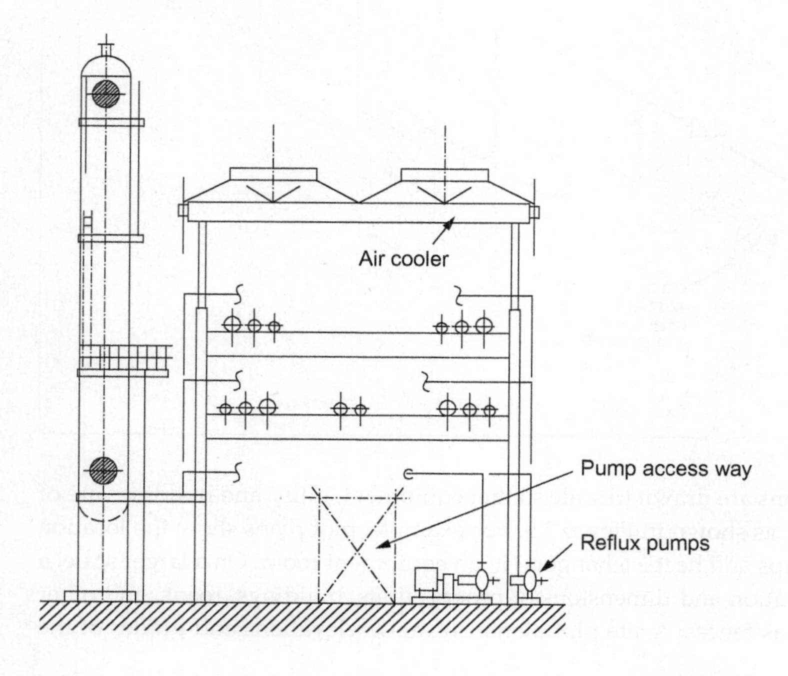

- *Equipment location diagrams*—show the relationship of units and equipment to facility boundaries.
- *Loop diagrams*—show all components and connections between instrumentation and a control room. For instance, a loop diagram might depict a control loop composed of a flow control valve that is reset by a liquid level controller. Some loop diagrams include all of the process information required to design the loop, which includes the service, the flow rate range, the calibration parameters for the flow control instrument, and other pertinent information. Most companies have their own loop diagram conventions; however, there are industry standard loop drawing software programs.
- *Foundation diagrams*—used by the construction crew pouring the footers, beams, and foundations. Concrete and steel specifications are designed to support equipment, integrated underground piping, and provide support for exterior and interior walls. Process technicians do not typically use foundation drawings; however, the drawings are useful when questions arise about piping that disappears under the ground and when new equipment is being added.
- **Safety, health, environmental (SHE) diagrams**—show all safety, health, and environmental (SHE) equipment throughout the operating facility. For instance, the eyewash and safety shower stations are indicated by location on the plot plan. The equipment is available for used by process technicians and construction workers in the event of an emergency while in the field. Process technicians are required by OSHA to participate in annual training program to understand the equipment and use it properly.

Safety, health, environmental diagrams a visual layout of the emergency access, personnel safety equipment, fire protection systems, and environmental systems.

Application block main part of a drawing that contains symbols and defines elements such as relative position, types of materials, descriptions, and functions.

Legend section of a drawing that defines the information or symbols contained within the drawing.

Process Drawing Information

Process drawings contain consistent elements. The **application block** (Figure 3.11) will generally be the largest part of the drawing. It provides the connections, positions, and functions of the various parts of the drawing. The **legend** (shown in top left of Figure 3.11) is a means of translating to the reader what the different visual elements are. The **title block** (shown in bottom right of Figure 3.11) provides the unique identifiers of that particular drawing.

Title block section of a drawing that contains information such as drawing title, drawing number, revision number, sheet number, originator signature, and approval signatures.

Figure 3.11 Application block. Note also the legend at the top left and the title block at the bottom right.

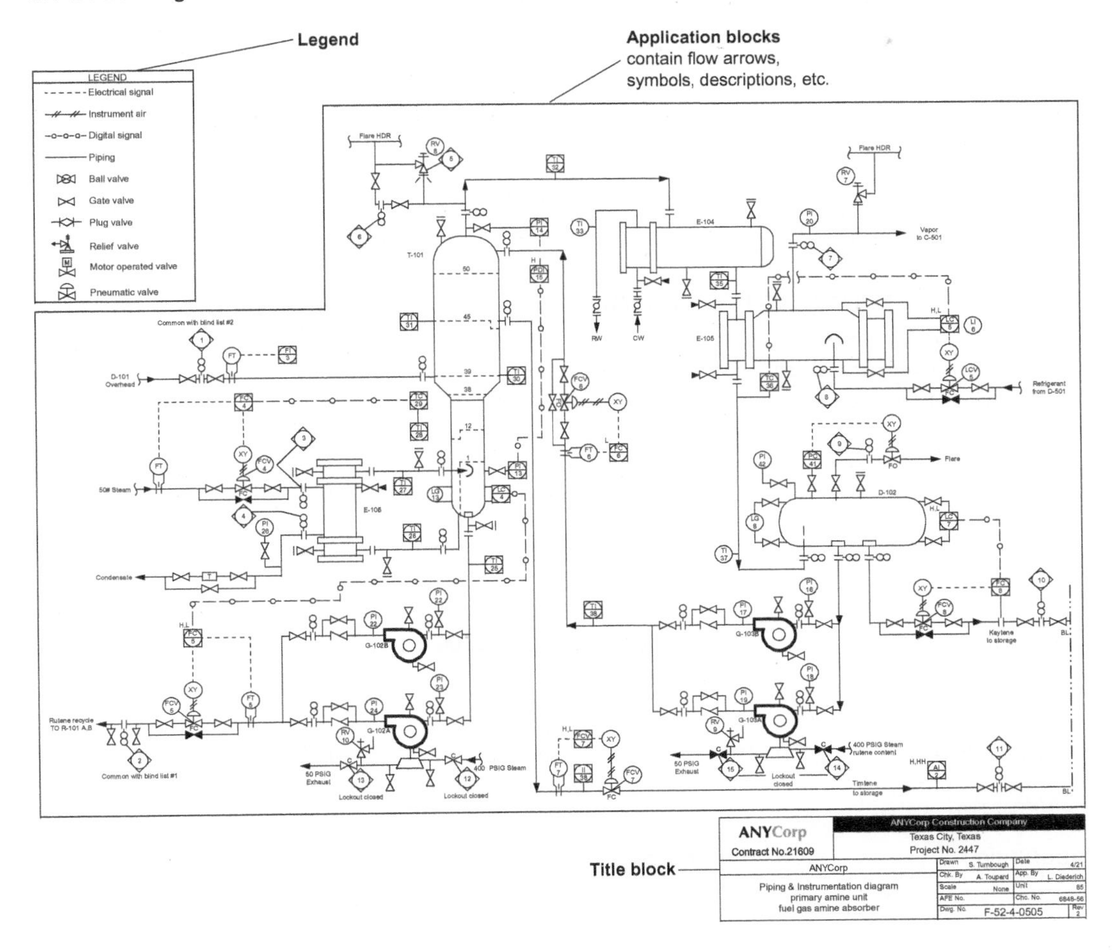

3.2 Symbols

Symbology

The use of **symbology** allows for generally standardized information across industrial drawings used all over the world. However, it is important to remember that there are few universally accepted standards; therefore, companies and organizations use slightly different symbologies. The most common control systems symbol standard is the ISA-5.1 standard (NOTE: ISA and instrumentation tag numbers are discussed later in this chapter). Process technicians should be able to recognize these symbols and any special lettering method used on a P&ID. Furthermore, technicians must be able to interpret process flows, as well as instrument and equipment designations.

Symbology various graphical representations used to identify equipment, lines, instrumentation, or process configurations.

Equipment symbols (see Figure 3.12) are figures used to identify types of equipment, instruments, and other devices on a PFD or P&ID. A set of common symbols has been developed to represent actual equipment, instrumentation, and other components. Some symbols may differ from facility to facility, while others may be specific to the individual process facility or company.

Figure 3.12 Common P&ID symbols.

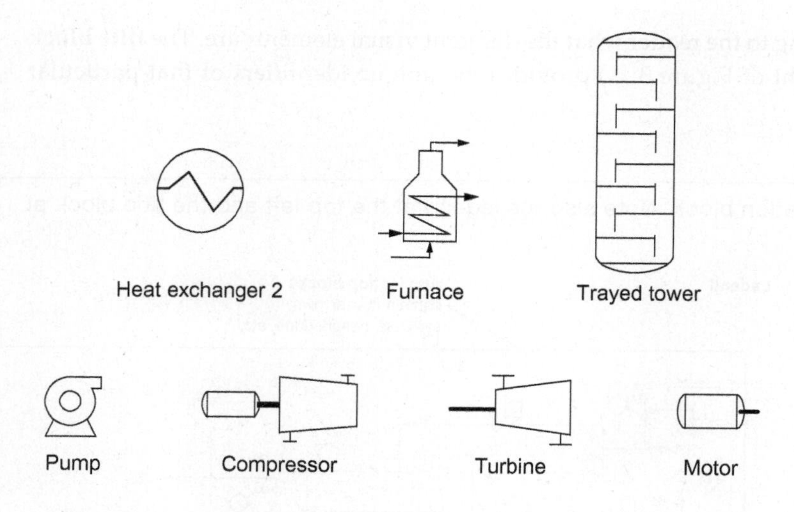

PIPING AND CONNECTION SYMBOLS Piping is a hollow tube used to transport process liquids and gases throughout a process facility. There are many symbols associated with piping. Although standards exist, symbols do sometimes vary slightly from facility to facility. Figure 3.13 shows examples of P&ID piping symbols.

Figure 3.13 P&ID piping symbols.

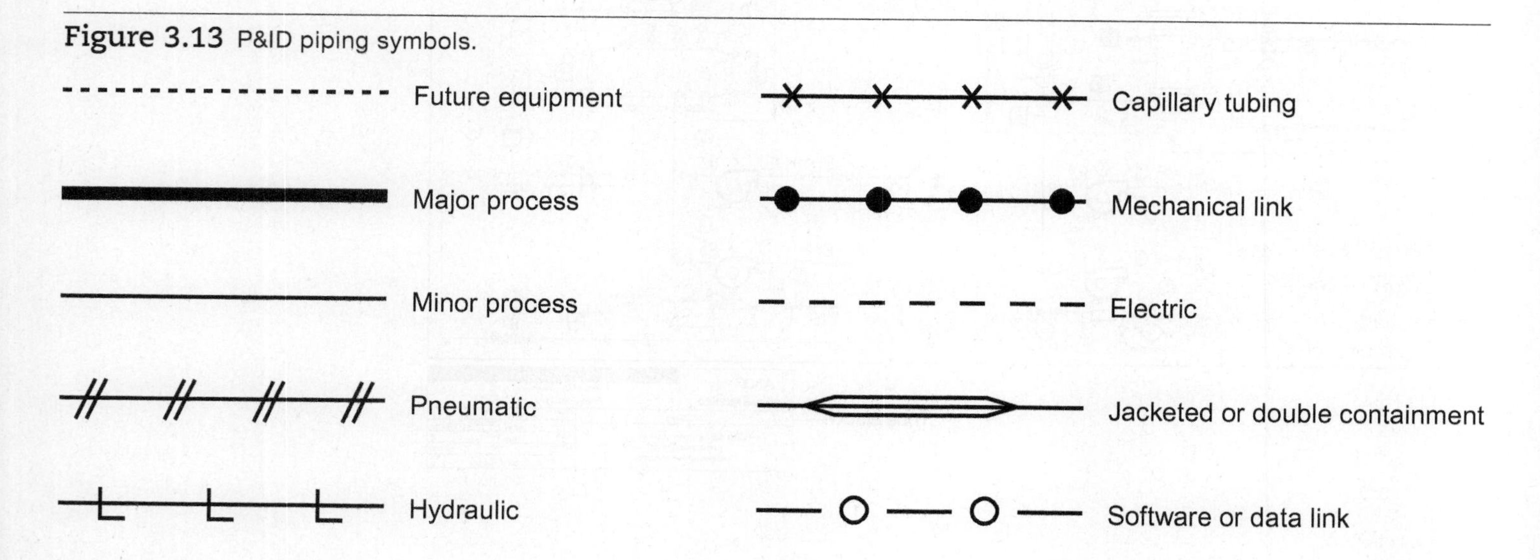

VALVE SYMBOLS Valves control the process flows through the unit piping. There are many types of valves used in the process industries. Each valve type has a unique symbol used to identify it on a process drawing. Figure 3.14 shows some examples of different valve symbols.

Figure 3.14 P&ID valve symbols.

Valve Type	Symbol	Valve Type	Symbol
Ball valve	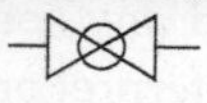	Needle valve	
Butterfly valve	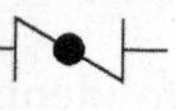	Nonreturn valve	 NR
Check valve	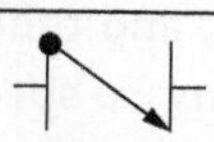	Piston valve	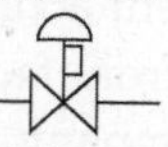

Valve Type	Symbol	Valve Type	Symbol
Actuator-operated valve		Plug valve	
Gate valve		Relief valve	
Globe valve		Stop-check valve	
Hand valve		Three-way valve	
Motor-operated valve			

ACTUATOR SYMBOLS Actuators are devices that convert electrical or pneumatic control signals to physical actions. Figure 3.15 shows examples of actuator symbols. However, these symbols vary from facility to facility.

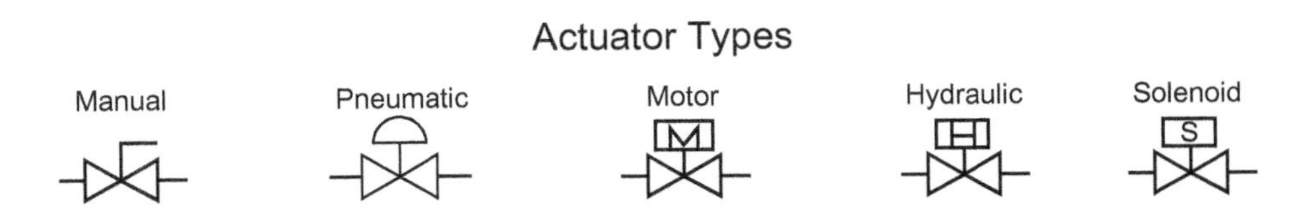

Figure 3.15 P&ID actuator symbols.

COMPRESSOR SYMBOLS Compressors increase the pressure of gases. In order to locate compressors on a P&ID, process technicians must be familiar with the different types of compressors and their symbols. Figure 3.16 shows examples of compressor symbols. However, these symbols may vary somewhat from facility to facility.

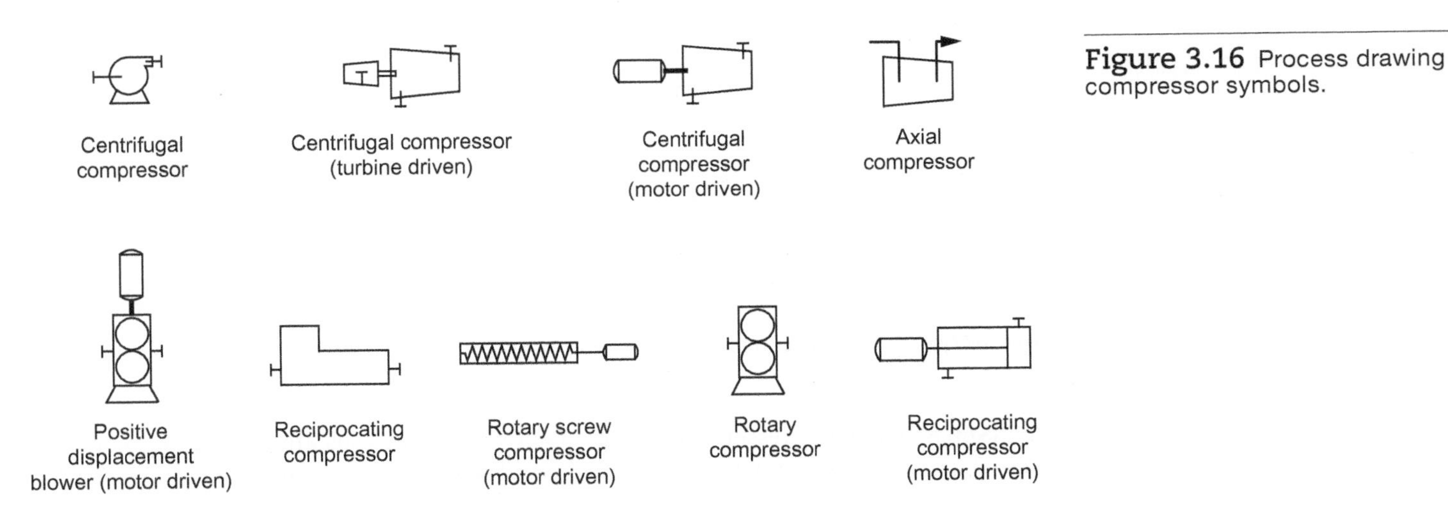

Figure 3.16 Process drawing compressor symbols.

PUMP SYMBOLS Pumps are used to move liquid materials through piping systems. The process industries use many different types of pumps. Each pump type has a unique symbol that appears on P&IDs. Figure 3.17 shows some examples of pump symbols.

Figure 3.17 P&ID pump symbols.

Pump Type	Symbol	Pump Type	Symbol
Centrifugal pump	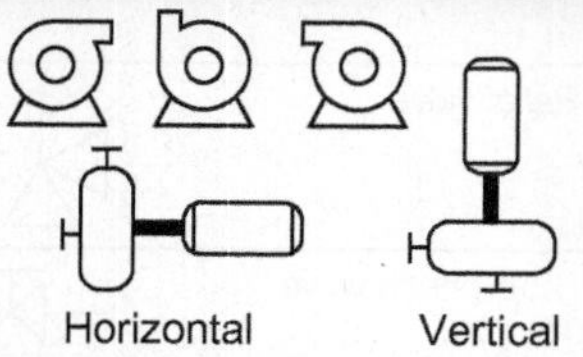 Horizontal Vertical	Rotary lobe pump	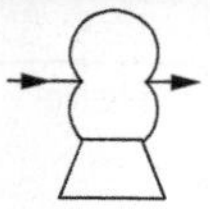
Positive displacement pump	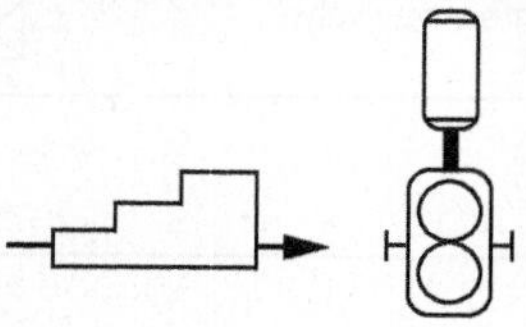	Turbine-driven equipment	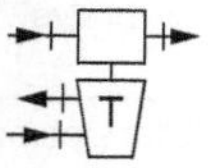 T
Electric-motor driven pump	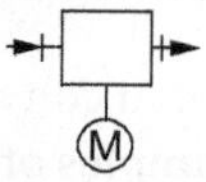 M	Electric-motor driven variable positive displacement pump	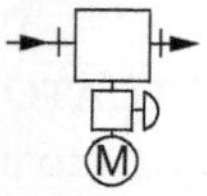 M
Piston pump	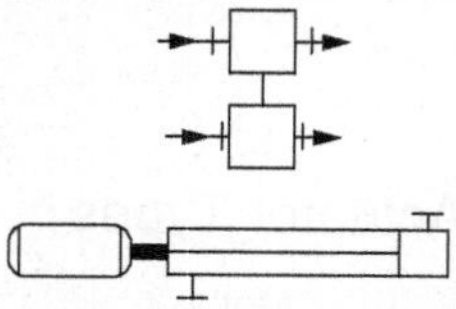	Vertical centrifugal or positive displacement pump	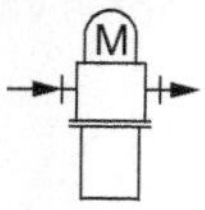 M

HEAT EXCHANGER SYMBOLS Heat exchangers transfer heat from one substance to another without the two substances physically contacting one another. The symbols shown in Figure 3.18 are examples of heat exchanger symbols a process technician might encounter on a P&ID.

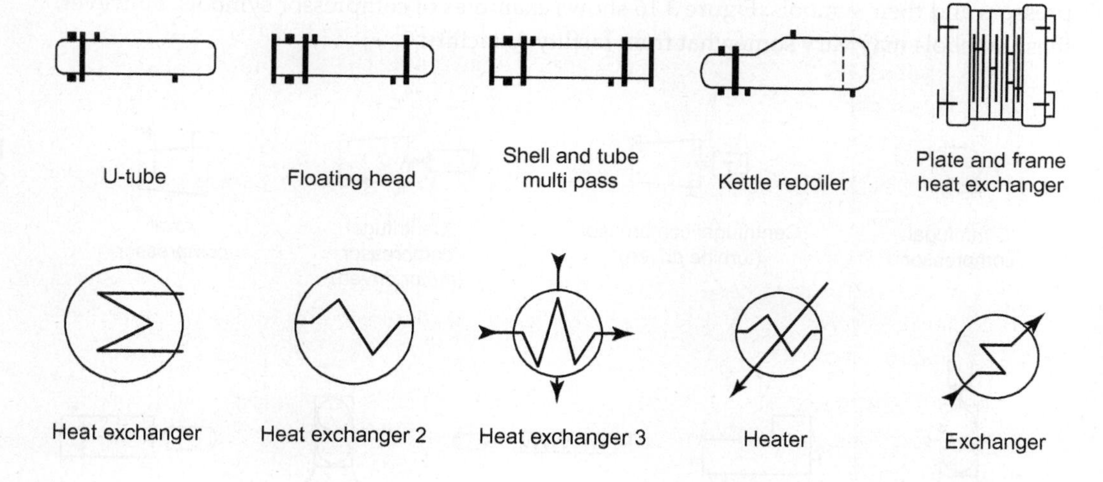

Figure 3.18 P&ID heat exchanger symbols.

VESSEL SYMBOLS Vessels are containers in which materials are processed, treated, or stored. In order to accurately locate vessels on a P&ID, process technicians must be familiar with the symbols. Figure 3.19 shows examples of some vessel symbols found on P&IDs.

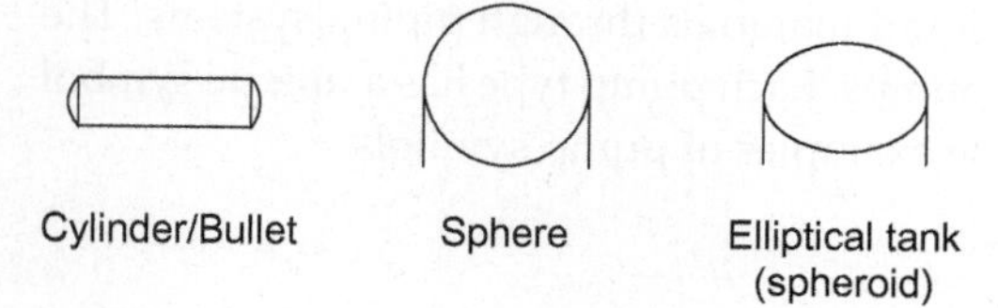

Figure 3.19 P&ID common vessel symbols.

COOLING TOWER SYMBOLS Cooling towers lower the temperature of water using latent heat of evaporation. In order to accurately locate cooling towers on a P&ID, process technicians must recognize the various cooling tower symbols. Figure 3.20 shows some of the symbols that indicate cooling towers.

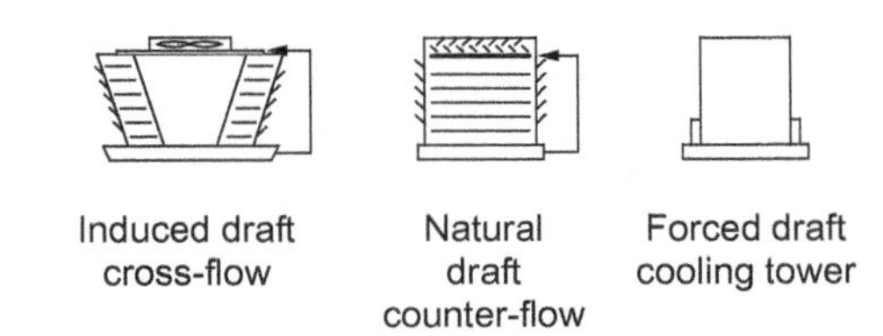

Figure 3.20 P&ID cooling tower symbols.

TURBINE SYMBOLS Turbines are used to provide the power necessary to drive equipment. In order to accurately locate turbines on a P&ID, process technicians must be familiar with turbine symbols. These symbols may vary from facility to facility. Figure 3.21 provides an example of a turbine symbol.

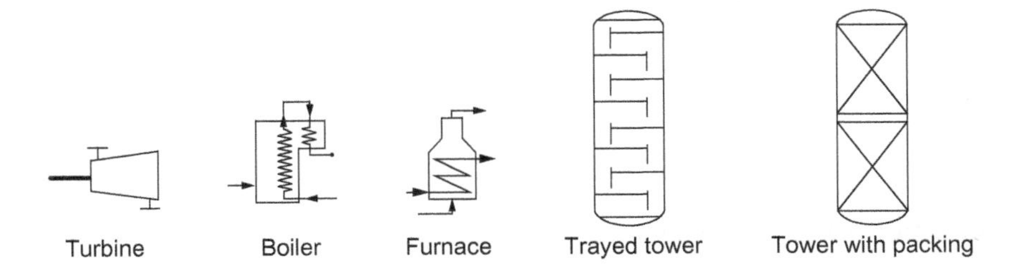

Figure 3.21 Turbine, boiler, furnace, trayed tower, and tower with packing symbols.

BOILER SYMBOL Boilers are devices that produce steam for various parts of a process. In order to accurately locate boilers on a P&ID, process technicians must be familiar with boiler symbols. These symbols may vary from facility to facility. Figure 3.21 provides an example of a boiler symbol.

FURNACE SYMBOL Furnaces are devices that produce heat for processes. In order to accurately locate furnaces on a P&ID, process technicians must be familiar with furnace symbols. These symbols may vary from facility to facility. Figure 3.21 provides an example of a furnace symbol.

REACTOR AND DISTILLATION COLUMN SYMBOLS Reactors are vessels in which chemical reactions are initiated and sustained. Distillation columns are devices used to separate liquid components by boiling point. In order to accurately locate columns and reactors on a P&ID, process technicians must recognize distillation column (tower) and reactor symbols. These symbols may vary from facility to facility. Figure 3.21 provides examples of tower and reactor symbols.

ELECTRICAL EQUIPMENT AND MOTOR SYMBOLS Electrical equipment can be used for a variety of functions. Each equipment type has a unique symbol that identifies it on process drawings. Examples of electrical equipment symbols and their descriptions are shown in Figure 3.22.

Figure 3.22 Electrical equipment and motor symbols.

Symbol	Name	Description
	Transducer	A device that converts one type of energy to another, such as electrical to pneumatic
	Motor driven	A symbol that Indicates a piece of equipment is motor driven (either AC or DC)

(Continued)

Figure 3.22 Electrical equipment and motor symbols. (*Continued*)

Symbol	Name	Description
	Current transformer	A device that can provide circuit control and current measurement
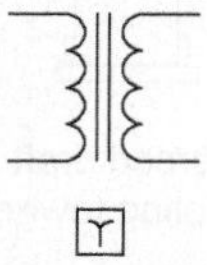	Transformer	A device that can either step up or step down the voltage of AC electricity
- - - - - -	Electrical signal	A signal that indicates voltage or current
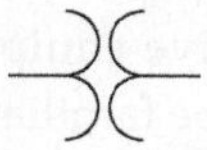	Potential transformer	A device that monitors power line voltages for power metering
	Inductor	An electronic component consisting of a coil of wire
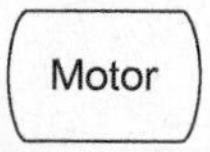 Motor	Motor	A device (either AC or DC) that converts electrical energy into mechanical energy
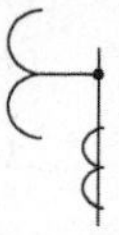	Outdoor meter device	A meter used to monitor electricity, such as a voltmeter (used to measure voltage) or ammeter (used to measure current)

INSTRUMENTATION SYMBOLS Instruments are devices that measure, indicate, and control process flows, temperatures, levels, and pressures, and provide analytical data. Instrumentation symbols identify instrumentation throughout a facility. The symbols may or may not look like the physical device represented. A 7/19-inch diameter circle, called a *balloon*, is commonly used to represent many functionally different instruments.

Figure 3.23 shows an example of a boxed instrumentation balloon, which represents a distributed control system (DCS) instrument, whereas the instrumentation balloon in Figure 3.24 represents a remote panel mount and local instruments.

Figure 3.23 Symbols representing DCS instrument and remote panel and local instruments. Note that the circle with a line through it (PIC 100) indicates a remote instrument, and a circle without a line through it (LG 10) indicates a field location.

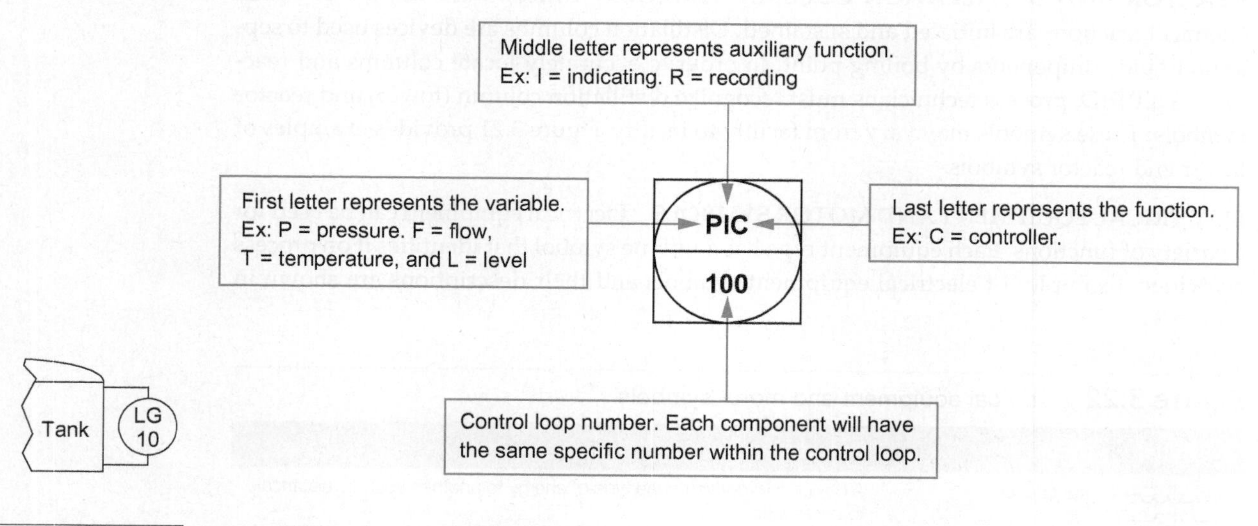

Field location

Remote location (panel board)

Figure 3.24 Symbols representing instrument in the field (left) and instrument on remote panel (right).

The only difference from one balloon to another is its unique alphanumeric tag number, explained in Figure 3.23. The tag number is the primary key to defining the instrument's function and control loop. Slight balloon modifications depict where the instrument is physically located. Considering the complexity of many control systems, this schematic approach works very well. Figure 3.25 provides some general instrument symbols.

GENERAL INSTRUMENT SYMBOLS - BALLOONS			
Location / Description	Control room instruments	Locally mounted instruments	Local board-mounted instruments
Discrete instrument			
Purpose / Description	Display	Function	
Computer system (FOX1A)			
Programmable logic controller			
Distributed control system			
Machinery health monitoring system (HP 10000 computer)	CH# MHMS		
Normally inaccessible or behind the panel devices, or functions may be depicted by using the same symbols but wih dashed horizontal bars: DCS			
Panel devices located on DCS controls			

Example of instrument bubble attribute
L, LL
Function description (where applicable)
Tag no.
Software alarms

Figure 3.25 General instrument balloon symbols.

ISA INSTRUMENT TAG NUMBERS Instrument tag numbers identify the measured variable, the function of the specific instrument, and the loop number. These tag numbers give the process technician an indication of what that instrument is monitoring or controlling. Letters and numbers on an ISA instrument tag describe the instrument (shown in Figure 3.26). The tag number should be unique since most process facilities use a global database to identify devices.

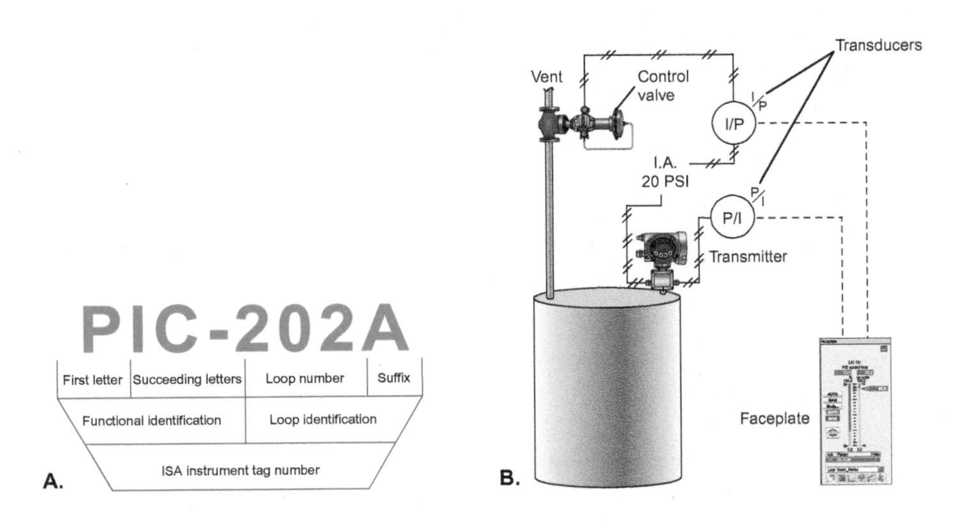

Figure 3.26 A. Instrument tag number. **B.** Tank showing local and remote instrumentation.

The first letter identifies the measured or initiating variable, and the following letters describe the function of the instrument. For example, in Figure 3.26, "P" stands for pressure, "I" for indicating, and "C" for controller. In other words, this instrument is a "pressure indicating controller." If the instrument is field mounted, it might have an identifier to say that it controls pressure and also might have an indicator on its faceplate.

The first letters (with possible modifiers) and the succeeding letters (the Readout or Passive Function column, Output Function column, and possible modifiers associated with the succeeding letters) can be found in ISA-5.1 Table 4.1, "Identification letters." Table 3.1 here is an example of an identification table based on data from the ISA-5.1 standard.

Did You Know?

The ISA-5.1 standard is a commonly used reference, but it is subject to change. For this reason, the most current standard should always be consulted for verification purposes.

The instrument tag examples listed in Table 3.2 can be interpreted using the functional identification information shown in Table 3.1.

Table 3.1 Functional Identification Table

	First Letters		Succeeding Letters		
Letter	Measured or Initiating Variable	Modifier	Readout or Passive Function	Output Function	Modifier
C	User's choice (any control device)			Control	
F	Flow rate	Ratio (fraction)			
H	Hand				High
I	Current (electric)		Indicate		
L	Level		Light		Low
P	Pressure, vacuum		Point (test) connection		
R	Radiation		Record		
T	Temperature			Transmit	
V	Vibration, mechanical analysis			Valve, damper, louver	

Table 3.2 Instrument Tag Number Functional Identification Examples

Letters	Functional Interpretation
C	Controller
CV	Control valve
E	Element
F	Flow
I	Indicator
L	Level
P	Pressure
T	Temperature
FFIC	A flow (ratio) indicating controller
FRC	Flow recording controller
LI	Level indicator
LV	Level valve (preferred way of identifying a control valve in a loop; may also be expressed as PV, FV, TV)
PC	Pressure controller (since this controller does not have an indicator or a recorder function, it would probably be behind the panel out of the sight of the operator)
PT	Pressure transmitter
TE	Temperature element (e.g., could be a thermocouple, RTD, or filled thermal system)
TT	Temperature transmitter

Equipment Standards

With the development of process drawings and equipment standards, a system of symbols has been utilized to depict the various drawings that are used to describe how both equipment and its associated instrumentation are interconnected. Various engineering and chemical companies have created a symbols system in their own facilities or companies, but there have been other groups that have formed organizations or societies to address issues across the various process industries:

- **ISA** (formerly known as the Instrumentation, Systems, and Automation Society) is the dominant source for instrumentation symbology under the ISA-5.1 standard (Figure 3.27).

ISA a global, nonprofit technical society that develops standards for automation, instrumentation, control, and measurement.

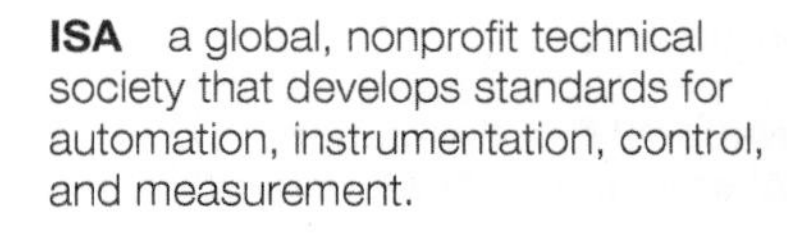

Figure 3.27 ISA logo.

CREDIT: Courtesy of ISA.

The ISA-5.1 standard is comprised of both specific symbols and a coded system built on the letters of the alphabet that depicts functionality. Although many, if not most, large companies have moved toward adopting the ISA-5.1 standard in its entirety; other preferred symbols may be kept in their inventory. Anyone who uses a drawing should not assume that the ISA-5.1 standard is used. All symbols, standard and nonstandard alike, should be identified in the legend of each drawing.

- The **American National Standards Institute (ANSI)** is an organization dedicated to supporting the competitiveness of U.S. business and quality of life by providing a framework for fair standards development. ANSI is successfully facing the standardization challenges of a global economy while addressing key issues such as safety and the environment. Although ANSI itself does not develop standards, it provides interested U.S. parties with a neutral venue to come together and work toward common agreements. The process to create these voluntary standards is guided by the Institute's cardinal principles of consensus, due process, and openness, and it depends heavily upon data gathering and compromises among a diverse range of stakeholders. ANSI accreditation is used as a baseline or "backbone" for standardization in various industries.

American National Standards Institute (ANSI) organization that oversees and coordinates the voluntary standards in the United States. ANSI develops and approves norms and guidelines that impact many business sectors. The coordination of U.S. standards with international standards allows American products to be used worldwide.

- **The American Petroleum Institute (API)** is a trade association that speaks on behalf of the petroleum industry to the public and the various government branches. The association sponsors and researches economic analyses and provides statistical indications to the public. API is a leader in the development of the petroleum and petrochemical industries for equipment and operating standards. Currently, API maintains over 700 standards and recommendations and has a certification program for the inspection of industry equipment. API also has various education programs including seminars, workshops, and conferences available to industry for ongoing education.

American Petroleum Institute (API) trade association that represents the oil and natural gas industry in the areas of advocacy, research, standards, certification, and education.

American Society of Mechanical Engineers (ASME) organization that specifies requirements and standards for pressure vessels, piping, and their fabrication.

National Electric Code (NEC) a standard that specifies electrical cable sizing requirements and installation practices.

National Fire Protection Association (NFPA) international organization that specifies fire codes including building construction codes, fire suppression systems, and firefighting capabilities required at facilities.

- **The American Society of Mechanical Engineers (ASME)** is a nonprofit organization focused on collaboration and knowledge sharing; it also provides professional development and learning opportunities for engineers across all disciplines. ASME assists in standards development across a range of topics including pressure technology, power plants, elevators, construction equipment, piping, nuclear components, and more.
- **The National Electric Code (NEC)** was established by the **National Fire Protection Association (NFPA)** to regulate electrical standards.
- The Occupational Safety and Health Administration (OSHA), a U.S. government agency, was created to establish and enforce workplace safety and health standards, conduct workplace inspections, propose penalties for noncompliance, and investigate serious workplace incidents.

Summary

There are many different types of drawings within the process industries. Each drawing type represents different aspects of the process and various levels of detail. Studying combinations of these drawings provides a more complete picture of the processes at a facility.

Process drawings provide viewers with visual descriptions and explanations of processes, equipment, and other important items in a facility. Process facilities use process drawings to assist with operations, modifications, and maintenance. The information contained within process drawings includes a legend, title block, and application block.

Examples of process drawings include block flow diagrams (BFDs), process flow diagrams (PFDs), piping and instrumentation diagrams (P&IDs), and plot plans.

Block flow diagrams (BFDs) are the simplest drawings used in the process industry. They provide a general overview of the process, but they contain few specifics. Block flow diagrams include the feed, product location, intermediate streams, recycle, and storage.

Process flow diagrams (PFDs) are basic drawings that use symbols and direction arrows to show the primary flows through a process. PFDs show the actual process and include design flow rates, temperatures, pressures, pump capacities, heat exchangers, equipment symbols, equipment designations, reactor catalyst data, cooling water flows, and symbol charts. Process streams are typically numbered for reference to material balance sheets containing stream compositions and other details.

Piping and instrumentation diagrams (P&IDs) are similar to process flow diagrams, but show more detailed process information such as equipment numbers, piping specifications, instrumentation, and other detailed information. In some cases, detailed equipment drawings may replace equipment symbols.

Utility drawings depict the utility systems found in the plant. They may include utilities such as steam, air, nitrogen, cooling water, and potable (drinking) water.

Electrical drawings are composed of at least two types: one-line (or single line) drawings that offer an overview of the whole electrical system within the plant or the unit and schematic drawings that show the actual wiring connections between the components.

Isometric drawings offer a three-dimensional perspective view. Piping isometric drawings show the routing of a line in three dimensions, making it easier for the process technician to follow in the field, or for the pipe fabricator to produce.

Plot plans are scale drawings that show the layout of equipment, units, and buildings. They are drawn to scale so that everything is of the correct relative size and shows proper dimensions.

Symbols are figures used to represent types of equipment. Examples of symbols representing many different types of equipment and instrumentation have been shown throughout the chapter. Different industry organizations develop and publish different symbology standards, so there is no universal standard followed by every company. Therefore, it is important for the process technician to be familiar with the symbols used by his or her company.

Checking Your Knowledge

1. Define the following key terms
   a. American National Standards Institute (ANSI)
   b. American Petroleum Institute (API)
   c. American Society of Mechanical Engineers (ASME)
   d. Application block
   e. Block flow diagrams (BFDs)
   f. Electrical diagrams
   g. Equipment symbols
   h. ISA
   i. Isometric drawings (Isoms)
   j. Legend
   k. National Electric Code (NEC)
   l. National Fire Protection Association (NFPA)
   m. One-line diagram
   n. Piping and instrumentation diagrams (P&IDs)
   o. Plot plans
   p. Process drawings
   q. Process flow diagrams (PFDs)
   r. Safety, health, environmental (SHE) diagrams
   s. Symbology
   t. Symbols
   u. Title block
   v. Utility flow diagrams (UFD)

2. Which of the following items are located on a process flow diagram? (Select all that apply.)
   a. Pump capacities
   b. Equipment symbols
   c. Internal mixer in a reactor
   d. Control valve

3. (*True or False*) P&IDs show less detail than PFDs regarding materials of construction, insulation, and equipment inner workings.

4. Process flow diagrams are typically read from _____.
   a. right to left
   b. left to right
   c. bottom right to top left
   d. bottom left to top right

5. A single-line diagram _____.
   a. consists of blocks representing processes or subsections of a process connected by straight lines
   b. contains symbols that represent the major pieces of equipment and piping systems used in the process
   c. shows the entire electrical system of interconnecting generators, transformers, transmission and distribution lines, and so on
   d. presents object images drawn at a 30-degree angle and may also contain cutaway views to show an object's inner workings

6. Match the following symbols to their descriptions.

Symbol	Description
I.	a. P&ID actuator hydraulic
II.	b. P&ID actuator pneumatic
III.	c. P&ID butterfly valve
IV.	d. P&ID needle valve

7. Match each of the following symbols to its correct meaning.

Symbol	Meaning
I.	a. Pump
II.	b. Heat exchanger
III.	c. Compressor
IV.	d. Trayed tower

8. The following is a valve symbol representing the _____ _____.

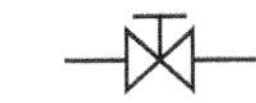

9. On the following ISA tag, what does the first letter "F" stand for'?

   FIC-202A

   CREDIT: Courtesy of ISA.

   a. Frequency
   b. Flow
   c. Force
   d. Function

10. Which society develops standards for automation, instrumentation, control, measurement symbols?
    a. API
    b. NFPA
    c. ISA
    d. ASME

NOTE: Answers to Checking Your Knowledge questions are in the Appendix.

Student Activities

1. Develop a block flow diagram for any familiar process, and present it to the class.
2. Write three to five paragraphs describing the purpose of a flow, temperature, level, and pressure control loop as found on the P&ID provided by your instructor.
3. In teams of two students, describe the location of various pieces of equipment based on their location in a drawing. (e.g., the valve is located downstream from the take—off to the reactor, but upstream of the analyzer).
4. Hand sketch a plot plan using the details provided by your instructor.
5. Complete the following chart, writing a 3 to 5 sentence description of each drawing, and how it is used:

Drawing Type	Description and Use
Block flow diagram (BFD)	
Process flow diagram (PFD)	
Piping and instrumentation diagram (P&ID)	
Plot plan	

6. Divide the students into teams for a troubleshooting competition on control loop scenarios (e.g., what happens if you lose instrument air to your control loops? Or what caused the FIC loop flow to increase or decrease?).

Chapter 4

Safety, Health, and Environmental Policy Compliance

Objectives

After completing this chapter, you will be able to:

4.1 Provide examples of safety, health, and environmental policies and their purpose. (NAPTA Operations, Equipment Maintenance: SH&E Impact 1 and Normal Operations Housekeeping and Complying with SH&E Policies 5*) p. 47

4.2 Describe the process technician's role in the execution of safety, health, and environmental policies. (NAPTA Operations, Normal Operations: Housekeeping and Complying with SH&E Policies 8) p. 49

4.3 Define *housekeeping* in process industries terms. (NAPTA Operations, Normal Operations: Housekeeping and Complying with SH&E Policies 1, 2, 3) p. 50

4.4 Describe the common types of equipment and procedures used to support safety, health, and environmental policies. (NAPTA Operations, Equipment Maintenance: SH&E Impact 2 and Normal Operations Housekeeping and Complying with SH&E Policies 6 and 7) p. 53

4.5 Describe safety and environmental hazards that safety, health, and environmental policies are utilized to mitigate. (NAPTA Operations, Equipment Maintenance: SH&E Impact 3 and Normal Operations Housekeeping and Complying with SH&E Policies 5) p. 55

*North American Process Technology Alliance (NAPTA) developed curriculum to ensure that Process Technology courses will produce knowledgeable graduates to become entry-level employees in process technology. Objectives from that curriculum are named here in abbreviated form. For example, "(NAPTA Operations, Equipment Maintenance: SH&E Impact 1 and Normal Operations: Housekeeping and Complying with SH&E Policies 5) means that this chapter's objective 1 relates to objective 1 of NAPTA's curriculum about equipment maintenance and SH&E impact and objective 5 of NAPTA's curriculum about housekeeping and safety, health, and environment (SH&E) policies during normal operations.

4.6 Provide examples of other possible hazards related to equipment maintenance. (NAPTA Operations, Equipment Maintenance: SH&E Impact 4 and 6) p. 56

Key Terms

Audio, visual, olfactory (AVO)—method used by process technicians to monitor the sounds, sights, and smells of a process unit or area during unit walkthrough inspections, **p. 49**

Blinding—the process and procedure to isolate equipment for hot work, cold work (like changing out a seal or bearings), or specific activities that require equipment removal, **p. 47**

Body harness—fall protection device worn while working at heights, **p. 53**

Bunker gear—protective clothing worn for firefighting, **p. 51**

Confined space entry (CSE)—policy that defines the process and procedure for entering confined spaces such as equipment, storage tanks, and excavations below grade, **p. 47**

Control of work (COW)—work practice that identifies the means of safely controlling maintenance, demolition, remediation, construction, operating tasks, and similar work, **p. 49**

Environmental Protection Agency (EPA)—a federal agency charged with authority to make and enforce the national environmental policy, **p. 47**

Fire retardant clothing (FRC)—wearing apparel for use in situations where there is a risk of arc, flash, or thermal burns that is regulated by NFPA-70E, ASTM, and OSHA, **p. 52**

Housekeeping—act of keeping a work area and equipment in a safe, clean, usable condition, **p. 49**

Immediately dangerous to life and health (IDLH)—condition from which serious injury or death to personnel can occur in a short amount of time, **p. 52**

Management of change (MOC)—method of managing and communicating changes to a process, changes in equipment, changes in technology, changes in personnel, or other changes that will impact the safety and health of employees, **p. 49**

National Emissions Standards for Hazardous Air Pollutants (NESHAP)—emissions standards set by the Environmental Protection Agency (EPA) for air pollutants that may cause fatalities or serious, irreversible, or incapacitating illness if not regulated, **p. 56**

Operations procedures—unit specific procedures used for the purpose of equipment and system startup, shutdown, normal operation, as well as emergency situations, **p. 49**

Resource Conservation and Recovery Act (RCRA)—primary federal law whose purpose to protect human health and the environment and to conserve natural resources. It completes this goal by regulating all aspects of hazardous waste management; generation, storage, treatment, and disposal, **p. 52**

Safety, health, and environmental (SHE) Policies—policies implemented by process facilities to minimize or prevent risks and/or hazards associated with the process industry and to ensure that the facility is in compliance with applicable regulatory agencies, **p. 47**

Self-contained breathing apparatus (SCBA)—independent breathing device worn by rescue workers, firefighters, process technicians, and others to provide breathable air in a hostile environment, **p. 50**

Turnaround (TAR)—the shutdown period for an operation unit, usually for mechanical reconditioning. The period from the end of one run to the beginning of the next, that is, the offstream to onstream period, **p. 47**

4.1 Introduction

This chapter provides an overview of various **safety, health, and environmental (SHE) policies** used within process industry facilities. SHE policies are overseen by agencies such as the following:

- The Occupational Safety and Health Administration (OSHA), a U.S. government agency created to establish and enforce workplace safety and health standards, conduct workplace inspections and propose penalties for noncompliance, and investigate serious workplace incidents
- The **Environmental Protection Agency (EPA)**, a federal agency that oversees environmental protection and pollution control

Safety, health, and environmental (SHE) Policies policies implemented by process facilities to minimize or prevent risks and/or hazards associated with the process industry and to ensure that the facility is in compliance with applicable regulatory agencies.

Environmental Protection Agency (EPA) a federal agency charged with authority to make and enforce the national environmental policy.

These regulatory authorities govern work process and procedures for process facilities and the safety rules and regulations that provide for the safety of workers within process facilities. The intention of operating within these guidelines is to ensure the safety and health of employees, the community, and the environment. In addition to these two federal agencies, individual state agencies exist to issue permits and enforce compliance with state rules and regulations.

Throughout this textbook, the term *safety, health, and environmental policy* is used to define policies and procedures developed by each process company in support of, and compliance with, the rules set forth by the various regulatory agencies. All process facilities that process hydrocarbons into various products contain similar types of chemicals and process equipment that can cause injury to personnel and damage to the environment. The policies and procedures discussed in this chapter help minimize the potential for these risks when applied. Processes and products vary across each process facility, and safety, health, and environmental policies are applicable to each process unit in much the same way.

Safety, Health, and Environmental (SHE) Policies and Practices

Good process safety management techniques will ensure that the process technicians, maintenance workers, and technical personnel are involved with the development and implementation of facility safety, health, and environmental policies. Many companies now have rules in place that, if violated, will result in the termination of employment. The following list provides an example of many of the typical safety, health, and environmental policies found in process industry facilities:

- **Blinding**—method for isolating equipment safely (Figure 4.1).

Blinding the process and procedure to isolate equipment for hot work, cold work (like changing out a seal or bearings), or specific activities that require equipment removal.

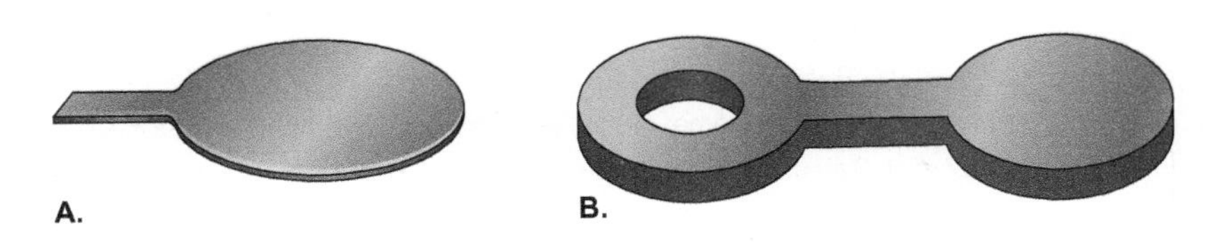

Figure 4.1 **A.** Paddle blind. **B.** Spectacle or figure-8 blind.

- **Confined space entry (CSE)**—a type of safety policy. Many companies utilize a single document that covers both CSE and hot work permits. An example of a confined space entry form is shown in Figure 4.2.
- Employee health monitoring—policy that defines the requirements for employee health monitoring while activities are conducted in hazardous areas, during hazardous chemical sampling, or where prolonged exposure to hazardous chemicals can occur, such as during **turnarounds (TARs)**.
- Equipment inspection and monitoring—policy that defines inspection and monitoring frequencies for both fixed and rotating equipment for the purpose of managing equipment reliability and mechanical integrity.

Confined space entry (CSE) policy that defines the process and procedure for entering confined spaces such as equipment, storage tanks, and excavations below grade.

Turnaround (TAR) the shutdown period for an operation unit, usually for mechanical reconditioning. The period from the end of one run to the beginning of the next, that is, the offstream to onstream period.

Figure 4.2 Example of a confined space entry form.

CONFINED SPACE ENTRY PERMIT

Confined Space Location/Description/ID Number **Date:**

Purpose of Entry

Time In: ________ **Permit Canceled Time:** ________
Time Out: **Reason Permit Canceled:** ________

Supervisor: ________

Rescue and Emergency Services

Hazards of Confined Space	Yes	No	Special Requirements	Yes	No
Oxygen deficiency			Hot work permit required		
Combustible gas/vapor			Lockout/tagout		
Combustible dust			Lines broken, capped, or blanked		
Carbon monoxide			Purge–flush and vent		
Hydrogen sulfide			Secure area–post and flag		
Toxic gas/vapor			Ventilation		
Toxic fumes			Other–list		
Skin–chemical hazards			**Special Equipment**		
Electrical hazard			Breathing apparatus–respirator		
Mechanical hazard			Escape harness required		
Engulfment hazard			Tripod emergency escape unit		
Entrapment hazard			Lifelines		
Thermal hazard			Lighting (explosion proof, low voltage)		
Slip or fall hazard			PPE–goggles, gloves, clothing etc.		
			Fire extinguisher		

Communication Procedures:

DO NOT ENTER IF PERMISSIBLE ENTRY LEVELS ARE EXCEEDED		Test Start and Stop Time: Start	Stop
	Permissible Entry Level		
% of Oxygen	19.5% to 23.5%		
% of LEL	Less than 10%		
Carbon monoxide	35 PPM (8 hr.)		
Hydrogen sulfide	10 PPM (8 hr.)		
Other			

Name(s) or person(s) testing: ________

Test Instrument(s) used–Include name, model, serial number, and date last calibrated:

CFM–ventilation	Size–cubic feet	Pre-entry time	❒ Central notified before entrance	Time notified:
			❒ Central notified after entrance	Time notified:

Authorized Entrants

Authorized Attendants

PERMIT AUTHORIZATION **I certify that all actions and conditions necessary for safe entry have been performed.**	
Name–(print):	
Signature:	
Date:	Time:

- Hot work—policy that defines the process and procedure for conducting hot work such as welding, grinding, or vehicle entry in or around process equipment.
- **Housekeeping**—policies for maintaining clean, orderly, and safe working conditions.
- Job Safety Analysis (JSA)—a procedure used to integrate accepted safety and health principles and practices into a particular task or job. In a JSA, each basic step of the job is analyzed to identify potential hazards and to recommend the safest way to do the job. A JSA is also sometimes called a job hazard analysis (JHA).
- Lockout/tagout (LOTO)—a procedure used in industry to isolate energy sources from a piece of equipment. Note: Chapter 5, *Lockout/Tagout* discusses lockout/tagout in detail.
- **Control of work (COW)**—a type of work practice focused on controlling for safety.
- **Management of change (MOC)**—method of ensuring that changes that will affect the safety and health of employees will be communicated to everyone involved.
- Material release reporting—policy that defines reporting requirements of regulatory authorities, such as the EPA or applicable state agencies, when venting, purging, or draining equipment or in the event of a material release.
- **Operations procedures**—detailed sets of steps for performing startup, shutdown, normal operations, and emergency actions safely and consistently.
- Personal protective equipment (PPE)—specialized gear that provides a barrier between hazards and the worker using the PPE.
- Process hazard analysis (PHA)—a systematic assessment of the potential hazards associated with an industrial process, taking into consideration specific hazards and locations of highest potential for exposure.
- Process safety management (PSM)—OSHA standard that contains the requirements for management of hazards associated with process using highly hazardous materials.
- Process safety information—policy that defines the type of documentation that is considered process safety information in support of the OSHA PSM regulation; including but not limited to operating procedures, inspection and maintenance procedures, operating and training material, process drawings, electrical one-line diagrams, instrument loop drawings, and electrical classification drawings.
- Vehicle entry—policy that defines the process and procedure for vehicle entry into process areas.

Housekeeping act of keeping a work area and equipment in a safe, clean, usable condition.

Control of work (COW) work practice that identifies the means of safely controlling maintenance, demolition, remediation, construction, operating tasks, and similar work.

Management of change (MOC) method of managing and communicating changes to a process, changes in equipment, changes in technology, changes in personnel, or other changes that will impact the safety and health of employees.

Operations procedures unit specific procedures used for the purpose of equipment and system startup, shutdown, normal operation, as well as emergency situations.

4.2 Process Technician's Role in Safety, Health, and Environmental Policies

Process technicians have the responsibility of knowing and understanding the inner workings of each process. This includes process technology and design criteria; process equipment and interconnecting piping, valves, and safety and control systems; as well as process specific hazards. This knowledge—along with the ability to utilize the proper safety, health, and environmental policies and applicable procedures to manage process operations—minimize the risks and hazards associated with the process industry.

Other common skills that the process technician must develop that are essential for managing process industry risks include:

- Following operations and equipment preparation procedures
- Using personal protective equipment (PPE) correctly. This equipment includes hard hats, safety glasses, goggles, protective footwear, hearing protection, protective gloves, protective clothing, and other equipment utilized to prevent exposure to, or injury from, hazardous materials or environments.
- Maintaining **audio, visual, olfactory (AVO)** equipment monitoring awareness

Audio, visual, olfactory (AVO) method used by process technicians to monitor the sounds, sights, and smells of a process unit or area during unit walkthrough inspections.

Self-contained breathing apparatus (SCBA) independent breathing device worn by rescue workers, firefighters, process technicians, and others to provide breathable air in a hostile environment.

- Staying familiar with safety/emergency equipment location, including fire and toxic gas alarms, fire extinguishers, protective clothing, fire turrets, safety showers, **self-contained breathing apparatus (SCBA)** locations, fall protection devices, and emergency egress routes. A self-contained breathing apparatus (SCBA) is shown in Figure 4.3.

Figure 4.3 Technician with self-contained breathing apparatus.

CREDIT: JeedJaad/Shutterstock.

- Staying familiar with first responder roles and responsibilities, including hazard identification, radio communications, and emergency evacuation procedures.
- Participating in the development of policies and procedures, including safety, health, and environmental policies; unit training material; operating procedures; equipment preparation procedures; and so on.

4.3 The Importance of Housekeeping

As mentioned, housekeeping defines activities that are completed in order to maintain a clean, orderly, safe facility. Housekeeping is often a combination of daily, weekly, and monthly activities.

Most process facilities or process units have housekeeping tasks that are common across the industry. Additionally, each process unit has additional "unit specific" tasks necessitated by the specific process. Maintaining a clean and orderly facility helps in the following ways:

- Minimizes or eliminates common risks to personnel
- Eliminates slipping and tripping hazards
- Eliminates potential for chemical exposure
- Eliminates environmental hazards
- Improves operations and maintenance efficiency
- Keeps tools and equipment in the proper place for quick access
- Maintains tool and equipment integrity and reliability.

Here are some common housekeeping examples:

- Monitor and maintain general area cleanliness at all times. Focus on area cleanup after maintenance and repair activities. A strong commitment to proper cleanup after the completion of every maintenance and repair activity ensures that tools and equipment are properly stored, and all trash and debris generated from the activity is removed (Figure 4.4).
- Maintain clean and orderly control rooms, offices, reference libraries, lab and sample rooms, storage and tool rooms, locker rooms, kitchens, and common areas.

Figure 4.4 Maintain general area cleanliness and proper storage at all times.

CREDIT: BigBlueStudio/Shutterstock.

- Maintain proper storage and maintenance of essential firefighting equipment, as shown in Figure 4.5, including fire hoses, hose nozzles, nozzle wrenches, extinguishers, fire turrets and turret nozzles, fire hydrants, foam addition equipment, and **bunker gear**.

Bunker gear protective clothing worn for firefighting.

Figure 4.5 Bunker gear in firefighting equipment closet.

CREDIT: Chaiwat boonsawat/Shutterstock.

- Maintain proper storage and maintenance of equipment essential for unit operations, including utility hoses (air, steam, nitrogen, water, and chemical), ladders (step ladders, extension ladders, and jack up platforms), valve wrenches, pipe wrenches, chain operators on manual isolation valves, and grease guns.
- Maintain proper storage and segregation of utility hose fittings (air, steam, nitrogen, water, and chemical) and pipe fittings (couplings, unions, tees, elbows, and pipe plugs).
- Maintain clean and orderly motor control centers (MCCs).

Here are unit specific housekeeping examples:

- Maintain lab and sample room stock with the appropriate sampling equipment, sample containers, and personal protective equipment (PPE) dictated by the unit sample schedule.
- Maintain clean and orderly specialty equipment storage areas for unit specific tools and equipment.

Resource Conservation and Recovery Act (RCRA) primary federal law whose purpose to protect human health and the environment and to conserve natural resources. It completes this goal by regulating all aspects of hazardous waste management; generation, storage, treatment, and disposal.

- Properly maintain waste storage areas, disposal bins, and containers utilized for waste hydrocarbon disposal, as defined by the **Resource Conservation and Recovery Act (RCRA).** The RCRA (pronounced "wreck-ruh") was enacted into law in 1976 to protect people, the environment, and resources from "cradle to grave."
- Remove algae from cold or hot process areas.
- Remove dust and particulates.
- Remove oil; clean up around process pumps, compressors, and lube oil consoles.

Personal protective equipment, as seen in Figure 4.6, must be used while performing various housekeeping activities. Some of the more *common* housekeeping tasks require only minimum personal protective equipment, such as the following:

Fire retardant clothing (FRC) wearing apparel for use in situations where there is a risk of arc, flash, or thermal burns that is regulated by NFPA-70E, ASTM, and OSHA.

- Fire retardant clothing, hard hat, safety glasses, steel-toed shoes. **Fire retardant clothing (FRC)** is worn by personnel while working in areas where a chance of fire or extreme heat exists.
- Hearing protection and communications radio
- Face shields, cover goggles, protective gloves, protective clothing
- Work gloves specific for the task (leather, cloth knobby, welders, and rubber).

Figure 4.6 **A.** Personal protective equipment. **B.** Double hearing protection with ear plugs and ear muffs.

CREDIT: A. Sam Oaksey/Alamy Stock Photo. **B.** Loco/Shutterstock.

A.

B.

Some facilities have process-related housekeeping tasks that may require a higher level of PPE. Such tasks could require the following:

- Half-mask organic vapor respirators and full-face organic vapor respirators. Selecting the correct organic vapor cartridge for the task is critical and based on specific exposure hazards of the task. Using the correct respirator cartridge can eliminate exposure to a wide range of hydrocarbons and particulates.
- Self-contained breathing apparatus (SCBA) or a hose-supplied air respirator to supply fresh air
- Total enclosure suit ("Moon suit") typically worn with SCBA for entry and activities that are considered **immediately dangerous to life and health (IDLH)**

Immediately dangerous to life and health (IDLH) condition from which serious injury or death to personnel can occur in a short amount of time.

- Firefighting bunker gear or heat suits designed for entry and activities where extreme heat or exposure to open flames (such as beneath an operating furnace or within a fin fan shroud) exists
- Portable hydrocarbon, oxygen (O_2), hydrogen sulfide (H_2S), and other detectors are used when a housekeeping task may expose personnel to a hazardous environment
- **Body harnesses** to maintain safety when working at heights.

Body harness fall protection device worn while working at heights.

4.4 Safety Equipment

Process units incorporate several types of safety equipment in order to mitigate emergencies, decrease environmental hazards, and increase personnel protection. One of the most common types of safety equipment is the fire extinguisher. Process technicians acting in a first responder role for extinguishing small fires can use the 30-pound dry powder fire extinguisher. The 150-pound dry powder extinguishers are used for fighting larger fires. CO_2 extinguishers are found in motor control centers and computer rooms to combat electrical fires. Motor control centers are enclosures that house the feeder breakers, motor control units, variable frequency drives, programmable controllers, and metering devices needed to supply power safely to unit equipment.

Fire hydrants, fire turrets, and fire hose reel stations are found in process units and tank farms for fire protection. They also provide cooling for equipment in the event of a large fire.

Tank farms and storage facilities may incorporate foam addition systems for applying aqueous film forming foam (AFFF), as seen in Figure 4.7. When applied correctly, AFFF will cover pools of burning hydrocarbon liquid to remove the oxygen from the fire triangle, thereby smothering the fire.

Figure 4.7 Applying aqueous film forming foam (AFFF) in a tank fire.

Deluge or sprinkler systems are used in many process applications, such as enclosed compressor buildings or congested areas that contain large quantities of equipment. Hydrocarbon detectors are also found in these types of processing areas to alert process technicians of a hazardous material release. Figure 4.8 shows an example of a flammable gas detector.

Self-contained breathing apparatus, or SCBA, is typical safety equipment found in process units that can be used to supply fresh air in the event of a hazardous material release (see Figure 4.3). The ELSA 5, 10, and 15 minute escape packs are another type of fresh air supplying device that may also be in place for emergency egress.

Safety showers and eyewash stations are strategically placed in and around process units and tank farms for emergency use in the event that personnel come in contact with chemicals. These systems are supplied with potable water so they can be safely used to flush hazardous material from the eyes or skin. Some stations are equipped with alarms to alert others of an emergency and the location. Figure 4.9 shows an example of a safety shower and an eyewash station.

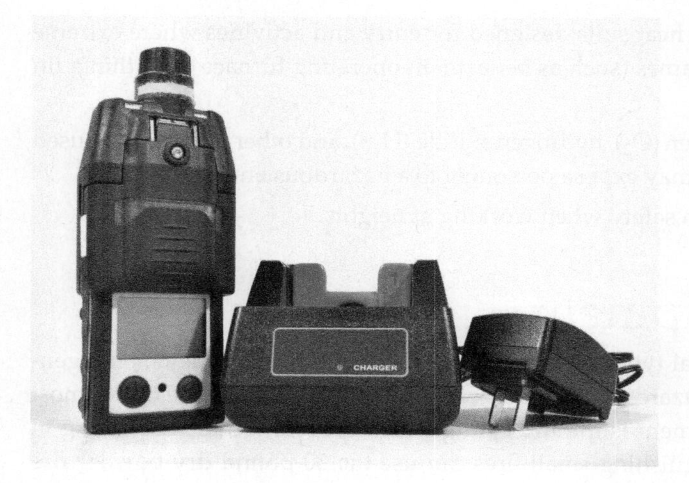

Figure 4.8 Flammable gas detector.

CREDIT: Tum ZzzzZ/Shutterstock.

Figure 4.9 A. Safety shower. B. Eyewash station.

CREDIT: A. coker/Alamy Stock Photo. B. fStop Images GmbH/Alamy Stock Photo.

A.

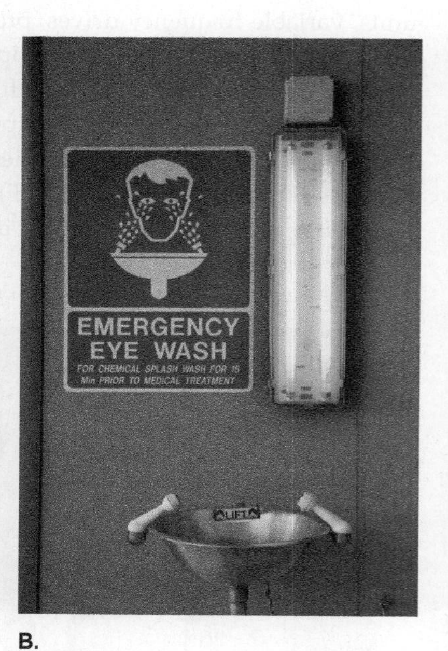

B.

Insulation on piping and equipment may also be considered safety equipment where it is installed to prevent exposure to extremely hot or cold temperatures.

Safety equipment inspection programs ensure the reliability and operability of the equipment. Process technicians are usually responsible for conducting safety equipment inspections. Inspection frequency and criteria vary by the type of equipment, the intended purpose, and, in some cases, equipment criticality, location, and exposure to the elements. For example, fire extinguishers should have fill data and seal tabs inspected on one frequency and the containers hydrostatically inspected on another frequency. (This approach also applies to SCBAs where the tank pressure, mask, and regulator condition are inspected on one frequency and the tank's hydrostatic integrity inspected on another.) Figure 4.10 shows an example of an inspection tag.

Given the complexity of deluge system operation, foam application systems, and hydrocarbon detector operation, these types of equipment are usually managed by system technicians using established preventive maintenance programs.

All safety equipment on a process unit should be inspected often. Safety showers and eyewash stations, as well as fire hydrants and fire turrets should be flushed periodically. Fire extinguishers should be refilled and replaced immediately after use. A partially used extinguisher must never be left in place.

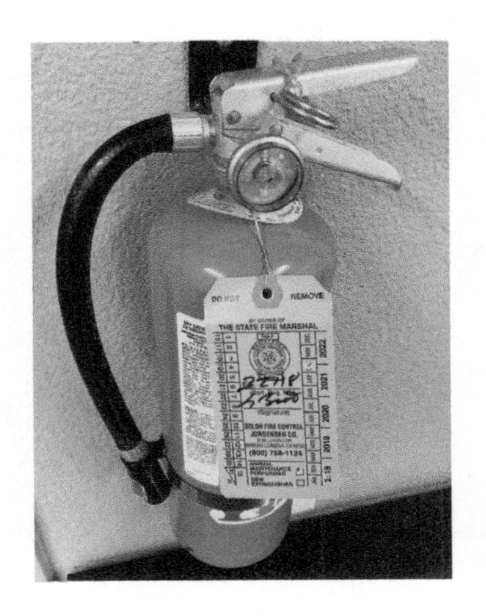

Figure 4.10 Inspection tag.
CREDIT: ZikG/Shutterstock.

The facility SHE personnel will typically conduct routine training exercises for operating personnel. During these exercises, personnel practice using safety equipment for personal protection, combating emergency situations, and using emergency egress.

4.5 Environmental Hazards

Safety, health, and environmental (SHE) policies are intended to mitigate many risks and hazards found within the process industry. Acute (short term), as well as chronic (long term) exposure to many chemicals found within the process industry can cause severe health problems. The risk of fire and explosion is always present due to the volatility of most hydrocarbons and can cause severe injury to personnel and damage to the environment and process equipment. A hydrocarbon release to the atmosphere, also known as a "material release," can affect the air quality in the entire facility and possibly surrounding communities. A material release may harm ground water, waterways, wildlife, and communities.

A material release can occur in many ways due to the vast quantities of piping and equipment within process facilities. Some of the most common are pipe flange leaks due to either improper gasket installation or rapid thermal expansion if the proper heat up or cool down procedures are not followed. Pump and compressor seal leaks are another example of how a material release can occur. Mechanical seals can leak to the atmosphere and cause a material release. Many of these seal systems are tandem seal or dry gas seal type. When damaged, they vent or leak to a closed system like a flare. Depending on the material being released, even venting to a flare system can be considered a material release. In addition, external and internal corrosion of piping and equipment can occur and lead to a release if a proper inspection program is not in place and followed.

Flare and vent systems, such as that shown in Figure 4.11, are common in process industries. They are designed to manage excessive pressure from equipment and relief valves and flare off-spec material. This includes unit emergencies, overpressure scenarios, and planned unit shutdowns or startups that may lead to excessive hydrocarbon flaring or atmospheric venting. In most cases, atmospheric releases and excessive flare venting are considered reportable material releases.

Controlled releases that occur when equipment is functioning properly are considered an environmental incident. It takes about 30 seconds to a minute for the steam control valve to catch up to the amount of gas going to the flare, so it will always smoke for about 30 seconds to a minute even when functioning properly.

Process controls that utilize flare and vent systems can also be the source of an environmental release. Unlike the simplified example of a pump or compressor seal venting to a closed system, there can be advanced control systems within processes that are intended to vent to the flare to mitigate many different scenarios, such as high pressure. Advanced controls may shut down equipment or an entire process unit in order to prevent venting and potential environmental release.

A.

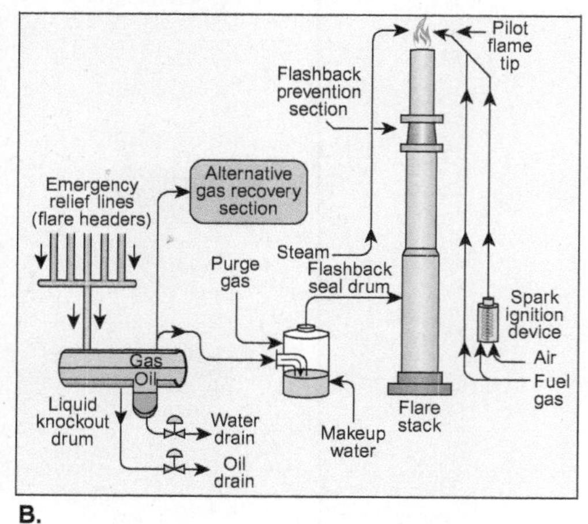

B.

Figure 4.11 Example of a flare and vent system.

CREDIT: A. photostock77/Shutterstock.

When a material release occurs, the proper agency must be notified, such as the EPA or the state environmental commission. Also, if a flare smokes for more than 5 minutes, this also must be reported to the EPA. The material released, known quantity, containment measures, and cleanup plans should be communicated to these agencies.

Safety, health, and environmental policies define procedures for material release reporting. The policies must be followed. Depending on the material and duration of release, environmental releases will need to be reported to the proper regulatory authority. Plants must follow the requirements of the **National Emissions Standards for Hazardous Air Pollutants (NESHAP)**.

National Emissions Standards for Hazardous Air Pollutants (NESHAP) emissions standards set by the Environmental Protection Agency (EPA) for air pollutants that may cause fatalities or serious, irreversible, or incapacitating illness if not regulated.

4.6 Other Potential Hazards

Isolating, de-energizing, draining, and purging process equipment for maintenance, inspection, or repair can also be the cause of environmental release. Many of the facility SHE policies address these activities due to their nonroutine nature.

Injury, illness, death, and damage to the environment are preventable by following policies and procedures in place for safely managing hazards. The following are examples of process unit activities and associated hazards that may occur without the use of proper safety, health, and environmental policies; energy isolation policy; and/or correct operation/maintenance procedure.

- *Confined space entry* (permit required)—exposure to hazardous materials, nitrogen, or oxygen-deficient atmosphere
- *Lockout/tagout and permitting for energy isolation*—exposure to hazardous materials, high voltage, material release
- *Hot work*—fire and explosion
- *Vehicle entry*—fire and explosion
- *Sampling*—hazardous material exposure

Improper equipment identification leading to performing repair or maintenance on the wrong piece of equipment is preventable by following policies and procedures. Policy and procedures identify the required safety measures to minimize the hazards surrounding equipment maintenance. The following are examples of such types of policy and procedure.

- Proper communication between process technicians to ensure that the correct equipment is selected for shutdown and isolation in preparation for maintenance. This is extremely important where the shutdown sequence for specific equipment is critical to other parts of the process that are intended to remain in operation due to the complex integration of equipment in most process units.

- Proper procedures to isolate equipment and energy sources in a safe, environmentally friendly manner. These isolation procedures typically include steps for draining or purging to containment or flare systems.
- Proper procedures to confirm that equipment maintenance is permitted and work executed in a safe manner as various craftspeople prepare the equipment, perform maintenance, and return the equipment to service. This also includes fall protection and height protection such as barricades.
- Proper use of PPE, including organic vapor respirators, impervious gloves and clothing, face shields, heat or flash suits worn while working near high voltage, and fall protection worn while working at heights (Figure 4.12).

Figure 4.12 Proper personal protective equipment, such as this flash suit, can save lives.

CREDIT: Mohd Nasri Bin Mohd Zain/ Shutterstock.

- Proper response to emergencies. Emergency response procedures define roles and responsibilities for individuals and groups for dealing with a medical emergency, fire/explosion, or material release. It is common that the process technician is the first responder in these events.
- Proper methods for the disposal of waste materials, as well as mitigation steps required in the event of a spill or material release.

Isolation Scenario

The following is a list of the primary SHE policies that direct the actions taken when preparing process equipment for removal and replacement:

- *Personal protective equipment*—policy that defines equipment that must be worn by personnel when working in process areas or when conducting specific activities such as sampling hazardous materials, entering hazardous areas, or opening process equipment
- *Operations procedures*—unit specific procedures used for the purpose of equipment and system startup or shutdown and normal operation, as well as emergency situations
- *Lockout/tagout*—procedure used in industry to isolate energy sources from a piece of equipment
- *Blinding*—policy that defines the process and procedure for equipment isolation for hot work, cold work (like changing out a seal or bearings), or specific activities that require equipment removal
- *Housekeeping*—policy that defines activities that must be completed in order to maintain the facility in a clean, orderly, and safe condition
- *Material release reporting*—policy that defines the reporting requirements of regulatory authorities such as the EPA and state environmental commission when venting or purging equipment, or in the event of a material release.

Routine Maintenance and Inspection

The following routine maintenance and inspection activities are also considered vital to successful SHE practices:

- External and internal inspection of equipment and piping in high-pressure, hazardous, or corrosive material service
- Preventive maintenance program for inspection/repair/replacement of insulation throughout a process unit
- Inspection of pressure relief devices at regular intervals.

Summary

Thorough safety, health, and environmental policies and procedures, coupled with the ability of the process technician to understand and properly execute these policies and procedures, are major controlling factors for minimizing and preventing the risks and/or hazards within the process industry. Safety, health, and environmental policies are in place to ensure that the facility is operated within strict guidelines provided by applicable regulatory agencies such as OSHA and the EPA. Process facilities would not be given a license to operate without a safety, health, and environmental policy in place to define work practices intended to ensure the safety and health of employees, the community, and the environment.

Safety, health, and environmental policies and procedures are used throughout the process industry, so it is important for a process technician to have a clear understanding of the policies and their intended purpose, as well as when and how they are applied. The process technician's role in safety, health, and environmental policy development and execution is equally important to their primary responsibility of understanding of the inner workings of each process.

The ability to access and use all available reference material is also an important skill needed by the process technician in order to be able to respond quickly and effectively in a potentially hazardous situation.

Most activities in a process facility are governed by a specific policy or procedure. This applies especially to activities that take place infrequently because the effect of a poor decision or work practice can cause injury, illness, or damage to the community and environment.

Checking Your Knowledge

1. Define the following terms:
   a. Audio, visual, olfactory (AVO)
   b. Blinding
   c. Body harness
   d. Bunker gear
   e. Control of work (COW)
   f. Confined space entry (CSE)
   g. Environmental Protection Agency (EPA)
   h. Fire retardant clothing (FRC)
   i. Housekeeping
   j. Immediately dangerous to life and health (IDLH)
   k. Management of change (MOC)
   l. National Emissions Standards for Hazardous Air Pollutants (NESHAP)
   m. Operations procedures
   n. Resource Conservation and Recovery Act (RCRA)
   o. Safety, health, and environmental (SHE) Policies
   p. Self-contained breathing apparatus (SCBA)
   q. Turnaround (TAR)
2. Which policies are most likely to vary from one company to another in the US?
   a. DHS
   b. SHE
   c. EPA
   d. OSHA
3. Many companies use a single document to cover both confined space entry (CSE) and __________ permits.
   a. Process hazard analysis (PHA)
   b. Process safety management (PSM)
   c. Hot work
   d. Turnaround (TAR)

4. Process technicians have a primary responsibility to know and understand the inner workings of each process in order to minimize the ___________ and ___________ associated with the process industry.
5. What is one of the key control points to facilitate a good overall housekeeping program?
   a. Thorough cleaning and organizing in preparation for customer visits
   b. Area cleanup after maintenance and repair activities
   c. Maintaining lab and sample room stock
   d. Maintaining clean and orderly specialty equipment storage areas
6. Which is one of the most common types of safety equipment found within a processing unit?
   a. 30-pound dry powder fire extinguishers
   b. 40-pound dry powder fire extinguisher
   c. 75-pound dry powder extinguishers
   d. 90-pound dry powder extinguishers
7. Which of these pieces of firefighting equipment would be used to extinguish a large fire in a tank farm storage facility?
   a. CO2 extinguisher
   b. Fire blanket
   c. Dry powder fire extinguisher
   d. A foam addition system for applying AFFF (aqueous film forming foam)
8. Excessive hydrocarbon venting to the atmosphere can be caused by: (Select all that apply)
   a. Blinding
   b. Overpressure scenarios
   c. Maintaining audio, visual, olfactory (AVO) equipment
   d. Planned unit startups
9. When a material release is reported, information to be given includes ___________: (Select all that apply.)
   a. material released
   b. known quantity
   c. containment measures
   d. cleanup plans
10. Match each of these key activities with the correct description.

Key activity	Description
I. Confined space entry	a. Exposure to hazardous materials, high voltage, material release
II. Lockout/tagout	b. Exposure to hazardous materials, nitrogen, or oxygen-deficient atmosphere
III. Sampling	c. Fire and explosion
IV. Vehicle entry	d. Hazardous material exposure

11. (*True or False*) The ability to access and use all available reference material is an important skill for the process technician to have in order to respond quickly and effectively in a potentially hazardous situation.

NOTE: Answers to Checking Your Knowledge questions are in the Appendix.

Student Activities

1. Select a piece of process equipment and list the primary SHE policies that define how to:
   - isolate the equipment from the process for safe shutdown.
   - define the PPE that must be worn when opening process equipment.
   - LOTO for equipment and energy isolation.
   - satisfy the necessary reporting requirements when venting/purging/draining equipment.
   - control the work activities surrounding the process equipment while maintenance is performed.
   - return the equipment to service.
   - clean up.
2. Conduct an online search of OSHA's 29 CFR 1910.119 Process Safety Management of Highly Hazardous Chemicals and write a two- to three-page report detailing the requirements of management of change and why they are important to process technicians.
3. Fill out a job safety analysis (JSA) for a tire change.
4. Write a short paragraph on the PPE that would be needed to inspect equipment.
5. Take a tour of the school building, and list all safety, health, and environmental risks or hazards that can be found and discuss ways to minimize these risks or hazards.
6. Write a one-page paper on why maintenance on equipment poses additional risks and hazards.
7. Write a one-page paper on why SH&E policies and procedures are helpful in minimizing or preventing risks.
8. Conduct an inspection on a piece of safety equipment.
9. Write a one-page paper describing how information found on an SDS is used when writing a permit.
10. In small groups of students, decide how to prepare a vessel for CSE. Discuss your methods in the class.

Chapter 5

Lockout/Tagout

Objectives

After completing this chapter, you will be able to:

5.1 Discuss the Occupational Safety and Health Administration (OSHA) standard for Control of Hazardous Energy (Lockout/Tagout). (NAPTA Operations, Normal Startup; Removal of Energy Isolation Devices 1*) p. 61

5.2 List the types of energy and various methods and devices that can be used to isolate equipment from the various types of energy:

- Lock
- Line break
- Tag
- Disconnect
- Blind
- Double block and bleed
- Switchgear. (NAPTA Operations, Normal Startup; Removal of Energy Isolation Devices 2, 3) p. 64

5.3 Describe the steps to follow when removing lockout/tagout devices. (NAPTA Operations, Normal Startup; Removal of Energy Isolation Devices 5, 6) p. 69

Key Terms

Affected employee—process technician or other employee whose job requirement is to operate or use a machine or piece of equipment that is being serviced or maintained under lockout or tagout conditions, or whose job requires them to work in an area in which servicing or maintenance is being performed, **p. 64**

*North American Process Technology Alliance (NAPTA) developed curriculum to ensure that Process Technology courses will produce knowledgeable graduates to become entry-level employees in process technology. Objectives from that curriculum are named here in abbreviated form. For example, "(NAPTA Operations, Normal Startup: Removal of Energy Isolation Devices 1)" means that this chapter's objective 1 relates to objective 1 of NAPTA's curriculum regarding removal of energy isolation devices during a normal startup.

Authorized employee—process technician or other employee who locks out or tags out a piece of equipment for required service or maintenance on that particular piece of equipment, **p. 63**

Capable of being locked out—a piece of equipment that has a multiple padlock attachment or other means of attachment to which, or through which, multiple locks can be affixed, **p. 62**

Energized—connected to an energy source; containing residual or stored energy, **p. 61**

Energy-isolating device—mechanical device that physically prevents the transmission or release of energy, **p. 61**

Energy sources—any source of electrical, mechanical, hydraulic, pneumatic, chemical, thermal, or other energy, **p. 61**

Interim test—test of equipment requiring removal of lockout/tagout devices prior to completion of maintenance or repair of equipment, **p. 70**

Lockbox—safety device ensuring no lockout/tagout (LOTO) devices are removed while work is performed. Lockboxes have multiple locks into which all keys and/or tags from the LOTO devices securing the equipment are inserted, and a single authorized employee using a LOTO device and a job lock during multishift operations then secures the box, **p. 64**

Lockout—a safety term used to describe the isolation of equipment for maintenance; a federally mandated safety precaution, **p. 61**

Lockout device—a device that utilizes a positive means such as keyed or combination type lock to hold equipment in a zero-energy state, **p. 64**

Multiple padlock attachment—clamp-like device used to install multiple locks on a lockout device, **p. 62**

Tagout—procedure of tagging valves, breakers, and so on, in preparation of equipment for maintenance; placement of a tag to indicate that the energy-isolating device and the equipment being controlled may not be operated until the tagout device is removed, **p. 62**

Tagout device—prominent warning device, such as a tag and a means of attachment, which can be securely fastened to an energy-isolating device in accordance with an established procedure, to indicate that the energy-isolating device and the equipment being controlled may not be operated until the tagout device is removed, **p. 65**

Zero-energy state—the state of equipment following specific process isolation and clearing procedures, followed by isolating all hazardous energy sources using lockout/tagout devices, **p. 61**

5.1 Introduction

The control of hazardous energy is defined in the Occupational Safety and Health Administration (OSHA) Standard 29 CFR.1910.147 and is commonly referred to as *lockout/tagout (LOTO)*. Lockout/tagout is a procedure used in industry to isolate energy sources from a piece of equipment. It is used when there is a need to inspect, repair, or replace process equipment. LOTO provides a mechanism to ensure that equipment that is to be worked on is and will remain in a **zero-energy state** prior to and during the repair process. **Lockout** uses an **energy-isolating device** to put the equipment into a zero-energy state.

In order to prevent injury, this OSHA standard requires employers to establish a program utilizing procedures for affixing the appropriate lockout or tagout devices to energy-isolating devices, and otherwise disable machines or equipment to prevent the equipment from being re-energized, being started up, or having an unexpected release of stored energy.

The employer should provide specific guidelines to take an **energized** device and isolate, de-energize, and secure the hazardous **energy sources**. The employer should have specific lockout/tagout procedures for the process technician to shut down and prepare the equipment for maintenance, perform LOTO, remove LOTO, and return the equipment to service.

Zero-energy state the state of equipment following specific process isolation and clearing procedures, followed by isolating all hazardous energy sources using lockout/tagout devices.

Lockout a safety term used to describe the isolation of equipment for maintenance; a federally mandated safety precaution.

Energy-isolating device mechanical device that physically prevents the transmission or release of energy.

Energized connected to an energy source; containing residual or stored energy.

Energy sources any source of electrical, mechanical, hydraulic, pneumatic, chemical, thermal, or other energy.

All hazardous energy sources of a machine, equipment, or piping system must be isolated, deenergized, secured, and verified in a safe position before and during service or maintenance activities. Attempting to remove locks or tags, operating, or otherwise tampering with equipment under lockout is prohibited.

Lockout/Tagout

Capable of being locked out a piece of equipment that has a multiple padlock attachment or other means of attachment to which, or through which, multiple locks can be affixed.

Multiple padlock attachment clamp-like device used to install multiple locks on a lockout device.

Lockout/tagout (LOTO) is a safety procedure that is used in many industries to ensure that equipment, lines, vessels, and machines are properly isolated, deenergized, and remain in a zero-energy state while maintenance or service work is occurring. The procedure also requires that hazardous power sources be isolated, and the equipment made inoperable before any repair procedure is started. LOTO is performed on equipment that is **capable of being locked out** and works in conjunction with a lock, usually locking the device or the power source with a **multiple padlock attachment** and placing it in such a position that no hazardous power sources can be turned on. The procedure requires that a tag be affixed, as shown in Figure 5.1, to the locked device, indicating who locked out the equipment and stating that it should not be turned on.

Figure 5.1 Lockout/tagout device.

CREDIT: Robeo/Panther Media GmbH/ Alamy Stock Photo.

Tagout procedure of tagging valves, breakers, and so on, in preparation of equipment for maintenance; placement of a tag to indicate that the energy-isolating device and the equipment being controlled may not be operated until the tagout device is removed.

In circumstances when equipment design prevents affixing a lock or multiple padlock attachment (or other device) for energy isolation or when the equipment has a locking mechanism built into it, employers may utilize a **tagout** system only. Examples of facilities that would use a tagout system are those with equipment predating the 1990s. However, whenever the equipment undergoes a major repair, renovation, or modification, the newly installed components are required to be designed to accept a lockout device.

If there is no valve to lock or tag out, flanges are disconnected and a blind is placed in the line between them to keep fluids away from the equipment being locked out. The blind is placed on the process (energized) side of the pipe to isolate the piece of equipment that is broken open or temporarily removed to be worked on (blinding is discussed later). This procedure may be called *breaking the line.*

OSHA policy CFR 1910.147, The Control of Hazardous Energy (Lockout/Tagout), establishes the minimum performance requirements for the control of such hazardous energy. Each process facility has developed its own policies to ensure compliance with the standard, and each facility must provide training to its employees to ensure understanding and compliance with the standard.

The lockout/tagout process starts with a prescribed procedure. The procedure is intended to assist the process technician in preparing the equipment for maintenance or servicing, getting the equipment to a zero-energy state, performing LOTO, and then, returning the equipment to service after the completion of work. Figure 5.2 shows an example of a generic lockout/tagout procedure.

P-101A Pump Lockout/Tagout	Time	Tech
Lock Box No. ____________		
1. Start up P-101B		
2. Allow flow to stabilize, then shutdown P-101A		
3. **Close** and **Lock** P-101A 4" discharge valve; tag do not operate		
4. **Close** and **Lock** P-101A 4" suction; tag do not operate		
5. **Close** and **Lock** ¾" seal oil supply valve; tag do not operate		
6. **Close** and **Lock** ¾" seal oil return valve; tag do not operate		
7. **De-energize** and **Lock** P-101A 480V breaker		
8. Depress the start button on the start/stop button on P-101A and check to see if pump starts. Tag start button do not operate.		
9. **Open** the ¾" case drain valve on P-101A to the closed drain system		
10. **Open** the ¾" checkvalve bypass valve on P-101 discharge; tag do not operate		
11. Install a nitrogen fitting and hose on the high point bleeder on P-101A discharge line.		
12. **Open** the ¾" high point bleeder valve and start a nitrogen purge to the closed drain system.		
13. Purge P-101A to the closed drain system for 20 min.		
14. **Close** the ¾" high point bleeder valve; close and depressurize the nitrogen hose.		
15. **Close** the ¾" case drain valve on P-101A to the closed drain system		
16. **Open** the ¾" suction bleeder valve and check for liquid. If pump and lines are dry, allow maintenance to perform scheduled work on P-101A; otherwise repeat steps 7-13 until pump and lines are dry. Tag bleeder valve do not operate.		
Process Technician Verifying Pump is Energy Free		
Maintenance Technician Verifying Pump is Energy Free		

Figure 5.2 Example of a section of a generic lockout/tagout procedure.

A LOTO procedure contains the following sections:

- Procedure unit identification nomenclature
- Procedure number
- Procedure name
- Purpose
- Lists of potential hazards associated with the equipment
- Instructions related to safety items
- Instruction related to handling deviations in the procedure
- Prerequisites
- Signature area to record those working on the procedure
- Notes or definitions related to the procedure
- Overview
- Detailed steps in taking the equipment out of service, preparing for maintenance, tagging and locking out
- Signature area for completion
- Area for comments
- Signatures of reviewers and procedure approver

The intent of the LOTO procedure is to ensure that an **authorized employee** refers to the company procedure to identify the type and magnitude of energy that the machine or equipment utilizes, understands the hazards, and knows the methods to control the energy. In some facilities, all process technicians have received the required training and may serve

Authorized employee process technician or other employee who locks out or tags out a piece of equipment for required service or maintenance on that particular piece of equipment.

as authorized employees, when needed. In other facilities, the authorized employee is designated by the employer based on knowledge and training on the company LOTO program and the unit process. The LOTO procedure also allows **affected employees** working in or near the area to be aware that work is taking place.

Affected employee process technician or other employee whose job requirement is to operate or use a machine or piece of equipment that is being serviced or maintained under lockout or tagout conditions, or whose job requires them to work in an area in which servicing or maintenance is being performed.

5.2 Types of Energy Requiring Isolation

Several types of energy must be isolated and purged from the system prior to allowing work to begin. These include the following:

- *Mechanical*—sum of kinetic and potential energy of a mechanical system; for example, a compressor or pump
- *Pneumatic*—power transmission system that uses the force of flowing gases to transmit power; for example, an industrial complex that produces its own instrument and plant air systems through compression
- *Hydraulic*—system that uses pressurized fluids as a means of generation, control, and transmission of power; for example, mobile cranes
- *Electrical*—energy made available by the flow of electrons through a conductor; for example, motors
- *Chemical*—energy stored in chemicals (compounds) and energy released or absorbed in chemical reactions; for example, flammables or corrosives
- *Thermal*—excessive heat or cold that cause injury; for example, heat tracing
- *Nuclear*—devices that emit ionizing radiation; for example, x-ray level measuring devices

Lockout and Isolating Devices

Locks and Lockboxes

Lockout device a device that utilizes a positive means such as keyed or combination type lock to hold equipment in a zero-energy state.

Lockbox safety device ensuring no lockout/tagout (LOTO) devices are removed while work is performed. Lockboxes have multiple locks into which all keys and/or tags from the LOTO devices securing the equipment are inserted, and a single authorized employee using a LOTO device and a job lock during multishift operations then secures the box.

A **lockout device** is used to hold equipment in a zero-energy state. Lockboxes have multiple locks into which all keys and/or tags from the LOTO devices securing the equipment are inserted. A single authorized employee using a LOTO device and a job lock during multishift operations then secures the box. Much of the equipment today, such as valves, start/stop stations, and electrical switchgear, has been designed with a built-in locking mechanism. When multiple locks are needed, a multiple padlock attachment, shown in Figure 5.3, is placed into the locking mechanism and then locks are attached. Keyed or combination locks are stored in the **lockbox** when not in use.

Figure 5.3 Example of a lock and multiple padlock attachments.

CREDIT: SocoXbreed/Shutterstock.

Lockboxes are used in situations involving a large number of workers and equipment and have advantages over using only a multiple padlock attachment. Lockboxes, like the one shown in Figure 5.4, contain keyed alike locks and are used when locking out larger pieces of equipment. The number of locks placed inside a lockbox varies and can be up to 50. Generally, for larger systems requiring more than 50 locks, a second lockbox can be used, and the lockboxes are connected together.

Figure 5.4 Example of a lockbox.

CREDIT: Somnuek saelim/Shutterstock.

When two or more contractors are working on different parts of a larger overall system, the locked-out device is first secured with a multiple padlock attachment that has many holes for multiple people attaching locks. Lockboxes and multiple padlock attachments can be used for contractors in this situation and can greatly reduce the number of locks that have to be used. Generally, the operations lock is placed first and removed last. Each subcontractor applies his or her own lock to the clamp. The lockout device cannot be removed until all workers have signed off on their portion of the work and removed their lock from the multiple padlock attachment.

Generally, the locks are different in color, such as red, blue, or orange, or the shape/size is used to designate different crafts, functions, and contractors of a facility. Group locks may be used in conjunction with personal locks. No two keys or locks should ever be identical when utilizing personal locks. A personal lock and tag must not be removed by anyone other than the individual who installed the lock and tag, unless removal is accomplished under the direction of the employer. Employer procedures and training for such removal must be developed, documented, and incorporated into the energy control program.

Chains/Cables

Chains and locking cables are also utilized as lockout devices. Chains are generally utilized with nonlocking valves and are placed through the valve handle and secured with a lock. Figure 5.5 shows an example of using chains to secure a valve. Locking cables are used to secure multiple valves with one lock, such as the valves to or from a seal oil pot with a valve leading to the flare.

Tags

The control of hazardous energy standard defines the requirement for **tagout devices**. Tags are to be substantial enough to prevent inadvertent or accidental removal, be attachable by hand, have the strength of no less than 50 pounds, and be weather resistant. Tags can be affixed using a variety of methods, but for safety purposes they are generally affixed

Tagout device prominent warning device, such as a tag and a means of attachment, which can be securely fastened to an energy-isolating device in accordance with an established procedure, to indicate that the energy-isolating device and the equipment being controlled may not be operated until the tagout device is removed.

Figure 5.5 Example of using chains to secure a valve.

CREDIT: audioscience/Shutterstock.

using a cable tie. Some of the tag types currently in use in the industry include the following:

- DO NOT OPEN
- DO NOT CLOSE
- DO NOT ENERGIZE
- DO NOT START
- DO NOT OPERATE.

The process technician may be required to sign his or her name to the appropriate tag (Figure 5.6). In a maintenance situation, the craftsperson completing the maintenance may also sign the tag.

Figure 5.6 Example of tagout tags.

CREDIT: SocoXbreed/Shutterstock.

Blinds

Blinds are installed as energy-isolating devices, a mechanical device that physically prevents the transmission or release of energy, but in certain instances can be considered lockout devices. Blinds cannot be installed in systems that are still energized. Blinds come in different sizes, ratings, and styles. The most common style used in the process industry is the figure 8 blind (Figure 5.7). This is sometimes called a spectacle blind for its resemblance to

Figure 5.7 **A.** Illustration of a figure 8 or spectacle blind. **B.** Photo of a spectacle blind.

CREDIT: B. Kingtony/Shutterstock.

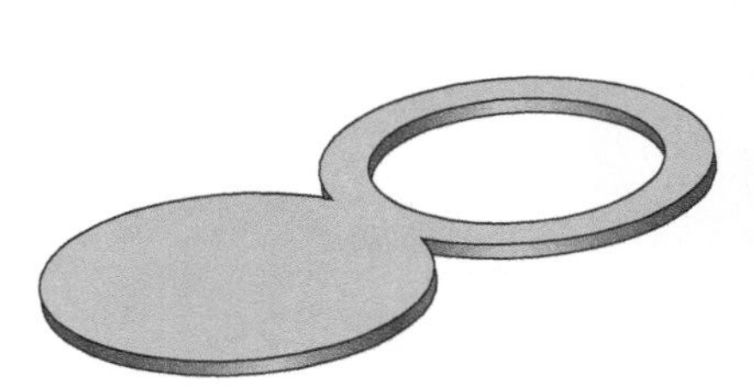

A.

B.

reading glasses. These blinds are always in the piping but are rotated 180 degrees during a shutdown or outage for safety reasons to prevent the possibility of flow through the pipe.

In addition to the figure 8 blind, there are paddle blinds (shown in Figure 5.8), also known as *T-handle blinds, line blinds,* and *pancake* blinds. These blinds are only installed during a shutdown or turnaround and are removed when equipment is prepared for restart.

Figure 5.8 **A.** Illustration of a paddle blind. **B.** A storage rack of paddle blinds for use during turnaround.

CREDIT: B. Red_Shadow/Shutterstock.

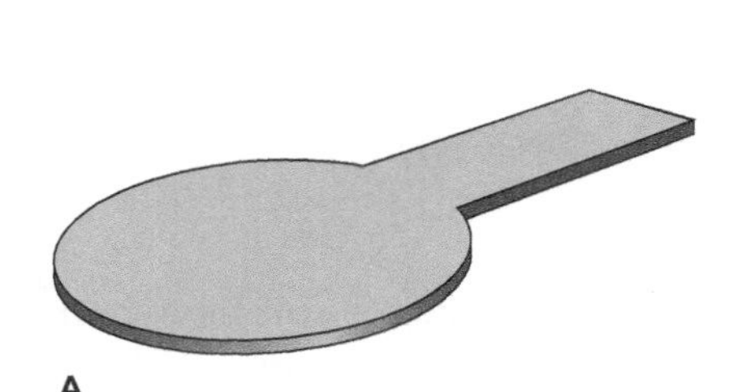

A.

B.

Blind flanges (Figure 5.9) are typically used at the end of a pipe to cap it off. Sometimes, during a turnaround for example, pipes need to be opened for inspection or to connect temporary piping for cleaning. Blind flanges are in place during operation and only removed when the associated equipment or unit is shutdown. These are sometimes called *blank flanges.*

Did You Know?

Even though handles protruding at the flange from an orifice plate and a paddle blind look very similar, the orifice plate handle will have a hole drilled through it, and there will be information etched into the handle displaying pipe size, orifice bore hole size, orifice composite information, and which side of the orifice points upstream.

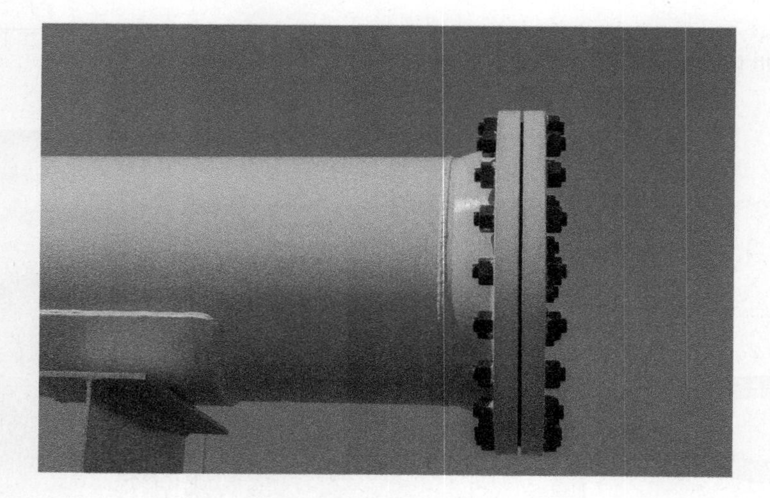

Figure 5.9 Blind flange.

CREDIT: Yama Tamito/Shutterstock.

A blind is considered a lockout device when it is the primary means of isolation of a system. If a LOTO requires a larger system to be isolated from a smaller system, then a line break is used, requiring a blind to be installed at the break location. The blind is then locked into place as shown in Figure 5.10.

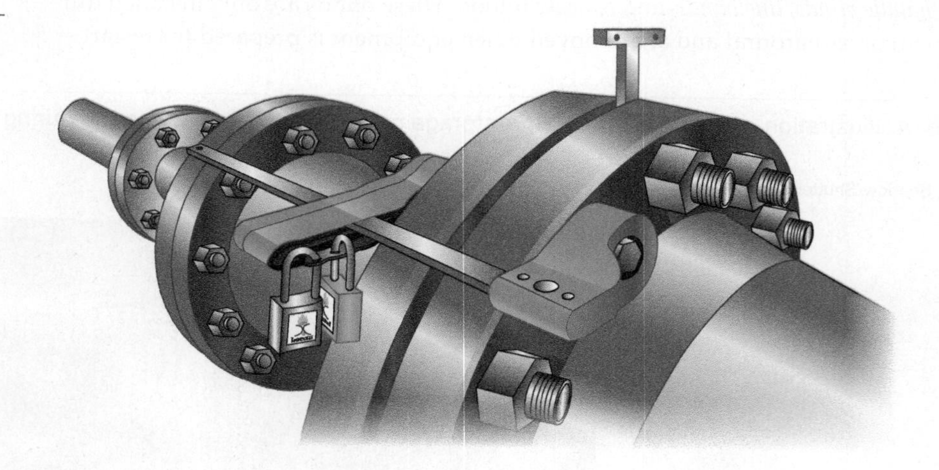

Figure 5.10 Example of a locked blind.

Double Blocks and Bleeds

Double blocks and bleeds consist of a valve configuration in which a full-flow vent valve is installed in a pipeline between two shutoff valves to provide a means of relieving pressure between them. Double blocks and bleeds, in the simplest configuration, consist of two gate valves on either end of a short run of pipe with a bleeder located on the pipe run. The valves are locked and tagged DO NOT OPEN or DO NOT OPERATE, with the bleeder valve in the open position tagged DO NOT CLOSE or DO NOT OPERATE. These configurations with the approval of the safety department, can sometimes be used in place of a blind. The simplest configuration of a double block and bleed is shown in Figure 5.11.

The double block and bleed is frequently used when working on analyzer systems. It may also be used when isolating and working on nonhazardous processes such as firewater and potable water systems. Double blocks and bleeds may be utilized on hazardous systems if certain criteria have been met.

Figure 5.11 **A.** Symbols for double block and bleed valves. **B.** Illustration of a double block and bleed.

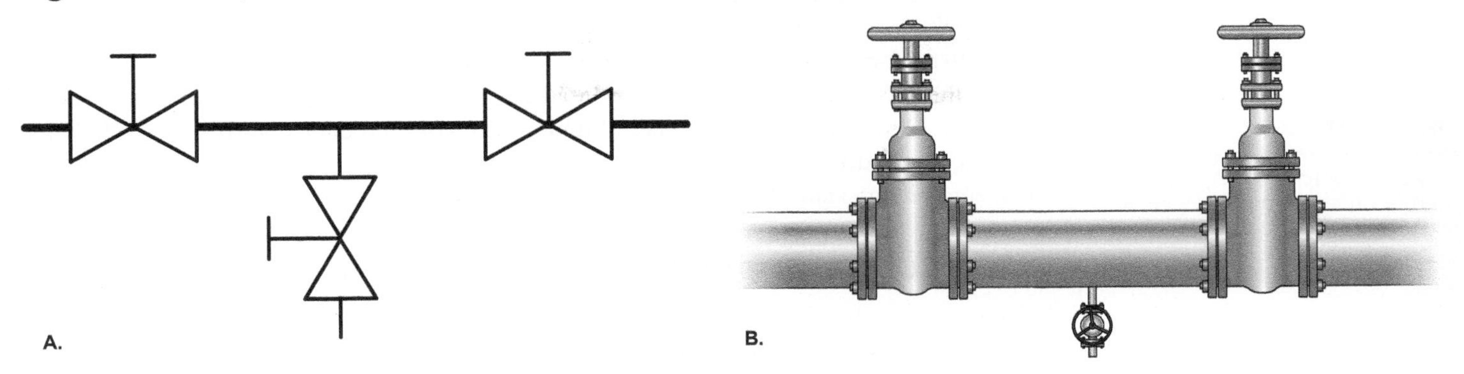

Switchgear

The electrical components of equipment must also be locked and tagged. For example, to LOTO a pump, the main breaker is de-energized and tagged DO NOT OPERATE with a multiple padlock attachment and lock placed through the eyelet of the breaker handle. The process technician then returns to the pump in the field and depresses the start button to ensure that the electrical components are de-energized. Figure 5.12 shows an example of a typical lockout/tagout for a pump breaker.

Figure 5.12 Example of typical lockout/tagout for a pump breaker.

CREDIT: Tap10/Shutterstock.

The process facility should issue specific requirements about who energizes and de-energizes electrical equipment. Some facilities allow the process technicians to energize and de-energize 480V breakers and below, whereas others require a qualified electrician to perform this function.

5.3 Removing Lockout/Tagout Devices

Lockout/tagout devices must only be removed by authorized employees. If work is finished prior to the end of the shift, the process technician locking and tagging the equipment removes the lockout/tagout devices and places the equipment in service. However, some maintenance jobs take longer and continue through shift change. In these cases, the oncoming process technician becomes the employee authorized to remove LOTO devices when the work is completed. This transfer of authority is given at shift change when passing on unit information from one technician to the other.

There are also times when equipment must be unlocked prior to work completion in order to perform an equipment test. An example would be to check for motor rotation. Only the authorized process technician, or the immediate supervisor, may remove any locks to permit the **interim test** to take place. Immediately after the interim test is concluded, and prior to any additional work commencing, the process technician must return the equipment to a zero-energy state and relock the equipment.

Interim test test of equipment requiring removal of lockout/tagout devices prior to completion of maintenance or repair of equipment.

Steps for Removing Lockout/Tagout Devices

The process technician should always refer to the facility LOTO procedure before starting the LOTO removal process. The following basic steps should be completed prior to returning the equipment to active service:

1. Ensure that the equipment has been returned to normal operating condition after maintenance, service, or repair, by confirming the following:
   a. All blinds or blind flanges have been removed.
   b. All piping has been reinstalled.
   c. All guards or protective guards are reinstalled.
   d. All fuses have been replaced, if necessary.
   e. All maintenance locks or group locks have been removed.
2. Inspect the area for housekeeping; all nonessential items should be removed.
3. Remove locks and tags from valves, control panels, or other equipment.
4. Close bleeder valves and remove tags.
5. Perform pressure tests of equipment, if necessary.
6. Notify affected employees of impending startup.
7. Remove locks and tags from the circuit breakers.
8. Startup equipment.

Once the equipment is online, the process technician should check the equipment periodically for line leaks, unusual noises, higher-than-normal temperatures, or high vibration.

Summary

The control of hazardous energy is critical in the process industry. The process facility policies are guides for the process technician to ensure safety and compliance with OSHA regulation 1910.147. The process technician is the first line of defense in the prevention of an incident or accident stemming from an uncontrolled release of hazardous energy.

Training is essential in the establishment and continued success of a control of hazardous energy program. Training for the facility control of hazardous energy program should occur as employees are hired, and retraining should occur as policies or equipment changes are implemented.

The lockout/tagout procedure is the process technician's guide to controlling hazardous energy. The process technician should strictly follow each step of the LOTO procedure created by the facility, and any deviations to the procedure should be approved and documented.

Checking Your Knowledge

1. Define the following terms:
   a. Affected employee
   b. Authorized employee
   c. Capable of being locked out
   d. Energized
   e. Energy-isolating device
   f. Energy sources
   g. Interim test
   h. Lockbox
   i. Lockout
   j. Lockout device
   k. Multiple padlock attachment
   l. Tagout
   m. Tagout device
   n. Zero-energy state
2. Which of the following is NOT a section that an operations lockout/tagout procedure should contain?
   a. Procedure unit identification nomenclature
   b. Purpose
   c. Signature area to record those working on the procedure
   d. The combination for each lock
3. OSHA policy CFR 1910.147 establishes the minimum performance requirements for _____.
   a. factory floor housekeeping
   b. the control of hazardous energy
   c. employee records retention
   d. pressure testing of equipment
4. Securing nonlocking valves in a lockout/tagout (LOTO) requires the use of _____ or _____.
5. Which of these are tags currently used in industry. (Select all that apply.)
   a. DO NOT CLOSE
   b. DO NOT ENGAGE
   c. DO NOT START
   d. DO NOT OPERATE
6. A lockout device uses a positive means, such as a lock, to hold equipment in a _____ _____ state.
7. Lockout and tagout devices should be removed only by __________.
   a. the affected employee
   b. the facilities safety manager
   c. the authorized employee
   d. the floor supervisor
8. After the equipment has been returned to service, the process technician should check for: (Select all that apply.)
   a. Discoloration
   b. Line leaks
   c. Unusual noises
   d. Scratches
   e. High vibration
9. Which of the following should be done before starting the lockout/tagout (LOTO) removal process? (Select all that apply.)
   a. Make sure all blinds or blind flanges are firmly in place.
   b. Perform pressure tests of equipment if necessary.
   c. Notify affected employees of impending startup.
   d. Remove all guards or protective guards.
   e. Open bleeder valves.
   f. Remove locks and tags from the circuit breakers.

NOTE: Answers to Checking Your Knowledge questions are in the Appendix.

Student Activities

1. Read OSHA regulation 1910.147 Control of Hazardous Energy, and create a presentation explaining the regulation.
2. Research previous incidents where control of hazardous energy guidelines were not followed per 1910.147. Select one incident and submit a two-page report on what failed in the process, and what may prevent similar incidents from occurring.
3. Research the double block and bleed system. Prepare a one-page report on whether you think this type of isolation and lockout is as safe as the installation of blinds.
4. Using the schematic that follows, write a LOTO procedure that places the A pump in active service and then stops and prepares the B pump for maintenance. *Hint:* The B pump will be removed for a seal replacement. The A and B pumps are equipped with local hand/off/auto (HOA) switches for start/stop and electrical switchgear in the MCC.

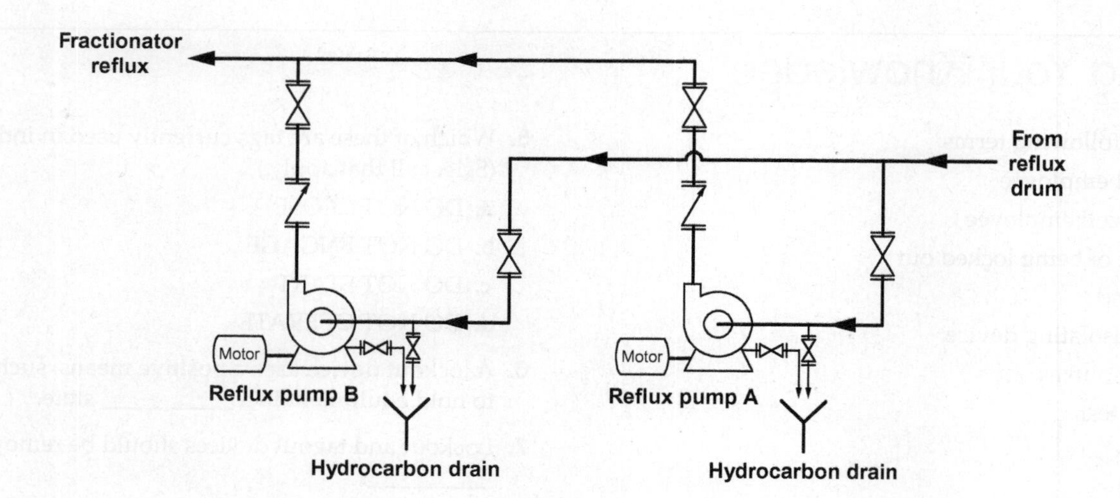

5. Perform the following tests and then discuss the following:
   a. Using a model, lock out a piece of equipment and then check to see if residual energy remains.
   b. Using a model, tag out a piece of equipment and then test to see if energy remains. Write a one-page paper on why lockout is preferred over tagout.

6. Divide into small groups and install a blind in a flange. As a class, discuss blind composition, pressure ratings, and torqueing procedures.

Chapter 6

Communication: Verbal, Nonverbal, and Written

Objectives

After completing this chapter, you will be able to:

6.1 List verbal communication components and techniques used to request or provide information:

- basic components associated with effective verbal communication: sender, receiver, message, interference, and feedback
- key obstacles that prevent effective verbal communication. (NAPTA Operations, Normal Operations: Verbal Communication 3*) p. 75

6.2 Explain important guidelines to follow when preparing written communication. (NAPTA Operations, Normal Operations: Written Communication 5–7) p. 77

6.3 Explain the verbal and nonverbal communication methods used in noisy operating areas. (NAPTA Operations, Normal Operations: Verbal Communication 7) p. 79

6.4 List the different types of electronic communication devices used in the process industry today and:

- Explain proper protocol when using the different types of electronic communication devices.
- Explain the features and functions that should be tested for operability prior to using the electronic communication device.

*North American Process Technology Alliance (NAPTA) developed curriculum to ensure that Process Technology courses will produce knowledgeable graduates to become entry-level employees in process technology. Objectives from that curriculum are named here in abbreviated form. For example, "(NAPTA Operations, Normal Operations: Verbal Communication 3)" means that this chapter's objective 1 relates to objective 3 of NAPTA's curriculum about verbal communication during normal operations.

- Demonstrate how to test the electronic communication device for operability. (NAPTA Operations, Normal Operations: Verbal Communication 9–11) p. 79

6.5 Explain the communication responsibilities of the process technician during startups and shutdowns:

- List all departments and personnel involved in or affected by the unit startup and shutdown.
- List the types of information that need to be communicated regarding unit startup and shutdowns.
- Explain the communication methods that may be used at different points during the process of starting up or shutting down the unit. (NAPTA Operations, Normal Operations: Verbal Communication 4, 5, Normal Startup: Overview and Communication 5, and Normal Shutdown: Overview and Communication 6, 7) p. 83

6.6 Identify the types of information communicated and communication methods utilized during the preparation or performance of routine maintenance. (NAPTA Operations, Normal Operations: Verbal Communication 6) p. 85

Key Terms

Blinding/unblinding permit—work permit that allows equipment isolation via the installation of blinds and blind flanges, **p. 85**

Cell phone—long-range electronic device used for mobile communication, text messaging, or data transmission across a cellular network of specialized base stations known as cell sites, **p. 82**

Communication—verbal, nonverbal, or written transfer of information between people, **p. 75**

Confined space entry (CSE) permit—permit that allows human entry and work within an OSHA-defined confined space, the issuance of which indicates all regulated and pertinent safety measures have been taken and/or are active, **p. 85**

General work permit—permit that allows work activity other than blinding/unblinding, hot work, lockout/tagout, and confined space, **p. 85**

Hot work permit—permit that allows hot work, such as welding, grinding, or vehicle entry in or around process equipment, **p. 85**

Intercom—stand-alone electronic communication system intended for limited or private conversation, **p. 81**

Intrinsically safe devices—electronic devices certified safe for use in explosive atmospheres, **p. 79**

Logbook—typically, hardbound ledgers used to handwrite significant activities that have occurred during the shift, **p. 77**

Nonverbal communication (NVC)—unspoken communication, such as gesture, expression, or body language, **p. 79**

Public address system (PA system)—system that reinforces and distributes a given sound throughout a venue, **p. 82**

Sound-powered phones—phones containing electromechanical transducers that convert voice directly into electrical energy, **p. 82**

Trunked radio system—complex type of computer-controlled radio system, **p. 79**

Two-way radio—radio that can transmit and receive voice communication, **p. 79**

Verbal communication—dialogue or conversation between two or more people for transferring information, **p. 75**
Written communication—communication by means of written or printed symbols or letters, **p. 77**

6.1 Introduction

Effective communication is critical to any successful business operation. In the process industries, effective **communication** is vital for safe and efficient operations. The process technician must effectively communicate with other process technicians, with maintenance technicians and supervisors, as well as with members of operations support groups such as safety, health, and environmental, engineering, emergency response personnel, and members of management.

Communication verbal, nonverbal, or written transfer of information between people.

A key component for successful workplace communication is to be direct, to include all important facts and information, and to leave out peripheral points. This is especially applicable during shift change (verbal) and when writing the shift log. When communicating about unit operations, all employees should learn to convey the information quickly and directly when talking or writing.

Verbal Communication

Process technicians must be able to use **verbal communication** with each other daily to keep the work flowing in a progressive, orderly fashion (Figure 6.1). The communication of critical operating information is essential for the process technician. There must be a two-way dialogue between the process technicians at shift change in order to maintain an uninterrupted workflow process. Many process technicians today utilize handwritten notes as reminders to facilitate their face-to-face discussion with other process technicians. Misunderstandings can lead to delays and may affect the productivity of the unit.

Verbal communication dialogue or conversation between two or more people for transferring information.

Figure 6.1 Process technicians observe and discuss equipment operation.

CREDIT: BASF North America.

Every time verbal communication occurs effectively, people follow a process:

- A sender conveys an original message either face-to-face or by telecom (phone, radio, etc.).
- A receiver hears the message and compares content to his or her perception of fact and his or her attitudes and values.
- The receiver responds to the sender by conveying his or her understanding of the original message, by asking clarifying questions, or by providing additional information about the original message.
- The sender responds with additional facts and information.

This communication process is repeated until the sender and receiver have complete understanding of the intent of the original message.

Various obstacles can prevent effective verbal communication, such as the following:

- Noisy or distracting environments
- Conflict between the sender and receiver
- Difference in culture and language with the sender and receiver
- Short-changing the communication process due to time pressures, complacency, or impatience with the dialogue.

When communicating verbally with other process technicians or other members of the process facility team, the process technician should take into consideration communication needs and the target audience. Listed below are some tips for verbal communication:

- *Use appropriate volume*—use volume appropriate for the setting. If talking in a quiet office setting, use a softer voice. If speaking to a larger group of people, or out in the field, speak louder.
- *Speak clearly*—enunciate words and avoid mumbling.
- *Pronounce words correctly*—know the correct pronunciation of words used.
- *Use the right words*—avoid words with unclear meaning.
- *Make eye contact*—maintain eye contact; it displays directness and may help to make a point.
- *Animate the voice*—avoid a monotone voice. Adjust voice pitch and learn to emphasize key points that require special attention.
- *Use gestures*—use movement, such as hand, arm, body, head, and facial expression to emphasize a point.
- *Do not send mixed messages*—avoid confusion by making tone and facial expressions match.
- *Slow down*—speak in a moderated voice to avoid the appearance of nervousness; instead, portray self-confidence.
- *Avoid ambiguity*—make instructions or other information clear by *not* using terms and jargon unrelated to the industry.
- *Listen*—listening is 50% of any verbal communication and requires focus.

Especially important with radio communication is to provide more than one point of information. For example, in order to ensure sender and receiver are communicating about the same equipment, each party should send and verify the equipment label and location (for example, FV-312, feed control valve to T-100). Providing multiple points of information helps reduce the risk of misunderstandings in the communication process.

The voice is an extremely powerful communication device for effectively passing along information during shift change. The process facility may have a list of required items that need to be communicated to the oncoming shift. The following information, at a minimum, should be verbally communicated at shift change:

- Safety information, such as malfunctions or out-of-service safety equipment
- List of personnel on previous shift and their areas of responsibility
- Changes in feed composition or rates
- Changes in product specifications or product destination
- Changes in equipment usage
- Any equipment failures or trips and/or instrument failures
- Any equipment or controls bypassed and why
- Abnormal situations in overall operation (feed, product flow, levels, pressures, temperatures, and composition)

- Active, previously active, or malfunctioning alarms
- List of the latest daily orders
- Procedures or other work currently in progress and continuing across shift.

When communicating this information, it is also helpful to communicate using a written checklist to ensure that all necessary information has been communicated. This is a helpful tool that the process technician going off shift can give the relief process technician coming on shift as a reference tool to be used during the relief process technician's shift.

6.2 Written Communication

During the shift change, communication between shifts is critical. Once shift change is over, **written communication**, usually by use of a log is the main source of information for any situations or activities that occurred during the previous shift. Other forms of written communication consist of operating procedures, maintenance procedures, night orders, and emails. When communicating through writing, process technicians must write clearly so that the reader correctly understands what the writer is trying to convey. As with verbal communication, poorly written communication can lead to misunderstandings, which may lead to delays that could ultimately affect personnel safety, the environment, or productivity of the process facility. Here are some tips for clearer, more concise written communication:

Written communication communication by means of written or printed symbols or letters.

- Avoid the use of inappropriate or slang words.
- Avoid the use of abbreviations (unless appropriately defined or commonly used).
- Avoid the use of symbols (such as ampersands [&]).
- Avoid or limit the use of clichés.
- Use correct spelling, especially the names of companies and people.
- Express numbers as words when the number is less than 10 or is used to start a sentence.
- Use quotation marks around any directly quoted speech or text and around titles of publications.
- Use technical and business writing principles:
  - Convey a single idea within a given sentence.
  - Create short sentences that are specific, using only the number of words needed to convey the message.
  - Provide a background for context in a short paragraph before writing the core message or requesting action or information.
  - Define any technical terms that may be unfamiliar to the receiver before writing the core message or requesting action or information.
  - Create the message with the most inexperienced person in the unit in mind.
  - If possible, identify members of the target audience to review the message and suggest needed revisions before it is sent.

The field and control room technicians are responsible for conveying written information between shifts. Improperly worded, incomplete, or unintelligible information can cause serious mistakes.

A **logbook** is a regularly kept record of activities, performance, and events. A unit logbook contains the information that process technicians must communicate to one another (Figure 6.2). Logbooks can be either electronic or hardbound and are considered legal records that can potentially be used as evidence in civil or criminal court. The information in logbooks provides a window to past operations and a basis to predict future events.

Logbook typically, hardbound ledgers used to handwrite significant activities that have occurred during the shift.

All information verbally communicated at shift change should also be part of the written or typed logbook. However, the process technician should clearly state the facts, as they

Figure 6.2 The process technician uses a logbook to record significant activities that have occurred during the shift.

CREDIT: Oil and Gas Photographer/ Shutterstock.

happened, over the course of her or his shift and should not include speculation or opinions. When handwriting entries into a hardbound logbook, the process technician should:

- Make legible log entries using an ink pen (do not use pencil).
- Make corrections using a red pen by striking through written text.
- Avoid cursive entries (print legibly).
- Make drawings in the logbook *only* if they are unit or process related.
- Log entries should be dated, and events/incidents should have a logged time.
- Initial a changed entry so that there is record of who corrected the entry.
- Proofread all shift entries for accuracy before shift change.

When making electronic logbook entries, use the software strikethrough feature to mark through the incorrect entry and insert the correct information on the next line below. Again, proofread all shift entries for accuracy before shift change.

Neatness, accuracy, and completeness are essential when making log entries. For example, an entry such as "Don't use the pump" is useless to the reader because the entry does not state:

- Which pump is affected
- Why the pump cannot be used
- If maintenance has been notified that the pump needs repair or replacement
- How long the pump will be out of service
- If the pump can be used in an emergency
- If the pump is locked out and tagged-out according to procedure or which procedure was used.

A correct logbook entry would read:

> "P-101A T-1 reflux pump out of service due to rough bearings. Maintenance has been notified and planning is in process to schedule the pump for repair. If the "B" pump becomes unusable, "A" can be used. P-101A is locked out and tagged per procedure number P101A_Maint."

The process technician also uses the process facility's email and communicates to other technicians, engineering personnel, maintenance personnel, safety personnel, or others. Expressing precise ideas through technical writing is necessary so that the receiver has a clear picture of what the sender of the email is trying to communicate.

6.3 Communicating in Noisy Areas

Nonverbal communication (NVC) is the process of communicating by sending and receiving wordless messages. Nonverbal communication includes all messages encoded without using written or spoken language.

Nonverbal communication (NVC) unspoken communication, such as gesture, expression, or body language.

Within the process industry, nonverbal communication is primarily used in high noise areas where radio communication and face-to-face discussions are hard to hear. In these types of areas, hand signals are the preferred method of communicating. Hand signals should be worked out between process technicians prior to performing work in high noise areas. The most common hand signals used in the industry are used in association with performing lifts with cranes. Operators should become familiar with the hand signals used by their company. Figure 6.3 shows the hand signals associated with communication between the spotter directing the lift and a crane operator. There are many other types of hand signals used for communication such as driving, motorcycling, bicycling, and signing for the deaf.

Figure 6.3 **A.** Crane operation hand signals. **B.** Technician signaling crane operator.

CREDIT: B. King Ropes Access/Shutterstock.

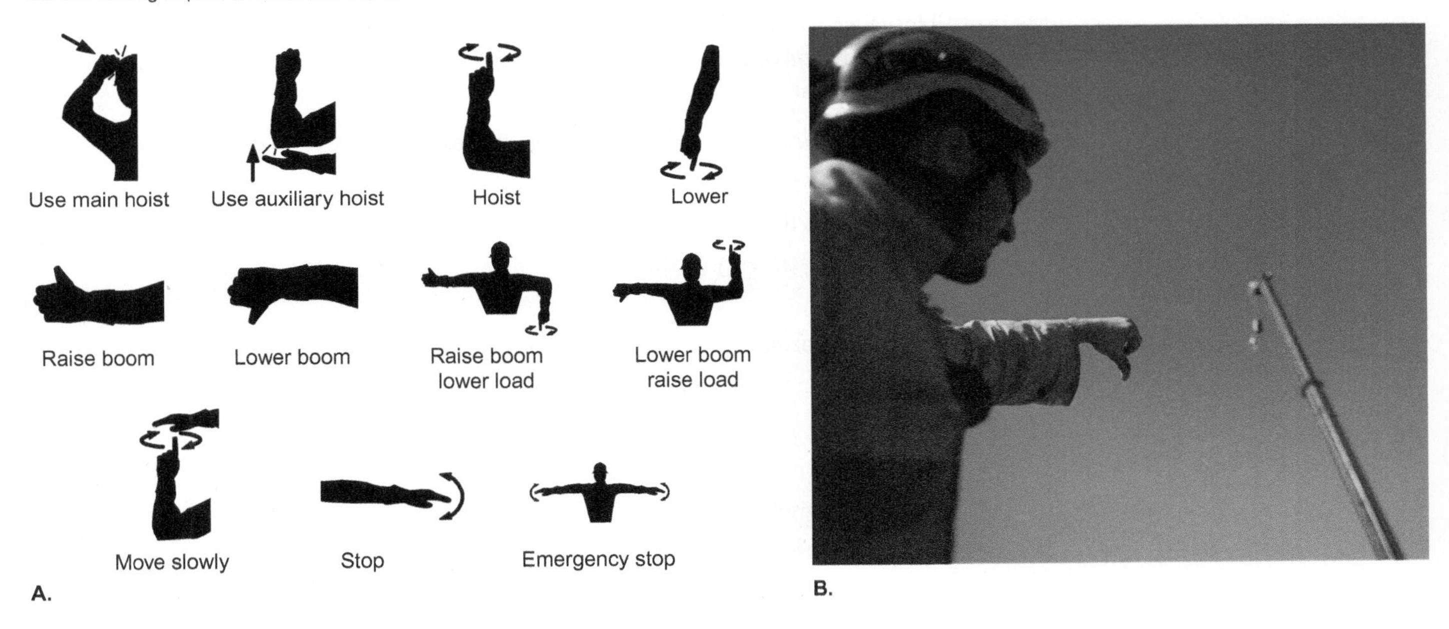

6.4 Electronic Communication Devices

Electronic communication includes devices such as two-way radio communication, intercoms, public address systems, and cell phones. Plant communication systems are selected for many reasons, and obviously safety is a major consideration.

Two-Way Radios

Two-way radios, as seen in Figure 6.4, are **intrinsically safe devices**. An intrinsically safe electronic device is designed to be incapable of releasing enough electrical or thermal energy to ignite flammable gases, vapors, or combustible dust. The device cannot light the fuel and cause fire or explosion.

Two-way radio radio that can transmit and receive voice communication.

Intrinsically safe devices electronic devices certified safe for use in explosive atmospheres.

Trunked radio system complex type of computer-controlled radio system.

Many of the radio systems within the process industry are **trunked radio systems**, which use fewer frequencies and are more efficient. They also provide the ability to divide the facility into groups, limiting the amount of nonessential conversations heard by all personnel. Trunked radio systems differ from conventional radio systems because a

Figure 6.4 **A.** Two-way radio. **B.** Two-way radio with headset.

CREDIT: A. goodluz/Shutterstock. **B.** ADAM GAULT/Science Photo Library/Alamy Stock Photo.

A.

B.

conventional radio system uses a dedicated channel, or frequency, for each individual group of users. However, trunked radio systems use a pool of channels that are available for a great many different groups of users. For example, a process facility can divide its personnel into the following groups:

- Operations channel (divided into smaller subgroups)
  - CAT 1 (Catalytic Cracking Unit 1)
  - CAT 2 (Catalytic Cracking Unit 2)
  - SRU (Sulfur Recovery Unit)
  - Utilities, etc.
- Maintenance channel (divided by subgroups for each craft)
  - Machinists
  - Pipefitters
  - Welders
  - Heavy equipment
  - Electricians, etc.
- Engineering channel (divided into engineering disciplines or by unit)
- Emergency response channel (divided by applicable subgroups)
  - Rescue
  - Fire
  - Emergency medical technicians, etc.
- Inspection channel
- Safety, health, and environmental channel
- Security channel

Trunked radio systems work on the probability that users do not need channel access all at once. Therefore, with a given number of users, fewer separate radio channels are required. However, trunked systems may experience delays if a major emergency occurs because of the tendency for users to overwhelm the system as they attempt to find out information about the emergency. Users must limit the traffic on the system during these times to be able to prevent overloading the system. These radios are designed to scan other channels to seek information in times of emergency.

The two-way radio is the primary communication device for the process technician when she or he is out in the process units. The radio should be inspected by the process technician for proper operation prior to entering the unit including a functioning mic and a fully charged battery. Many two-way radios are intrinsically safe, but to verify that a particular radio or battery is intrinsically safe, look for the following on the back panel of the radio or the inside panel of the battery:

FM APPROVED
NONINCENDIVE APPARATUS
CLASS I DIVISION 2 GROUPS A, B, C, D
CLASS II DIVISION 2 GROUPS F, G

In some operating facilities, a two-way radio is issued to each employee. In other operating facilities, each operating unit is allotted a certain number of radios that are shared by those on shift.

Radios are provided by the process facility for communication and reporting purposes. They are also provided to the process technician as a safety device. In the event of an emergency, a radio provides the process technician with a quick way to summon help from other unit process technicians or emergency responders. Many of the trunked radio systems have a feature called All Call that, when activated, transmits to all radio users on all channels. If available, All Call is normally reserved for emergency responder notification and communication for the entire facility site.

Note that radio systems use the public airwaves and are licensed by the Federal Communications Commission (FCC). Improper usage of the radio or inappropriate language could cause the company to lose its radio operating license.

Intercoms

Intercoms are distributed amplifier systems, and most of these systems consist of strategically placed indoor and outdoor stations that contain a power amplifier for driving speakers. The outside stations are intrinsically safe and are generally housed in an explosion-proof, wall-mounted container. Inside units are stationed near the control board for use by the control board technician.

Intercom stand-alone electronic communication system intended for limited or private conversation.

The inside unit is generally desk-mounted and not housed in a container. The intercom system is simple to use. In the field, the field process technician opens the explosion-proof door of the unit, depresses the page button, and speaks into the handset. The responding individual lifts the handset of his or her nearest station for telephone-type communication over the party line. Several individuals may engage in conversation on each intercom system. Figure 6.5 shows an example of a field-mount intercom.

Figure 6.5 Field-mount intercom.

CREDIT: manine99/Shutterstock.

PA Systems

Public address system (PA system) system that reinforces and distributes a given sound throughout a venue.

Public address systems (PA systems) are often used in small venues such as school auditoriums, churches, and restaurants. Public address systems with a larger number of speakers are widely used in institutional and commercial buildings to read announcements or declare states of emergency.

In processing facilities, public address systems, as seen in Figure 6.6, are generally used for paging personnel or for emergency notifications to the facility site. The system consists of a primary control panel, user interfaces as required, and amplifiers and speakers placed strategically throughout the process facility. These systems are designed as failsafe, or unable to fail. They are capable of programmable monitoring to detect fault status in order to ensure maximum availability at all times. These systems can perform general paging functions, emergency broadcasts, and alarm tones. Alarms can be automatically initiated so that alarm tones or prerecorded voice messages can be broadcast. The alarm tones may be set up in a series of tones or beeps that indicate a certain area or unit that may have an emergency. The process technician is required to learn the appropriate tones at his or her process facility.

Figure 6.6 Public address system.

CREDIT: Lee Ramsden/Alamy Stock Photo.

Cell Phones

Cell phone long-range electronic device used for mobile communication, text messaging, or data transmission across a cellular network of specialized base stations known as cell sites.

Although not prohibited, in some process industries, such as the refining and petrochemical industry, **cell phone** use is restricted. A process technician should not take a personal cell phone into an operating process unit. Cell phones should not be used in process units unless they are intrinsically safe.

Cell phones that are intrinsically safe may be carried into operating process units if allowed by the organization. These types of cell phones are generally bought by the process facility for use by emergency responders, shift supervisors, and chief operators. It is important that process operators become aware of and follow the procedures of their employers regarding cell phone use in process areas.

Sound-Powered Phones

Sound-powered phones phones containing electromechanical transducers that convert voice directly into electrical energy.

Sound-powered phones, using either headsets, handsets, or a combination of the two, can be used in a process facility. Typical use includes crane operations, in-facility maintenance, wire pulling by electricians, or even during rescue operations. Keeping this type of communication on dedicated devices ensures the two-way radios are used for plant personnel only. Sound-powered devices contain electromechanical transducers that convert voice directly into electrical energy. A headset microphone transducer converts sound pressure from a user's voice into a minute electrical current, which is then converted back to sound by a transducer at the other end. Sound-powered telephones are useful in any situation where two or more people must communicate who cannot hear or see one another, and when clear, reliable communication is imperative. They are useful in other industries, such as concrete pumping, sports spotter systems, communication underground (for example in train/subway tunnels and in tanks), production line balancing, and many others.

6.5 Communication during Startups or Shutdowns

Communication is essential for a successful startup or shutdown, which are critical times in the operation of a process unit.

Startups

During a unit startup, there must be communication between many different departments and personnel within the facility. Many departments and personnel are affected by a unit startup, and the process technician must communicate detailed information to them. During a unit startup, the process technician is expected to communicate with the following work groups:

- *Operations*—including other process technicians, supervisors, other process units, and management. Examples of the information the process technician is expected to communicate during the unit startup includes, but is not limited to:
  - Current process status
  - Work permit status
  - Current equipment status, including lockout/tagout status of equipment
  - Procedure status
  - Maintenance status of remaining work or work in progress
  - Any abnormal situations
  - Notifications of major equipment startups that have the potential to affect the entire facility site or neighboring units.
- *Maintenance*—including maintenance technicians assigned to the startup and maintenance management. The information the process technician supplies to maintenance during startup includes:
  - Maintenance work that may be needed
  - Work permit status
  - Status of ongoing maintenance work
  - Current startup status.
- *Engineering*—including process, mechanical, and electrical engineers. The information that a process technician communicates includes:
  - Current startup status
  - Rotating equipment problems the unit may have or is experiencing
  - Any electrical issues the unit may be experiencing
  - Process related issues such as temperatures, pressures, or catalyst activation.
- *Safety*—including industrial hygienist, safety representatives, safety engineers, and safety management. The information that is communicated includes:
  - Current startup status

- Any safety related issues that are ongoing
- Any startup related issues that may be industrial hygienist related, such as leaks or spills.

Shutdowns

As with startup, communication is important during a planned unit shutdown. Many departments and personnel are involved, and communication begins months prior to the actual shutdown date. During a unit shutdown, a process technician is expected to communicate with several work groups.

- *Operations*—during the unit shutdown, the process technician may communicate with other process technicians, supervisors, other process units, and management. Some of the information the process technician is expected to communicate includes:
  - Current shutdown status
  - Work permit status
  - Current equipment status, including lockout/tagout status of equipment
  - Procedure status
  - Maintenance status of remaining work or work in progress
  - Any abnormal situations that were encountered during the shutdown process.

  Prior to the shutdown, the process technician is expected to communicate some of the following:
  - Input for the turnaround work list
  - Request for turnaround supplies
  - Input on procedures.
- *Maintenance*—this group includes maintenance technicians assigned to the shutdown and maintenance management. The information the process technician supplies to maintenance includes:
  - Maintenance work that may be needed during the shutdown
  - Work permit status
  - Status of ongoing maintenance work related to preshutdown work
  - Current shutdown status.
- *Engineering*—the process technician will communicate with process, mechanical, and electrical engineers. The information that the process technician communicates includes:
  - Current shutdown status
  - Rotating equipment problems experienced during the shutdown of the unit
  - Any electrical issues encountered during the shutdown
  - Any abnormal process-related issues encountered during the shutdown.
- *Safety*—the safety personnel the process technician will encounter include industrial hygienists, safety representatives, safety engineers, and safety management. The information communicated includes:
  - Current shutdown status
  - Any safety related issues that are ongoing
  - Any hygienist related issues, such as leaks, spills, etc.

Communication to these groups is generally verbal. However, written documentation is always encouraged as a guide to dialogue. In addition, written documentation serves as a record that pertinent information was passed on. A process technician must also realize that communication is a two-way street, and so the technician should also ask questions that are relevant to her or his job or task during discussion with each of the groups.

6.6 Communication during Routine Maintenance

During routine maintenance, the process technician provides the maintenance technician with verbal and written communication. Verbal communication can consist of discussion regarding the permits associated with the work to be performed. A written communication tool, the permit, serves as a guide to this discussion. A process technician is responsible for all work taking place in the unit. Control of work (COW) is a work practice that identifies the methods of safely controlling maintenance, demolition, remediation, construction, operating tasks, and similar work activities. In most operating facilities, the process technician is required to fill out and discuss the work permit with the maintenance technician or work crews prior to the work starting.

The permit is the process technician's key to controlling the maintenance activities in his or her unit. Depending on the work, the process technician may be required to fill out the following permits:

- **Confined space entry (CSE) permit**—a permit that allows entry and work within an OSHA-defined confined space once safety measures are in place. The process technician is required to follow an isolation and lockout/tagout procedure to prepare the confined space for entry.
- **Lockout/tagout (LOTO)**—a procedure to isolate energy sources from a piece of equipment. It is a general term used to refer to the control of hazardous energy as defined by OSHA standard 29 CFR 1910.147. Implementation is accomplished by identification and isolation of hazardous energy sources and hazardous substances. The process technician, along with the maintenance technician who is to perform the work, must verify that the isolation has been completed prior to work beginning.
- **Hot work permit**—permit for tasks such as welding or grinding. The process technician, along with the technician to perform the work, must verify the procedure has been completed correctly and have a discussion centering on the permit conditions.
- **Blinding/unblinding permit**—work permit for equipment isolation using blinds and blind flanges. This may also be referred to as a *line breaking permit*. The process technician, together with the technicians performing the work, should verify the LOTO has been completed correctly, verify associated equipment is energy free, and have a discussion centered on the permit conditions.
- **General work permit**—a broader permit for work activity that does not require a specific permit. Different operating facilities refer to this permit differently. The process technician, together with the technicians or maintenance personnel performing the work, should verify the LOTO procedure has been completed correctly (if applicable), verify the equipment is energy free (if applicable), and have a discussion centered on the permit conditions.

Confined space entry (CSE) permit permit that allows human entry and work within an OSHA-defined confined space, the issuance of which indicates all regulated and pertinent safety measures have been taken and/or are active.

Hot work permit permit that allows hot work, such as welding, grinding, or vehicle entry in or around process equipment.

Blinding/unblinding permit work permit that allows equipment isolation via the installation of blinds and blind flanges.

General work permit permit that allows work activity other than blinding/unblinding, hot work, lockout/tagout, and confined space.

The process technician can control the work on the process unit by issuing and discussing the applicable permit for a given type of work with the personnel performing the work. These permits and face-to-face discussions help to ensure that all personnel remain safe.

Summary

In the process industries, effective communication is vital for safe and efficient operations. Communication is evolving into a skill that involves both conversation and electronic sharing of information. Within the process industry, face-to-face discussions and electronic sharing of information are equally important.

Communication is a process of transferring information between people in a verbal, nonverbal, or written way. The most important part of communication is for the information conveyed to be clearly understood. Partial understanding only creates problems. Competent process technicians know that clarity is an important component in both verbal and written communication. To support clear communication, they make sure to prepare, organize, and be ready to answer questions when communicating. They cover the most important factors first and avoid unimportant details. They support verbal communication with written notes.

Advances in information technology have dramatically increased the speed of communication. Instant communication with others has become easy; and fast access to information for decision making has become common. Intrinsically safe devices are used in plants to aid communication.

Startups and shutdowns require a high level of communication and awareness. Permits of various kinds are used to plan and to prepare for startups, shutdowns, and routine maintenance events.

Checking Your Knowledge

1. Define the following key terms:
   a. Blinding/unblinding permit
   b. Communication
   c. Confined space entry (CSE) permit
   d. General work permit
   e. Hot work permit
   f. Intercom
   g. Intrinsically safe devices
   h. Logbook
   i. Nonverbal communication (NVC)
   j. Public address system (PA system)
   k. Sound-powered phones
   l. Trunked radio system
   m. Two-way radio
   n. Verbal communication
   o. Written communication
2. Which of these is NOT true about effective verbal communication?
   a. A sender conveys an original message.
   b. A receiver hears the message and compares content to his or her perceptions.
   c. The receiver responds to the sender his or her understanding.
   d. The sender interrupts the receiver's questions.
   e. The sender responds with additional information.
3. List five tips for good verbal communication skills.
4. List three principles specific to technical and business writing.
5. Which of the following are valid tips for good business and technical written communication? (Select all that apply.)
   a. Avoid or limit the use of clichés.
   b. To save space, always use abbreviations.
   c. For brevity, use symbols such as ampersands [&] whenever possible.
   d. Avoid the use of inappropriate or slang words.
   e. Write all numbers as numerals unless used to start a sentence.
   f. Use correct spelling, especially the names of companies and people.
   g. Use square brackets around any directly quoted speech or text and around titles of publications.
   h. Use technical and business writing principles.
6. The process of communicating by sending and receiving wordless messages is known as _________ communication.
7. In high noise areas where radio communication and face-to-face discussions are hard to hear, _________ are the preferred method of communicating.
   a. cell phone text messages
   b. hand signals
   c. flares
   d. flags
8. Name the four work groups with whom the process technician is expected to communicate.
9. During a shutdown, the process technician would communicate any hygiene related issues such as leaks or spills to the work _________ group?
10. _________ are stand-alone electronic communication systems intended for limited or private conversation.
11. A public address system is used in the refining and process industry for _________ and _________.
    a. calling home / playing music
    b. paging personnel / emergency notification
    c. public speaking / voicing an opinion
    d. none of the above
12. Match each the following work groups who would receive communications from the process technician during a startup to an example of the information they would receive.

Work Group	Example of Information
I. Engineering	a. Any startup related issues that may be industrial hygiene related, such as leaks or spills.
II. Maintenance	b. Current equipment status, including lockout/tagout status of equipment
III. Operations	c. Process related issues such as temperatures, pressures, or catalyst activation.
IV. Safety	d. Status of any ongoing repairs and other housekeeping work

13. What duties must the process technician perform before allowing entry and work within an OSHA-defined confined space? (Select all that apply.)
    a. Obtain permission from upper management.
    b. Issue a confined space entry (CSE) permit
    c. Follow an isolation and lockout tagout procedure to prepare the confined space for entry.
    d. Clean up the confined space area after the work is completed.

14. The ________ is required to issue all permits on a process unit.
    a. operations supervisor
    b. maintenance technician
    c. process technician
    d. operations superintendent

NOTE: Answers to Checking Your Knowledge questions are in the Appendix.

Student Activities

1. Using the information described, create a written logbook providing necessary detail, including date and time:

   Scenario: You are working the day shift during which you engage in the following activities:

    a. You prepare P607A (pump with leaking seal) for maintenance using procedure P607_Maint, which is a lockout/tagout procedure.
    b. You fill out a general work permit for a scaffold to be erected for work on P-607A.
    c. You unload a truck containing 500 gallons of sodium hypochlorite for your unit cooling tower.
    d. You add oil to the following pumps: P-607B, P-601A, P-602A, P-603A, and P610.
    e. P-615A T-6 bottoms pump shuts down, causing a low flow through H-6 furnace and tripping the furnace offline. You immediately start the B pump and reestablished bottoms flow. Fortunately, the furnace restarts within a few minutes after starting P-615B, and the unit upset is minimal. You call the electricians to check out P-615A after troubleshooting.
    f. You and the other process technicians attend a morning safety meeting lasting 15 minutes.
    g. You find a problem with P-622A cooling water pump (leaking seal) and write a work order.
    h. You unloaded another truck of Nalco 657 inhibitor for the cooling tower.
    i. Maintenance blinds P-607A.
    j. Maintenance (pipefitter and machinist) working over to pull P-607A and taking to shop for machinist repair. (Machinist will work until pump repaired and back in the hole or 9:00 PM, whichever comes first.)
    k. You take routine readings and find no abnormalities.

2. Select a classmate and simulate a verbal shift change using the written log sheet from Activity 1.

Chapter 7
Procedure Writing

Objectives

After completing this chapter, you will be able to:

7.1 Define the function of operating procedures. (NAPTA Operations, Procedure Writing 1*) p. 89

7.2 Explain the steps for compiling necessary information to develop a safe and effective operating procedure. (NAPTA Operations, Procedure Writing 1) p. 91

7.3 Explain methods of organizing and presenting procedure information effectively. (NAPTA Operations, Procedure Writing 1) p. 92

7.4 Use writing techniques that help create a clear, easily followed procedure. (NAPTA Operations, Procedure Writing 1) p. 93

7.5 Apply the techniques and principles presented in this chapter. (NAPTA Operations, Procedure Writing 1) p. 95

Key Terms

Checklist—procedure written in a list format that requires the user to initial or check the completion of each step, **p. 95**

Internal procedure—company specific procedure, **p. 89**

International Organization for Standardization (ISO)—regulates safety and health standards internationally, **p. 89**

Occupational Safety and Health Administration (OSHA)—U.S. government agency created to establish and enforce workplace safety and health standards, conduct workplace inspections and propose penalties for noncompliance, and investigate serious workplace incidents, **p. 89**

Personal protective equipment (PPE)—specialized gear that provides a barrier between hazards and the worker, **p. 91**

Procedure owner—individual who is accountable for the accurate development and maintenance of a procedure, **p. 89**

*North American Process Technology Alliance (NAPTA) developed curriculum to ensure that Process Technology courses will produce knowledgeable graduates to become entry-level employees in process technology. Objectives from that curriculum are named here in abbreviated form. For example, "(NAPTA Procedure Writing 1)" means that this chapter's objectives relate to objective 1 of NAPTA's operations curriculum on procedure writing.

Procedure template—form or guide that accurately and effectively shapes procedure presentation and content, **p. 92**
Procedure user—process technician trained and qualified on the subject matter of the procedure prior to use, **p. 92**
Process safety management (PSM)—OSHA standard that contains the requirements for management of hazards associated with processes using highly hazardous materials, **p. 89**
Standard operating procedures (SOPs)—a set of directions or instructions that has a recognized and permanent value and that defines the particular steps to take when a certain situation or condition occurs, **p. 91**
Subject matter experts (SMEs)—individuals within an organization possessing a very high level of expertise regarding a particular job, task, or process, **p. 91**

7.1 Introduction

Operating procedures are perhaps the most important documents in industry because they guide personnel to perform tasks without safety or environmental incidents. Poorly written or improperly executed procedures have been identified as the root cause, or at least a contributing factor, in industrial accidents across the globe.

To promote safety and meet the demands of governmental regulations and industry standards, process facilities have developed internal standards that define *procedure development and use requirements* within their organizations. Governmental organizations, such as the **Occupational Safety and Health Administration (OSHA)**, create safety standards like the **process safety management (PSM)**. The **International Organization for Standardization (ISO)** is another standard-setting agency.

Process technicians create **internal procedures** that provide accurate, clear instructions to aid the user in safely completing a task within the facility. They write, maintain, and use operating procedures relevant to their area of responsibility. As the **procedure owner**, they are accountable for the accurate development and maintenance of the procedure. However, process technicians do not change (update/maintain) procedures without many other people being involved and approving any changes. When following a procedure step-by-step during a shutdown or other activity, any slight deviation must be approved by the supervisor (Figure 7.1). As an example, if the procedure says to wait until a temperature reaches 150 degrees Fahrenheit before moving forward, the supervisor might approve moving forward at 155 degrees Fahrenheit instead. Such deviations are infrequent.

Occupational Safety and Health Administration (OSHA) U.S. government agency created to establish and enforce workplace safety and health standards, conduct workplace inspections and propose penalties for noncompliance, and investigate serious workplace incidents.

Process safety management (PSM) OSHA standard that contains the requirements for management of hazards associated with processes using highly hazardous materials.

International Organization for Standardization (ISO) regulates safety and health standards internationally.

Internal procedure company specific procedure.

Procedure owner individual who is accountable for the accurate development and maintenance of a procedure.

Figure 7.1 The process technician must get approval for any deviation made to the procedure.

CREDIT: Chatchai.wa/Shutterstock.

A successfully written procedure will follow basic writing principles and techniques that will make it easily identifiable, accessible, executable, task specific, accurate, well organized, audience specific, and error preventing. Development of a successful procedure can be divided into stages. Three fundamental stages that are essential to develop any procedure include:

- Gathering information
- Organizing the information
- Presenting the information.

An example of a procedure template is shown in Figure 7.2.

Figure 7.2 Operating procedure template.

Operating Procedure Template

Title	**Name the procedure (be specific)** Give the procedure a title that tells the user exactly what completion of the procedure will produce. For organizational purposes and ease of locating the procedure, it is very helpful to start the title with the functional name of the equipment or process.
Purpose	**State the purpose of the procedure**
Before you begin	**List and explain anything the user should know or do before executing the procedure.**
Safety, health and environmental considerations	**This is an extremely important element that often requires input from specialists in each of these critical areas.** Describe all potential hazards associated with execution of the procedure and what actions are required to mitigate or prevent each hazard from occurring.
Tools or equipment needed	**Name any tools the user will need on hand.**
Procedure	**This is the How To Do It part of the procedure.** This section focuses on the main objective of the procedure. List the steps required to execute the procedure and achieve the desired result. Keep your instructions simple, clear and concise.
Procedure completion signatures	**This procedure was completed by:** ____________________ Process Technician Signature Date/time:__________
Procedure validation signature	**This procedure was validated by:** ____________________ Subject Matter Expert (SME) Signature Date/time:__________
Procedure approval	**This procedure was approved by:** ____________________ Authorized Approver Signature Date/time:__________
Revision history	**Update as revisions are made**

Date	Description of change	By

7.2 Procedure Writing–Gathering Information

A procedure may be relatively simple, such as describing how to change a flat tire, or long and complex, such as providing instructions for placing a gas fired turbine into service. No matter how simple or complex, the technical accuracy of the content is critical. Therefore, the information gathering process establishes a solid foundation for the procedure.

Process technicians often create **standard operating procedures (SOPs)**. The following list identifies many of the considerations necessary when gathering information prior to a draft of a specific procedure:

Standard operating procedures (SOPs) a set of directions or instructions that has a recognized and permanent value and that defines the particular steps to take when a certain situation or condition occurs.

- Purpose, such as what, when, how, and why
- Target audience: Who must understand and execute the procedure?
- Federal laws and regulations
- State laws and regulations
- Corporate requirements, such as internal policies and procedures
- Safety, health, and environmental (SHE) considerations
- Equipment manufacturer requirements and recommendations
- Impact of the procedure on the overall process
- Scheduling.

The text of the procedure must incorporate specialized information as well. For example, an SOP for startup or shutdown of a particular pump must address:

- Required **personal protective equipment (PPE)**
- Safety and environmental considerations
- Hazards associated with operation of the pump
- Hazards associated with the process material
- The function of the pump in the process.

Personal protective equipment (PPE) specialized gear that provides a barrier between hazards and the worker.

The procedure may incorporate additional information, such as:

- Physical construction of the pump, associated valves, and instruments
- Type of pump, such as positive displacement or centrifugal
- Type of service the pump is in, such as cold or hot.

Although the procedure writer may have extensive knowledge on the subject of a procedure, a procedure that meets all requirements often necessitates input from several resources. The knowledge and experience of other **subject matter experts (SMEs)** may help accurately address concerns such as safety, risk management, environmental protection, consequences of deviation, operating constraints or parameters, permit requirements, and more.

Subject matter experts (SMEs) individuals within an organization possessing a very high level of expertise regarding a particular job, task, or process.

In addition, a manufacturer representative may be interviewed, or a technical publication may be reviewed and/or referenced during information gathering. The manufacturer reference materials can help ensure that design specifications and recommendations are met. In certain situations, it may be suitable to use the manufacturer procedures without modification.

At times, facility procedures will differ from the manufacturer's literature. This is because, in many applications, the equipment is used in ways that vary from the manufacturer's design. For example, the equipment may run at a speed or load that is different from what the manufacturer planned.

Procedures are written for the operation of equipment in the environment for which it is being used. Process technicians should always follow the facility's procedure, even if it is different from the manufacturer's literature.

7.3 Procedure Writing–Organizing and Presenting Information

Organizing the Information

After the information has been gathered, it is then assembled into a logical order. There are several methods that can be used to organize information into a manageable format. Many organizations have developed a **procedure template** to provide an accurate and consistent format for procedure content. These procedure templates are specifically designed and formatted to meet exact requirements.

Procedure template form or guide that accurately and effectively shapes procedure presentation and content.

The use of a template to complete the entire process of developing a procedure offers several benefits. These benefits include:

- Directing the information gathering process and promoting logical organization
- Developing consistency that will continue into the presentation process
- Guarding against omission of critical steps or facts
- Promoting ease of execution.

The procedure template represented in Figure 7.2 demonstrates how information can be organized. The layout and grouping of information into specific components simplifies the writer's job and improves the user's ability to perform the procedure.

The suggested template is an example demonstrating how the complex components of a procedure can be logically structured. The order in which the components are listed is significant because the sequence in the template forms the foundation for the entire procedure.

A template that contains the most common components helps the writer avoid omitting critical information and serves to reduce user errors. In the example, each of the required components is listed in the left hand column with space available on the right to provide descriptive information necessary to carry out the procedure safely.

Presenting the Information

Once the gathering and organization of information is completed, the focus shifts to the presentation of the information. The presentation is the final written document. The procedure is incomplete until accurately written for the **procedure user**, or target audience.

Procedure user process technician trained and qualified on the subject matter of the procedure prior to use.

Taking into consideration the target audience, the written steps of a procedure should reflect the actual job tasks being performed in the field. It should enable the user to easily answer the following questions:

- WHAT is to be done?
- WHEN is it to be done?
- HOW is it to be done?
- WHY is it to be done?

Using these questions, a procedure writer can more effectively communicate action steps precisely. Through practice and experience, a procedure writer learns to expand these concepts and improve writing skills.

WHAT Start each step with an action word or a statement that clearly defines *what* to do in that step. For example:

- Open the feed control valve.
- Start the pump motor using the local start/stop switch.
- Stop the transfer when the level in the hold tank reaches 72%.

WHEN Describe *when* to execute a procedure, or certain steps within a procedure. This step is very important to safe, reliable operation. For example:

- When the temperature drops below 150 degrees Fahrenheit, block in the steam supply valve.
- After completion of the safety startup checklist, proceed to . . .
- Before warming any steam lines on the turbine, establish normal operation of the lube oil system.

HOW Use adjectives and expressions to define limitations or clarify *how* to perform the action. For example:

- Close the minimum flow recycle hand valve gradually while monitoring the discharge flow. If the forward flow does not increase, reopen the minimum flow hand valve.
- Position your body to avoid hazards. Hold down the valve lever and rotate the lever in one smooth motion so that the lever is pointing to the opposite filter.

WHY Explain *why* a step or an action is required. In some instances, this can be very important. Consider the following example from a column startup procedure:

- The distillation column can be fed two different feeds to produce two different products. When the column takes feed A, the column overhead is the product stream. When the column takes feed B, the column bottoms is the product stream.

In some instances, it is important to include notes or warnings describing important operating facts prior to an action step. Warnings may include operating limits, personal hazards, environmental details that provide information, and a chance to pause and think about the upcoming action.

7.4 Techniques for Effective Written Communication

Clear communication is necessary for successful procedure writing. The procedure user must be able to interpret and execute the instructions accurately, and must be aware of any potential hazards. A well written procedure results in high confidence for the technician using the procedure. The most effective writing techniques include being precise, using common names, avoiding subjective words, following company procedures, grouping and labeling, employing visual techniques, and practicing.

Be Precise

Use the authoritative voice for clarity and economy of words; do not leave room for interpretation. Unclear instructions may result in lost production, product contamination, environmental release, or worse—injury or loss of life.

For example, when describing an amount, such as the amount of salt to add to a recipe, and an exact amount is not necessary, then it is acceptable to say a "pinch." However, if 1/8 teaspoon is the correct amount, then the recipe must say 1/8 teaspoon. Other examples of specified measurements include the following:

- Do not allow the pressure to exceed 80 pounds per square inch gauge (PSIG).
- Add chemical A to the mixer, increasing the mixer level 1 inch.

Accurate, detailed information is required when referring to equipment or identifying valves. Examples of accurate equipment descriptions include:

- FC-5555, column feed valve flow controller
- PC-1111, reflux drum pressure valve controller.

Use Common Names

When writing procedures, use commonly accepted terminology. Although many different terms may be used to identify and describe a piece of equipment, use the common acceptable name and accurate, correct descriptions. Examples of common acceptable names include:

- Double valve and vent, or double block and bleed, or pad and depad
- Reboiler or calandria
- Suction pot or suction drum
- Heater or furnace
- Operator or technician.

Avoid Ambiguous Words

Ambiguous, unclear words create uncertainty. Avoid words like *could, should, may, might,* and *ought* when writing procedures. Also, avoid using the words *about* and *approximately*, as they indicate a lack of precision. These types of words are subjective and allow for a choice. Procedures are specific; they are not suggestions or recommendations.

Follow Company Procedures

Internal procedures should be designed to meet the standards and requirements of the specific facility. An employer may have an exact method of procedure writing that is designed to meet the organization's specific standards and requirements.

Group and Label Information

Grouping information into small manageable units will help define a logical sequence. Studies have shown that most people begin to have difficulty processing written instructions that exceed seven to nine steps. Complexity of the instructions can influence the number of steps required in a set of instructions.

The goal of procedure writing is to present a procedure that is specific, clear, and concise without overwhelming the user. Grouping and labeling the information into like groups is a method that is useful to break down a more complex set of steps into separate activities that work together to produce the desired result. Figure 7.3 contains examples of grouping and labeling information.

Figure 7.3 Examples of grouping and labeling.

Label: Warm up the Primary Lube Oil Pump Steam Driver

Item #	Initial each item as it is completed	Initial
1.	Action 1	
2.	Action 2	
3.	Action 3	
4.	Action 4	
5.	Action 5	
6.	Action 6	

Grouping: items 1–6

Label: Safety Checklist - Primary & Auxiliary Lube Oil Pumps

Item #	Initial each item as it is completed	Initial
1.	Action 1	
2.	Action 2	
3.	Action 3	
4.	Action 4	
5.	Action 5	

Grouping: items 1–5

Label: Place the Auxiliary Lube Oil Pump in the Standby Mode

Item #	Initial each item as it is completed	Initial
1.	Action 1	
2.	Action 2	
3.	Action 3	
4.	Action 4	

Grouping: items 1–4

Employ Visual Techniques

Visual techniques can be used to promote clarity and provide emphasis. Examples of common visual techniques include the following:

- Use underlined bolded text to emphasize a target value.

STEP	ACTION
4.	Hold the column pressure **below 50 PSIG**.

- Center, bold, or capitalize text on a line to call attention.

STEP	ACTION
6.	IMPORTANT! Notify the process technician before starting the reflux pump.

- Use numbers as opposed to text when stating values.

STEP	ACTION
2.	Start feed to the column at a rate between 25 and 35 gallons per minute (GPM).

- Present "if/then" statements in a table when a decision must be made.

STEP	ACTION	
1.	If the column pressure is:	Then:
	< 15.0 PSIG	Fully open the methane makeup hand valve.
	> 90.0 PSIG	Open the vent to flare hand valve until the local flow indicator reads 2 standard cubic feet per second (SCF/S).

- Use check/initial boxes to document step completion. This helps to ensure that each step is completed in the correct order. This technique is commonly called a **checklist**.

Checklist procedure written in a list format that requires the user to initial or check the completion of each step.

STEP	ACTION	CHECK
1.	Open the discharge isolation valve wide open.	________
2.	Start the pump using the local START/STOP switch.	________
3.	Verify discharge flow is > 30 pounds per hour (PPH).	________

As a rule, if a template or table is not used when writing a procedure, it is important to *write only one line per step* and to *leave plenty of white space.* Do not write paragraphs containing multiple steps. Figure 7.4 provides an example of an operating procedure that includes a checklist.

7.5 Practice Procedure Writing

Learning to write procedures correctly is important, and practice will enhance the writer's skill. A simple method to practice writing procedures is to choose a simple task and write a procedure. For example, begin by choosing a task involving a home or car with which you are very familiar. Write an in-depth procedure to perform the task.

Note: Keep in mind the procedure is intended for a skilled and qualified user.

Most procedures begin in the field. Here is an example of how to write a procedure to change a tire on a vehicle:

- Stand by the car with pad and pencil.
- Consider safety first and throughout the process of documenting the steps.
- Envision the skill and knowledge level of the person who will use the procedure (audience analysis).
- Envision the circumstances such as when and where the procedure will be used, and possible hazards created by location.
- Next, list the various major steps required to complete the task.

(Text continues on p. 99.)

Figure 7.4 Sample of an operating procedure.

OPERATING PROCEDURE SAMPLE

Title	**PRODUCT COLUMN FEED PUMP – STARTUP AFTER MAINTENANCE**
Purpose	This procedure provides instructions for the startup of a Product Column Feed Pump after maintenance.
Before you begin	Verify that the maintenance department has signed completion of any work performed. Visually inspect the pump and motor. Check for any signs that the equipment is not ready to be returned to normal service. **If you have any concerns, STOP!** *Contact the maintenance representative and resolve all concerns before proceeding.*

Safety, Health and Environmental Considerations

Possible Hazard	**Mitigation**
Personal exposure to cold product material	Wear proper PPE including leather gloves.
Damage to pump internals/pump seal possible release of hazardous material	Reduce the pump internal temperature at a rate less than 10 degrees F per hour
Personal exposure to rotating parts	Motor and pump shaft guards must be properly installed

Tools or Equipment Needed

- Printed copy of this procedure
- Personal protective equipment
- 6-inch Valve wrench
- 16-inch Valve wrench

Procedure Pre-Startup Checklist

- All items must be satisfactorily completed prior to any attempt to start the pump. Contact maintenance to resolve any problems.
- Initial each item as it is completed.

STEP	**ITEM DESCRIPTION**	**INITIAL**
1.	Verify proper reading on the local pump cavity temperature gauge. The pump has been opened to the air. If the reading is not ambient, have the gauge replaced.	______
2.	Visually confirm each of the following conditions	______

POSITION	***DESCRIPTION***	
CLOSED (HS-20)	Pump suction motor operated isolation valve	______
CLOSED (FC-21)	Pump minimum flow control valve (suction tank level control valve)	______
CLOSED	Pump suction hand valve	______
CLOSED	Pump discharge hand valve	______
CLOSED	Pump cavity vent to flare hand valve	______
CLOSED	Pump suction line vent to flare header hand valve	______
CLOSED	Pump discharge line vent to flare header hand valve	______
CLOSED	Pump seal vent to flare hand valve	______

page 1 of 4

Figure 7.4 *(Continued)*

OPERATING PROCEDURE SAMPLE

Procedure	STEP	ITEM DESCRIPTION	INITIAL
Pre-Startup Checklist	3.	Visually confirm the following valve positions	

POSITION	DESCRIPTION	
OPEN	Electrical breaker to pump motor	______
OPEN	Pump suction line process safety valve	______
OPEN	Pump discharge line process safety valve	______
OPEN	Pump cavity vent to air hand valve	______
OPEN	Pump suction line vent to air hand valve	______
OPEN	Pump discharge line vent to air hand valve	______
OPEN	Pump suction line drain valve	______
OPEN	Pump discharge line drain valve	______

4. Verify the console technician is reading a level between **40% and 60%** in the pump suction tank.

Do NOT attempt to start the pump motor if the level is < 40%. ______

5. Stroke the pump minimum flow control valve (FC-2221) from the control console. Visually confirm the valve travels full stroke. Escalate to maintenance if there are problems with proper operation. ______

Procedure	STEP	ACTION	INITIAL
N_2 Purge The Pump	1.	Connect a N_2 hose to the pump suction line drain valve.	______
	2.	Open the N_2 supply one full turn. Allow N_2 to purge through the pump,venting out the: - pump cavity vent to air - suction line vent to air - discharge line vent to air - discharge line drain valve	______
	3.	After 5 minutes, close all vents and drains. Begin pressure purging the pump 3 times to 50 PSIG. Blow down through the discharge line vent to air only. Leave all vents and drains closed after the last blow down.	______
	4.	Disconnect the N_2 hose and store properly.	______

Procedure	STEP	ACTION	INITIAL
Liquid Fill The Pump	1.	Close the electrical breaker that supplies power to the pump suction isolation valve.	______
	2.	Set the following controllers in the position described:	

Controller	Description	Position
HC-2220	Pump Suction Motor Operated Isolation Valve	FULL OPEN
FC-2221	Pump minimum flow controller /pump suction tank level controller	25% OPEN in MANUAL

______

3. Open the hand valve before and after the minimum flow control valve. ______

page 2 of 4

(Continued)

Figure 7.4 *(Continued)*

OPERATING PROCEDURE SAMPLE

Procedure
Liquid Fill The Pump

STEP	ACTION	INITIAL
4.	Open the vent to flare hand valve on the discharge line full open.	______
5.	Open the pump suction hand valve two full turns. Process liquid will begin to fill the pump suction line, the pump cavity, and the pump discharge line.	______
6.	Observe the pump discharge line vent to air tubing for signs of frost, which indicates flow of process material through the pump.	______
7.	Monitor the local temperature gauge and adjust the discharge line vent to air hand valve to limit the pump internal cooldown rate to < 10 degrees per hour.	______
8.	Close the discharge line vent to air hand valve when the pump cavity internal temperature is –40 degrees F.	______
9.	Hold the pump in this position until the product column is ready to receive feed.	______

Procedure –
Place the Pump
In Normal Service

When the Product Column is ready to receive feed, follow these steps to place the pump in normal service.

STEP	ACTION	INITIAL
1.	Visually confirm the following conditions.	______

POSITION	*DESCRIPTION*
CLOSED	All process vents and drains
FULL OPEN	Pump seal vent to flare
FULL OPEN	Pump suction hand valve
FULL OPEN	Pump discharge hand valve
FULL OPEN	Minimum flow hand valves before and after the control valve

______

STEP	ACTION	INITIAL
2.	Close the electrical breaker that supplies power to the pump motor.	______
3.	Set the pump minimum flow controller on **AUTO** at the normal control point of **50%**.	______
4.	Set the Product Column feed flow controller on **AUTO** at the startup set point of **30 GPH**.	______
5.	Start the pump motor using the local START/STOP/REMOTE switch.	______
6.	Observe the operation of the pump and motor locally for unusual noise or other signs of abnormal operation. **SHUT DOWN the pump if there is any reason for concern!**	______

page 3 of 4

Figure 7.4 *(Continued)*

OPERATING PROCEDURE SAMPLE

	STEP	ACTION	INITIAL
Procedure – Place the Pump In Normal Service	7.	Place the local START/STOP/REMOTE switch in the REMOTE position.	______
	8.	Refer to procedure 'Product Column Normal Operation'. Increase column feed rates per that procedure.	______

Procedure Completion Signatures — This procedure was completed by:

____
Process Technician Signature

Date/Time:__________________

Procedure Validation Signature — This procedure was validated by:

____
Subject Matter Expert (SME) Signature

Date/Time:__________________

Procedure Approval — This procedure was approved by:

____
Authorized Approver Signature

Date/Time:__________________

Revision History — **Update as revisions are made**

DATE	DESCRIPTION OF CHANGE	BY

page 4 of 4

- Then, break down the activities into steps, noting the tools, equipment, and other materials that are necessary for the task.
- When you have finished writing the procedure, go to the beginning and test the procedure several times until you are comfortable with your draft.
- Think critically and look for something missed or not previously considered.
- When you are satisfied with the steps for the procedure, transfer your notes into a format suitable for presentation and review.
- Have another person start at the beginning and test the procedure.

Remember that a good procedure meets the following criteria:

- Is easily identifiable
- Is easy to access and execute
- Is task specific

- Is accurate
- Is well organized
- Targets the intended audience
- Is not overwhelming
- Eliminates or reduces the potential for error.

The procedure is finally ready for a subject matter expert (SME) to review and evaluate critically. The SME may ask the following:

- Were all safety issues documented?
- Were any steps or elements missing?
- Is the procedure clear and concise with nothing left open to interpretation?
- Do the procedural steps follow a logical sequence?
- Were there any unanswered questions?

To ensure accuracy and clarity, request that another individual pilot test the procedure. Do not be discouraged if the procedure has flaws and must be reworked a few times to make it perfect.

Summary

The training and skills acquired as a process technician eventually qualify an individual to play a critical role in the development, creation, and use of operating procedures. Operating procedures are a very important document in the industry because they guide personnel to perform tasks successfully. Process technicians may also be asked to review existing procedures for accuracy or to update an existing procedure due to a process change. The role of procedure writer demands the same level of ownership and accountability as other responsibilities of the job.

The goal of procedure writing is to present a procedure that is specific, clear, and concise without overwhelming the user. With practice, a procedure writer will master the writing techniques that produce procedures that are both easy to follow and allow the process to be completed in an efficient, safe manner.

Several principles and techniques can help a procedure writer to write effective procedures. Development of a successful procedure can be divided into stages, information gathering, organization, and presentation. Gathering information includes considering requirements and references from subject matter experts. Procedure templates can be used to organize the information in logical manner. Presentation of the information includes the actual writing of the procedure, as well as answering the questions *what, when, how,* and *why.* Major techniques for effective procedure writing include being precise, using common names, avoiding subjective words, following company procedures, grouping and labeling, and employing visual techniques. Learning to write procedures correctly is important, but practice will enhance the writer's skill.

Checking Your Knowledge

1. Define the following terms
   a. Checklist
   b. Internal procedure
   c. International Organization for Standardization (ISO)
   d. Occupational Safety and Health Administration (OSHA)
   e. Personal protective equipment (PPE)
   f. Procedure owner
   g. Procedure template
   h. Procedure user
   i. Process safety management (PSM)
   j. Subject matter expert (SME)

2. Identify characteristics of a successfully written procedure. (Select all that apply.)
   a. Short, unspecific
   b. Executable, error preventing
   c. Audience specific, accurate
   d. Rambling, easily identifiable
   e. Well organized, clear
3. (*True or False*) Poorly written procedures have sometimes been identified as a root cause of industrial accidents around the globe.
4. Which of the following phrases are acceptable to use in a procedure? (Select all that apply.)
   a. Open the block valve 5 full turns.
   b. Open the block valve about a third.
   c. Open the block valve gradually until you observe condensate coming out of the downstream drain valve.
   d. Close the vent when the pressure is below 20 PSIG.
5. What is the purpose of having a spot for initialing the completion of the steps in a procedure?
   a. It helps identify the technician who did that particular step incorrectly.
   b. It helps ensure that each step is completed in the correct order.
   c. It helps determine who has completed the most steps in a given work day.
   d. It ensures that only authorized individuals have performed the procedure.
6. Which of the following is NOT a benefit of using a template?
   a. Directs the information gathering process and promotes logical organization
   b. Encourages creativity
   c. Guards against omission of critical steps or facts
   d. Promotes ease of execution
7. Name three words to avoid when writing procedure instructions.
8. Which of the following is NOT a visual technique used to promote clarity and provide emphasis.
   a. Center, bold, or capitalize text on a line to call attention.
   b. Use colorful, descriptive language
   c. Use check/initial boxes to document step completion.
   d. Use numbers as opposed to text when stating values.
9. ________ is always the first consideration when writing a procedure.
10. What should be step one in a procedure for cleaning the chuck on a drill press?
    a. Take out any drill bit still in the chuck.
    b. Clean out any big pieces of dirt still lodged in the chuck itself.
    c. Disconnect the power cable from the electrical source.
    d. Clean and oil the drill press thoroughly.
11. Which of these is not an *optional* part of a procedure?
    a. Valves and instruments associated with the equipment
    b. Description of equipment (like hot or cold)
    c. Personal protective equipment for the procedure
    d. Type of equipment (like rotary or displacement pump)

NOTE: Answers to Checking Your Knowledge questions are in the Appendix.

Student Activities

Using the *process scenario, current operating conditions,* and *process sketch* provided on the next page, write an operating procedure to prepare the north reflux pump for service, put it into service, and shutdown the south reflux pump. Your procedure will stop when the north pump is operating normally, and the south pump has been safely removed from service.

- Demonstrate the principles, concepts, and techniques introduced in this chapter.
- Ensure compliance with applicable safety, health, and environment (typical process facility policies) and OSHA regulations.

NOTE: For the purpose of this exercise, do not address preparation of the south reflux pump for seal replacement.

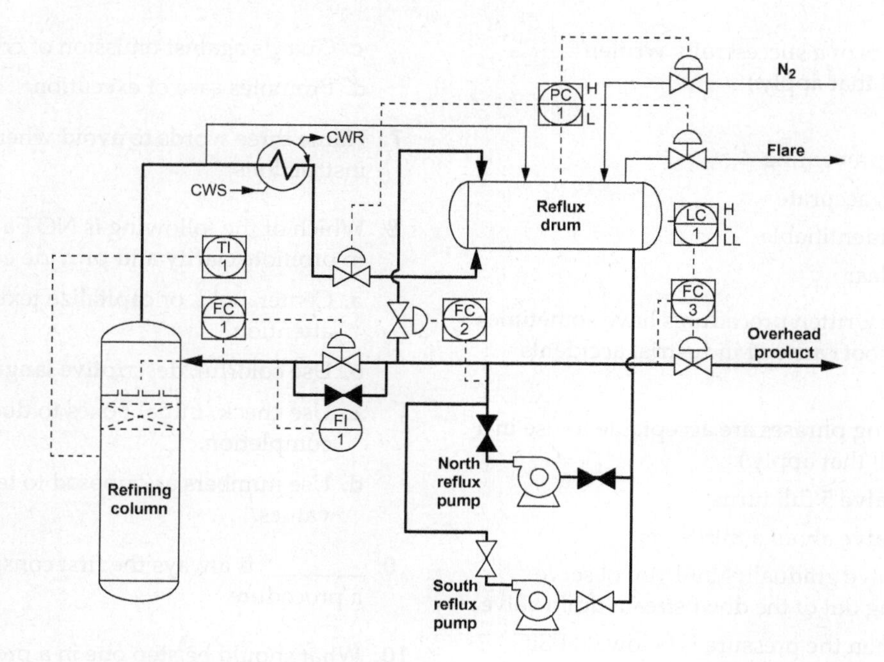

Process Scenario

- During normal operation, one reflux pump maintains reflux to the refining column and the second pump remains on standby.
- The refining column is operating normally and within limits; however, the south reflux pump has developed a seal leak. It must be shut down and prepared for seal replacement.
- There are no environmental or safety concerns because of the seal failure. Leakage across the seal of the south pump is venting to the flare through a dedicated vent line.
- The north reflux pump was recently out of service for maintenance and must be prepared and placed in service.
- Reflux flow must be maintained to avoid a disruption to the process and loss of production.

Current Operating Conditions

- All control parameters are within required limits and at set point.
- Isolation valves around the north pump are still closed because of the recent maintenance work on the pump.

Helpful Hints

- Remember, your procedure ends when the north pump is operating normally and the south pump has been safely removed from service.
- The current valve positions are identified on the sketch: white = Open, black = Closed.
- There are both local and remote START/STOP/REMOTE controls for each pump (not shown).

Chapter 8

Shift Change/Relief

Objectives

After completing this chapter, you will be able to:

8.1 Identify the types of information that need to be conveyed during shift change. (NAPTA Operations, Normal Operations Shift Change 1*) p. 104

8.2 Discuss the level of detail necessary to convey unit status information accurately. (NAPTA Operations, Normal Operations Shift Change 2) p. 105

8.3 Discuss the different methods used to make relief. (NAPTA Operations, Normal Operations Shift Change 3, 4) p. 106

8.4 Name the individuals who will be typically present during shift change. (NAPTA Operations, Normal Operations Shift Change 3) p. 110

8.5 Discuss the importance of making timely relief and establishing good relationships with members of your shift and other shifts. (NAPTA Operations, Normal Operations Shift Change 3, 5) p. 110

Key Terms

Effective communication—communication skills that help convey the intended meaning efficiently, **p. 109**

Electronic logbook (eLog)—computer based event logging program developed to assist the process technician record and report significant shift activities, **p. 106**

Maintenance activity—events performed by the maintenance personnel in a process facility, such as pump or compressor repair, pipe work, and routine general maintenance, **p. 105**

Nonoperating personnel—personnel other than process technicians who are performing work on the unit; these may include people visiting the unit, such as engineers, members of the management team, maintenance staff, and contractors, **p. 105**

*North American Process Technology Alliance (NAPTA) developed curriculum to ensure that Process Technology courses will produce knowledgeable graduates to become entry-level employees in process technology. Objectives from that curriculum are named here in abbreviated form. For example, "(NAPTA Operations, Normal Operations Shift Change 1)" means that this chapter's objective 1 relates to objective 1 of NAPTA's curriculum about shift change during normal operations.

Paraphrasing—summarizes the information received to clarify understanding, **p. 110**
Reflection—act of repeating what was communicated to confirm the information was understood correctly, **p. 110**
Shift change/relief—handing off the responsibility for operation and maintenance of a facility from one crew to another at a designated time; also known as *shift handover, shift pass-down, shift turnover, making relief,* and by other terms, **p. 104**
Unit status report—information gathered by the current operating shift for reporting to the oncoming shift during shift change, **p. 105**
Work permits—documents that allow individuals or groups to perform work on a process unit, **p. 105**

8.1 Introduction

In any continuous operation, maintaining continuity across shifts is vital. It is essential that process technicians, as well as their shift leaders, be able to communicate relevant information accurately so that the operation can continue to run safely and effectively. Oncoming personnel need a thorough and accurate understanding of facility status so that they can make correct decisions and perform appropriate actions (Figure 8.1).

Figure 8.1 Workers in process control room in oil blending factory.

CREDIT: Monty Rakusen/Cultura Creative RF/ Alamy Stock Photo.

Shift change is a critical activity and has been cited as a contributing factor in a number of major accidents, incidents, and process deviations. It is essential that organizations have the right tools and processes in place to execute effective shift changes.

Shift Change/Relief

Shift change/relief handing off the responsibility for operation and maintenance of a facility from one crew to another at a designated time; also known as *shift handover, shift pass-down, shift turnover, making relief,* and by other terms.

A **shift change/relief** is also known as *shift handover, shift pass-down, shift turnover, making relief,* and other terms. There are many items relevant to the operation and maintenance of the process unit that must be communicated at shift change. Important information includes:

- Safety issues
- Environmental issues
- Alarms and their current status
- Equipment conditions/problems
- Procedures in progress
- Process status
- Process trends
- Maintenance activities: completed, in progress, and planned

- Presence of nonoperating personnel
- Status of permits in force
- Product quality issues (off test or specification).

These items make up the **unit status report**. The status report is either written or verbal. A written status report can be used as a reference or guide during the verbal discussion process technicians have at shift change and as a tool for the relief technician during the coming shift.

Unit status report information gathered by the current operating shift for reporting to the oncoming shift during shift change.

8.2 Details About Unit Status

Both the field technician and the control board technician must provide an accurately detailed report of the current status of the process unit. The reports must include the following:

- Alarms—status of all current and previous alarms.
- Equipment conditions/problems—current equipment-related problems on the unit, including equipment placed in or taken out of service, any problems experienced with any of the equipment during the shift, as well as the results of the equipment inspection/monitoring performed during the current shift.
- Procedures in progress—operating and maintenance procedures in progress. The field and control board technicians will review the copy with their appropriate relief. Both the field and master copies of the procedure should be updated prior to shift change to provide the oncoming shift with accurate information so that procedure continuity across shifts is maintained and procedures can be safely and accurately completed. At the end of the shift, the field copy and the master copy should be identical.
- Process status—current and past process status, such as process safety management (PSM) limits exceeded and action(s) taken to return to normal, operational requirements exceeded and action(s) taken to return to normal, equipment being deinventoried, and abnormal situations on unit.
- Process trends—high levels, low levels, swings in temperature, or any other process-related variable that may be helpful to the oncoming shift, including nonroutine or unexplained events.
- **Maintenance activity**—should include completed, in progress, and/or planned, maintenance activity. It also includes maintenance activities that will be continuing across shift, the scope of the job, procedure information, and the number of personnel remaining on the unit. Activities planned for the next day requiring equipment preparation by the oncoming shift should also be communicated during the unit status report.
- Presence of **nonoperating personnel**—people remaining on the unit during or after shift change, including the number of personnel and their current and planned activities.
- Product quality issues—discussions of any off-specification process samples or analyzers.
- Status of **work permits**—work permit status needs to be discussed when covered work continues across the shift change (Figure 8.2). An updating of permit signatures may also be required. Shift change information includes a detailed status of each work permit that will require revalidation, the scope of the work, number of employees continuing to work, and current job status. The different types of work permits include hot work permits, confined space entry permits, and general work permits.
- Safety and environmental issues—any safety or environmental items, events, near misses, upsets, potential hazards, or potential upsets that occurred with emphasis on possibility of recurrence.

Maintenance activity events performed by the maintenance personnel in a process facility, such as pump or compressor repair, pipe work, and routine general maintenance.

Nonoperating personnel personnel other than process technicians who are performing work on the unit; these may include people visiting the unit, such as engineers, members of the management team, maintenance staff, and contractors.

Work permits documents that allow individuals or groups to perform work on a process unit.

Shift change/relief is part of a process technician's job requirement. If technicians do not make proper relief or leave without ensuring all important details have been clearly communicated, they may be subject to disciplinary action.

Figure 8.2 Two-person teams discuss the status of active work permits at shift change.

CREDIT: Hybrid Images/Image Source/Alamy Stock Photo.

8.3 Methods Used to Make Relief

There are several methods used for shift relief. The most common method in industry has been to communicate verbally to the appropriate oncoming technician (Figure 8.3). However, U.S. Chemical Safety Board (CSB) investigations have revealed deficiencies in shift changes, so many process facilities have opted to incorporate additional tools, such as a logbook or a shift pass-down sheet, to assist the technician in making a thorough relief. Coupled with effective verbal communication, these additional tools have proven to increase the transfer of relevant and critical information at shift relief.

Figure 8.3 Technicians making shift relief in the control room.

CREDIT: branislavpudar/Shutterstock.

The aforementioned methods of making relief are a regularly kept record of activities, performance, and events that occurred during each working shift. These forms of communications, whether electronic or handwritten, are considered legal documents that can potentially be used as evidence. The process technician should clearly state the facts as they occurred (for example, "P-101 broke down at 23:00 and was taken out of commission to be fixed.") and should not include speculation or opinions. This type of documentation can also be used for troubleshooting process systems. (For example, "How many times have we worked on P-101? Maybe it needs to be sent back to engineering.")

Electronic Logbooks

Electronic logbook (eLog) computer based event logging program developed to assist the process technician record and report significant shift activities.

Many facilities have added electronic logbooks to assist the technician in providing accurate shift change information. **Electronic logbooks (eLogs)** may be purchased or created in-house in an applicable worksheet or database. Figure 8.4 shows an example of an eLog with entries.

Figure 8.4 eLog entries.

Date	Time	Shift	Initial	Area	Type	Entry details
5/18/22	22:45	D	CC	Outside	Comment only	Made rounds Caught routine samples P-108B on line, P-108A down for PM P-104B on line, P-104A down for PM 2100 drained D-102 boot; 24" to 5" LOTO'd P-108A and P-104A for PM
5/19/22	4:31	D	ST	Outside	Comment only	[0400] Checked D-220 water boot 6" did not drain
	7:47	C	CC	Outside	Comment only	Flare verified lit with no visible signs of emissions
	12:13	C	CC	Outside	Comment only	Unloaded Toluene truck into D220
	14:39	C	CC	Outside	Comment only	Siemens came out and changed one of the water injection filters Took normal samples and made normal rounds
	16:27	C	CC	Outside	Comment only	Maintenance PM'd P104A, P108A, and P216 UnLOTO'd P104A, P108A, and P 216 and all are ready for service.
	16:28	C	CC	Outside	Comment only	[1430] LOTO'd P215 for PM
5/20/22	4:25	D	ST	Outside	Comment Only	Flare verified lit with no visible signs of emissions Routine rounds Routine samples Drained D-220 water boot 19" to 5"
	12:49	C	CC	Outside	Comment only	Completed Wednesday Audio/Visual Route Completed emergency horn testing (PASSED) URD switched from TK 105 to 106 at the docks for ship loading Flare verified lit with no visible signs of emissions Routine rounds
	13:26	C	CC	Outside	Comment only	Routine samples
5/21/22	11:47	C	CC	Outside	Comment only	11:00 lined the cooling tower back up to normal users UnLOTO'd cooling water service and return lines Returned hoses to proper location Chuck, Moe, and I filled the unit truck with gas
	15:30	C	CC	Outside	Comment only	Routine samples Routine rounds Flare verified lit with no visible signs of emissions
5/22/22	4:19	D	CC	Outside	Comment only	Flare lit with no visible signs of emissions Routine rounds Routine samples Tank farm good

Electronic logbooks can help ensure there is complete shift change information by providing areas where essential information is to be supplied.

Manual Logbooks

Logbooks are typically hardbound ledgers used to handwrite significant activities that have occurred during the shift. Figure 8.5 shows an example of a page of a handwritten hard-bound ledger. These are still used today in some locations, although the eLog is more widely used. Some drawbacks of the handwritten log include:

- Omitted information
- Legibility

Figure 8.5 Handwritten hardbound ledger entries.

Date	Time	Shift		Dwyer, Mendez, Turnbough
5/18/22	22:45	D	Outside	Made rounds
				Caught routine samples
				P-108B on line, P-108A down for PM
				P-104B on line, P-104A down for PM
				2100 drained D-102 boot 24" to 5"
				LOTO'd P-108A and P-104A for PM
5/19/22	4:31	D	Outside	[0400] Checked D-220 water boot; 6" did not drain
	7:47	C	Outside	Flare verified lit with no visible signs of emissions
	12:13		Outside	Unloaded Toluene truck into D220
	14:39		Outside	Siemens came out and changed one of the water injection filters
				Took normal samples and made normal rounds
	16:27		Outside	Maintenance PM'd P104A, P108A, and P216
				UnLOTO'd P104A, P108A, and P 216 and all are ready for service.
	16:28		Outside	[1430] LOTO'd P215 for PM
5/20/22	4:25	D	Outside	Flare verified lit with no visible signs of emissions
				Routine rounds
				Routine samples
				Drained D-220 water boot 19" to 5"
	12:49	C	Outside	Completed Wednesday Audio/Visual Route
				Completed emergency horn testing(PASSED)
				URD switched from TK 105 to 106 at the docks for ship loading
				Flare verified lit with no visible signs of emissions
				Routine rounds
	13:26		Outside	Routine samples

- Limited search ability
- Requirement to enter data in pen, making correction of incorrect entries difficult.

Shift Pass-Down Sheets

Shift pass-down sheets are generally preprinted or computer-generated forms with designated categories of information that facility management has deemed critical and relevant to facility or unit operation. These sheets are used for shift-to-shift pass-downs to provide an oncoming shift with a synopsis of facility operations. Figure 8.6 shows an example of a shift pass-down sheet.

Figure 8.6 Shift pass-down sheet.

Unit: NRU
Shift: Night Shift
Process technicians: Mendez-Board, Carpenter-outside, Turnbough-outside, Berry-Foreman

	Time	Comments
Flare status	1830	Verified lit by Carpenter and Turnbough
Sample status	1900	Samples caught and analysis showing no abnormalities, unit on-test at routine rates
Chemical inventory status	0000	Toluene in D-102: 50 % Toluene in D-220: 50 % Cooling tower sodium hypochlorite: 85 % Cooling tower sulphuric acid: 50 %
Tank levels	0000	TK-183: 51 %, Gauge: 15 FT 6 IN TK-184: 46 %, Gauge: 14 FT 8 IN
Maintenance activities	1830-2330	P-108A LOTO'd for maintenance PM on day shift tomorrow P-104A LOTO'd for maintenance PM on day shift tomorrow M-212A Had to place M-212B on line to clean "A" filter as delta P exceeded 15 PSIG limit. "B" filter back in standby after cleaning and placing "A" back on-line LV-102-had slight issue with valve sticking, Turnbough greased and have not had any problems since No other maintenance issues
AVO route	0300	Completed Audio/Visual/Olfactory route with no unusual observations
Process adjustments	1830-0600	D-102 Increased level to 55% when LV-102 was not tracking, see maintenance issue entry on this. Have since lowered drum level back to 50% No other process moves
Safety	1830-0600	No safety issues or concerns this shift
Miscellaneous	1830-0600	Running low on 3/4" bull plugs and sample bottles- please order tomorrow
Oncoming shift	0630	Cobb-Board, Dwyer-outside, Perry-outside, Duncan-Foreman

Report approved by: Jack Berry - Foreman, NRU

Verbal Communication

Verbal communication is extremely effective only if all necessary information is passed on to the oncoming technician. Occasionally, a process technician will remember pertinent information after leaving the facility. When this happens, it is important to call back to inform the relief of the missed information, especially if the information is critical to process performance.

If the technician is using only verbal communication to execute a proper shift pass-down, it is extremely important that each technician involved be trained in **effective communication**, which helps to convey the intended meaning more efficiently. Verbal

Effective communication communication skills that help convey the intended meaning efficiently.

communication, coupled with additional tools as mentioned earlier, can be very effective in executing a productive shift change.

8.4 Participants in the Shift Change

Shift change can occur in one of two ways: individually or in a group setting with both shifts attending. Each facility has its preferred manner for executing shift change. The most common is the individual, or technician-to-technician, method.

In the individual, technician-to-technician communication method, the current unit technician sits down with the oncoming technician and communicates, face-to-face, the shift's activities (Figure 8.7). The technician communicates verbally using handwritten notes, entries from an eLog, entries from a hardbound logbook, or shift pass-down sheets.

Figure 8.7 Shift change participants in the control room.

CREDIT: Dmitry Kalinovsky/Shutterstock; embedded image Simulation Solutions.

In the group communication method, both the current shift and the oncoming shift sit in a group setting and communicate the activities that occurred during the shift. In this type of setting, a member of the facilities management team may lead the meeting. Each technician may report activities that occurred during the current shift. Like with the individual method, this type of communication is done verbally, using handwritten notes, entries from an eLog, entries from a hardbound logbook, or shift pass-down sheets. Group shift change is often used for turnarounds, unit startups, unit shutdowns, or special projects.

An efficient tool to use, whether communicating in a technician-to-technician or in a group situation, is **paraphrasing** and **reflection**. An oncoming process technician should listen intently to the information being reported during shift change, and then should paraphrase what was heard, repeating the information back to verify that he or she understood the meaning of what was being conveyed. This ensures that the current shift has relayed all information and that the oncoming shift understands what is being shared. The difference between what was intended and what was understood becomes apparent.

Paraphrasing summarizes the information received to clarify understanding.

Reflection act of repeating what was communicated to confirm the information was understood correctly.

8.5 Timely Relief and Relationships among Staff

Facilities have an appointed time to make shift change/relief. Shift change/relief is an essential requirement of a process technician's job. If technicians do not make proper relief or leave without ensuring all important details have been clearly communicated, they may be subject to disciplinary action. The responsibility is shared by both the outgoing and oncoming shift. Outgoing technicians should be comprehensive and detailed with the communication of the current unit status. Oncoming technicians should make sure they understand all pertinent information. Asking questions, paraphrasing, and reflecting may take an extra moment, but this may make a difference in having the next shift run smoothly and efficiently.

It is extremely important that the shift change occur on time and be effective. If a person makes relief late, the knowledge transfer between the technicians is generally rushed, and information may be lost. In most 12-hour shift schedules, the person being relieved is the person most likely to make relief at the end of the next shift. A successful shift change will communicate all necessary information to the person coming on shift. The expectation is that that person will do the same when being relieved.

Establishing good relationships with members of one's own shift, as well as members of other shifts, is extremely important to the role of process technician. Soft skills (interpersonal skills) are critically needed to maintain good working relationships (Figure 8.8). Within an industrial environment, technicians will typically have to interact with one another to get a task completed. Without good interpersonal skills, their knowledge and occupational skills can become lost in the shuffle. Shift change is a time when communication and focus are especially important. Attentiveness, accepting feedback, dealing with difficult personalities, using persuasive and clear language, and showing respect are examples of soft skills that are vital to an effective and safe shift change.

Figure 8.8 Interaction with team member.

CREDIT: Hybrid Images/Image Source/ Alamy Stock Photo.

As one set of technicians leaves the site, the next set will need be fully aware of the state of the processes and equipment, as well as other facts specific to the site. If a person coming on shift does not appear to be open to hearing and learning about an event or a problem, then it is essential for the person going off shift to leave written documentation behind as part of the turnover.

When forming relationships with other technicians, some important behaviors to consider are the following:

- Be safe. Work safely, never place a coworker in harm, stop unsafe acts, and follow procedures.
- Be attentive. Listen to peers and especially technicians with experience; they are there to help.
- Be honest. Conduct yourself honestly in all things, and if you make an error, own up to it and learn from it. This will protect you in the long run, and it will also protect your coworkers and plant equipment.
- Be knowledgeable. Learn everything possible about your area of operating responsibility and help others to learn, too.
- Be prompt. Arrive at work on time and make good shift change/relief; these qualities are greatly appreciated by others.
- Be a team player. Help coworkers establish a positive working relationship with other shift technicians.
- Be willing. Listen, learn, help others, and volunteer for tougher job assignments.

Summary

The shift change/relief is critical in exchanging key information between the current operating shift and the relieving shift. Without a proper exchange of information between the shifts, the likelihood of an incident increases. Accidents, incidents, and process deviations can be related to poor shift changes.

Face-to-face communication is a good practice for shift change/relief that can be improved by the addition of structured written material, such as entries from a logbook, an electronic logbook (eLog), or a checklist, which reduce the risk of incomplete communication.

As a process technician, the goal is to provide sufficient information to the relief operator so that he or she has a complete understanding of the unit operations during the previous shift. A relief technician with good communication skills asks questions, paraphrases, and reflects what has been related. Paraphrasing and reflecting what was heard tells the other party what image her or his words painted. Differences between what was intended and what was understood then become apparent. Poorly chosen words frequently create incorrect mental images for the listener, but good communication skills help operating technicians obtain a complete understanding of the information being conveyed. Developing soft skills will enhance your ability to perform effective shift change/relief.

Checking Your Knowledge

1. Define the following terms:
   a. Effective communication
   b. Electronic logbook (eLog)
   c. Maintenance activity
   d. Nonoperating personnel
   e. Paraphrasing
   f. Reflection
   g. Shift change/relief
   h. Unit status report
   i. Work permits

2. Which of the following are included in a unit status report? (Select all that apply.)
   a. Alarms and their current status
   b. Procedures in progress
   c. Welcoming messages to new staff
   d. Process status
   e. Hours worked by process technicians
   f. Status of permits in force

3. A shift change/relief is also known as __________. (Select all that apply.)
   a. shift handover
   b. shift refresher
   c. shift passdown
   d. making relief

4. At the end of the shift the field and master control room copies of a working procedure should:
   a. add extra notations to the field copy to assist workers on the next shift
   b. be identical
   c. add greater detail to the master control room copy for the benefit of the archives
   d. edit the field copy to abbreviate and summarize steps to streamline work

5. Process status reporting includes which of the following? (Select all that apply.)
   a. Process safety management (PSM) limits exceeded
   b. Contractors remaining on the unit after shift change
   c. Equipment on order through purchasing
   d. Operational requirements exceeded
   e. Equipment to be deinventoried in the next fiscal year
   f. Abnormal situations on unit

6. Which of the following has been the most common method of communication to oncoming technicians during shift relief?
   a. Electronic logbooks (eLogs)
   b. Manual logbooks
   c. Shift pass-down sheets
   d. Verbal communication

7. Which of the following is a characteristic of manual logbooks?
   a. Easy to make corrections
   b. Excellent legibility
   c. Used when eLogs are not in use
   d. Unlimited search capability
8. The most common manner for executing shift change is the individual, or __________, method.
   a. Written memo
   b. Technician-to-technician
   c. All-hands
   d. Info passdown
9. Which of the following are efficient tools to use either in a typical technician-to-technician situation or in a group situation? (Select all that apply.)
   a. Refraction
   b. Paraphrasing
   c. Sign language
   d. Reflection
10. Which of the following are suggestions for forming a good relationship with other process technicians? (Select all that apply.)
   a. Be attentive.
   b. Insist on being the leader.
   c. Be a team player.
   d. Be willing.

NOTE: Answers to Checking Your Knowledge questions are in the Appendix.

Student Activities

1. This activity contains two parts:
   - Work with a classmate to perform a shift pass-down, developing logbook entries and verbal or written communicating shift and section activities to each other.
   - When acting as the oncoming relief technician, practice *paraphrasing and reflecting* what you are told for the off-going technician to compare your understanding with what he or she intended.
2. Write a one-page paper on what your expectations are in receiving and giving a shift pass-down to/from another process technician.
3. Perform research on electronic logbooks; select a program and write a one-page paper on how that program works and why you made that selection.

Chapter 9

Utility and Auxiliary Systems

Objectives

After completing this chapter, the student will be able to:

9.1 Describe a typical plant steam system, uses of the various steam pressures, as well as hazards and mitigating factors. (NAPTA Operations, Normal Startup Utilities and Auxiliaries 1, 2, 3*) p. 115

9.2 Discuss water treatment systems, as well as the hazards and mitigating factors that apply to them. (NAPTA Operations, Normal Startup Utilities and Auxiliaries 2, 3, 4) p. 120

9.3 Describe an aeration-type sanitary sewer system, as well as hazards and mitigating factors that apply. (NAPTA Operations, Normal Startup Utilities and Auxiliaries 1, 2, 3) p. 125

9.4 Describe the refrigeration cycle, its major components, hazards, and mitigating factors. (NAPTA Operations, Normal Startup Utilities and Auxiliaries 1, 2, 3) p. 127

9.5 Describe the types and function of cooling tower system, related hazards, and mitigating factors. (NAPTA Operations, Normal Startup Utilities and Auxiliaries 1, 2, 3) p. 129

9.6 Explain electrical systems and the importance of lockout/tagout while working on electrical equipment. (NAPTA Operations, Normal Startup Utilities and Auxiliaries 1, 2, 3) p. 131

9.7 Discuss air systems, hazards, and mitigating factors. (NAPTA Operations, Normal Startup Utilities and Auxiliaries 1, 2, 3) p. 132

*North American Process Technology Alliance (NAPTA) developed curriculum to ensure that Process Technology courses will produce knowledgeable graduates to become entry-level employees in process technology. Objectives from that curriculum are named here in abbreviated form. For example, "(NAPTA Operations, Normal Startup Utilities and Auxiliaries 1, 2, 3)" means that this chapter's objective 1 relates to objectives 1, 2, and 3 of NAPTA's curriculum about normal startup of utilities and auxiliaries.

9.8 Describe the basic operation of pressure relief and flare systems and associated hazards. (NAPTA Operations, Normal Startup Utilities and Auxiliaries 1, 2, 3) p. 134

9.9 Explain the use of nitrogen, and identify hazards associated with its use. (NAPTA Operations, Normal Startup Utilities and Auxiliaries 1, 2, 3) p. 136

9.10 Describe the use of natural gas and the differences between natural gas and process off-gas. (NAPTA Operations, Normal Startup Utilities and Auxiliaries 1) p. 137

Key Terms

Boilers—vessels in which water is boiled and converted into steam under controlled conditions, **p. 116**

Electric heat tracing—series of self-regulating heating cables designed to provide freeze protection and temperature maintenance to metallic and nonmetallic pipes, tanks, and equipment, **p. 124**

Firewater—water from plant firewater lines used for emergencies, **p. 122**

Flashback—situation in which gas vapors ignite and return to the source of the vapors, **p. 134**

Hydrogen sulfide (H_2S)—highly toxic, highly flammable, colorless gas with a very distinctive, rotten egg-like odor, **p. 131**

Natural gas—a flammable gas associated with gas and oil fields and consisting principally of methane and the lower saturated paraffin hydrocarbons. It may also include impurities such as water vapor, hydrogen sulfide, and carbon dioxide, **p. 137**

Nitrogen—an odorless, invisible, inert gas, forming approximately 80% of the atmosphere. An important purging and blanketing medium, **p. 136**

Potable water—water that is safe to drink and use for cooking, **p. 123**

Steam—water vapor, or water in its gaseous state. The gas or vapor into which water is converted when heated to the boiling point, **p. 116**

Steam clouds—tiny drops of water that have condensed from steam and are carried along by the invisible vapor, **p. 119**

Steam generators—any shell and tube exchanger or kettle type exchangers using boiler feedwater (BFW) to remove process heat, convert BFW to steam, and then pressure control that steam to a supply header, **p. 120**

Steam jets—essentially a steam nozzle that discharges a high velocity jet and used to create a particular pattern of spray steam, **p. 119**

Steam tracing—tubing that is installed adjacent to a pipeline and is enclosed with the pipeline by insulation. Steam is then passed through the tubing providing heat, **p. 119**

Steam turbines—(1) prime movers for the conversion of heat energy of steam into work on a revolving shaft, utilizing fluid acceleration principles in jet and vane machinery. (2) Turbines driven by the pressure of steam discharged at high velocity against the turbine vanes, **p. 116**

Water hammer—hydraulic action associated with a noncompressible fluid in a pipe, so named because it sounds like a pipe being hit with a hammer; the energy developed by the sudden stoppage of fluid in motion, **p. 119**

9.1 Introduction

Every process facility is dependent on one or more utility or auxiliary systems to produce products. Most facilities require energy, and most manage utility sections that specialize in steam generation, water systems (boiler feedwater, firewater, potable water, cooling water, and wastewater), and compressed gases (nitrogen, air, and natural gas). In addition, process facility auxiliary systems include electricity, flare systems, and refrigeration.

Process technicians are responsible for maintaining and operating these systems according to the operating guidelines of the process facility. Technicians ensure that the utility systems operate safely and efficiently just as they do for the main process systems in the plant.

Steam Generation and Distribution

Steam water vapor, or water in its gaseous state. The gas or vapor into which water is converted when heated to the boiling point.

Boilers vessels in which water is boiled and converted into steam under controlled conditions.

Steam in industrial settings drives turbines, operates pumps, provides process heating, warms heat tracing, and provides building heat. Steam is vaporized water at a temperature above the boiling point of water at sea level and atmospheric pressure (above 100 degrees Celsius [212 degrees Fahrenheit]).

A steam distribution system consists of valves, fittings, piping, and connections designed for the system's pressure and temperature. **Boilers** are equipment used to heat water to produce steam. They generate steam at a given design pressure to satisfy the process requirements. High pressure steam (250 to 600 PSIG) drives turbines on large generators, gas compressors, and pressure letdown stations. High pressure steam from boilers may be supplemented with high pressure superheated steam from various plant waste heat steam generating equipment.

Letdown stations are pressure control valves that reduce high pressure steam to a lower pressure supply header. Letdown stations vary by the type of process, but typically are 400 to 150 PSIG and 150 to 50 PSIG. Note, though, that plant steam system pressures vary widely and that process operators may be employed in industries that do not use steam energy.

A principal application of steam is as a source of heat in heat exchangers (a heat medium). The supplied steam on the tube side gives up its temperature to the process liquid on the shell side of the exchanger.

Condensate is another valuable source of heat at much lower pressures and temperatures. An advantage of condensate is that, because it is treated boiler feedwater, it can be sent back to the steam generation boiler feedwater system.

Steam turbines (1) prime movers for the conversion of heat energy of steam into work on a revolving shaft, utilizing fluid acceleration principles in jet and vane machinery. (2) Turbines driven by the pressure of steam discharged at high velocity against the turbine vanes.

Steam turbines (Figure 9.1) are rotating mechanical drivers powered by high velocity steam. They may power pumps, compressors, or generators. The work performed by the steam is increased by reducing the pressure of the turbine steam exhaust. In a process using a large steam driven turbine, the steam may exhaust from the turbine and condense in a water-cooled surface condenser, which creates a partial vacuum on the turbine discharge.

A surface condenser is normally a large horizontal shell and tube exchanger with cooling water flowing through the tubes. Condensate is level controlled to the BFW system, and the

Figure 9.1 **A.** Illustration of a steam turbine. **B.** High pressure to low pressure turbine connected to generator.

CREDIT: B. arrogant/Shutterstock.

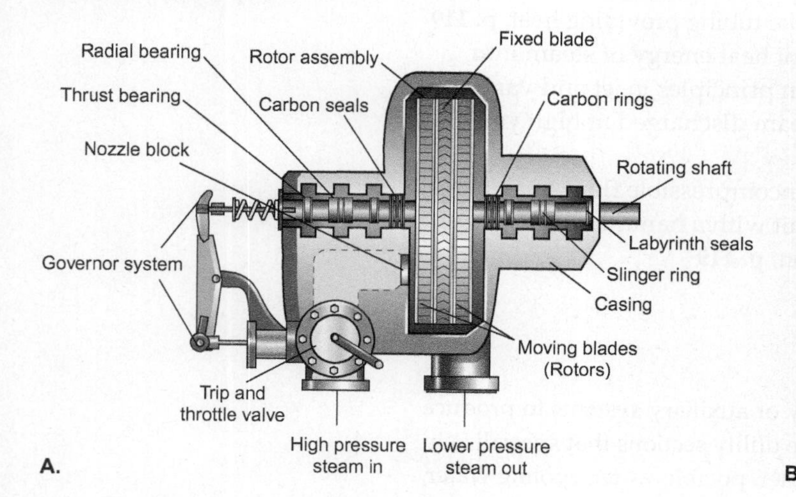

vessel operates under a partial vacuum. Some process units may supplement the vacuum on the condenser by utilizing steam ejectors.

TURBINE-SUPPLIED STEAM LETDOWN Steam turbines using high-pressure steam may exhaust directly into a lower pressure steam header and function as a pressure letdown system, such as a 150 PSIG steam header. Depending on design, turbines can exhaust at various pressures. However, the turbine is not the final control for the letdown station pressure.

If plant equipment consumes a larger volume of steam than the steam turbine exhaust produces, the steam header pressure falls. When the lower steam pressure falls below a set value, a pressure control valve lets down additional high pressure makeup steam into the lower pressure steam header (Figure 9.2). At times, the delicate balance of generated steam can fluctuate due to high or low demands. In that event, direct consumption of supply is suddenly reduced, which can cause rapid changes in steam pressure. To stabilize the system supply and adequately protect against overpressurizing, a vent to atmosphere is installed. This system helps control any imbalance in the system.

DESUPERHEATING When steam is let down to a lower pressure, it may become superheated, meaning that the temperature is higher than saturated steam of the same pressure. The temperature of the letdown steam can be desuperheated, or cooled, by adding atomized condensate. The degree of superheat is dependent on the facility design. The amount of desuperheating needed depends on the downstream demand for process heating that uses exchangers.

Heat exchangers function more efficiently when heating the process with saturated steam, which condenses in the heat exchange. Steam yields the greatest amount of heat when the phase change from steam to condensate occurs. Although superheated steam has a higher temperature, under most circumstances, it does not condense in a heat exchanger, so it yields less heat than saturated steam does.

POWER GENERATION WITH STEAM Every process facility requires electrical power to operate. Steam is often the preferred source of energy used to create electricity because of safety and economic reasons. Unlike hydrocarbons, steam is not explosive and often there is sufficient heat within a facility to create enough steam to use for electrical generation.

To produce electricity with steam turbines, pressure greater than typical process steam is necessary. Turbine inlet steam pressure is usually in the range of 250 to 850 PSIG at 700 degrees Fahrenheit (377 degrees Celsius) to 900 degrees Fahrenheit (482 degrees Celsius). The 600 PSIG steam turbine drives the generator and exhausts into a condensing system or a considerably lower pressure steam header, such as 50 PSIG.

Typically, high pressure steam enters the turbine casing, and since the casing is much larger than the steam supply line and is connected to the exhaust, almost immediately drops to near exhaust pressure. This is equal to either the partial vacuum of the condensing system or the letdown steam header pressure. Most turbine rotors (wheels attached to the turbine shaft) are equipped with blades or buckets. The steam velocity hitting the blades or buckets forces the rotor to turn.

Some turbines incorporate inlet nozzles, which increase the steam velocity to the rotor. Many turbines contain two types of blades, which act as nozzles and impart a velocity increase to the steam. Exhaust steam pressure varies depending on the type of turbine, such as single stage, multistage, condensing, or noncondensing. Whatever the type of turbine, the rotating turbine shaft drives the generator, which produces electric power.

Hazards and Mitigation

Superheated steam is invisible because the temperature is very high, and there are no condensing droplets to see in the atmosphere. Steam leaks are difficult to locate visually, but usually make excessive noise, alerting anyone in the area. Because a high pressure superheated steam leak necessitates wearing hearing protection, pinpointing the location of the leak may be difficult. However, personnel must wear all appropriate personal protection equipment (PPE) to avoid serious bodily injury.

Figure 9.2 Example of steam generation and distribution.

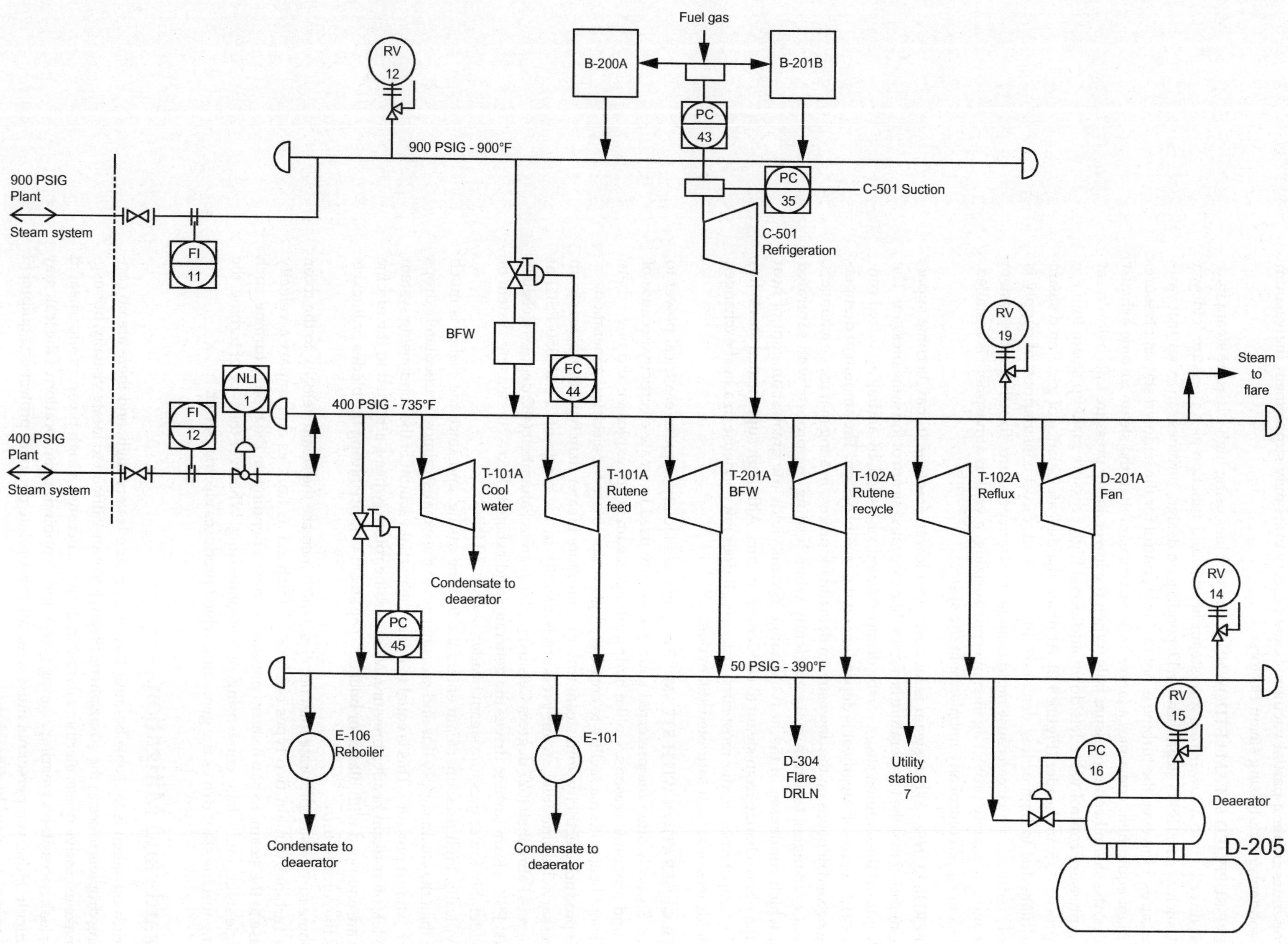

Because steam leaks are very costly, process technicians should tighten packing leaks on steam valves to eliminate waste. Large steam leaks can produce **steam clouds**, which can become hazardous, reducing visibility in an area and possibly creating trip hazards.

Steam clouds tiny drops of water that have condensed from steam and are carried along by the invisible vapor.

Condensing steam can cause damage to equipment and create other hazards. When equipment is purged with steam, a suitable gas, such as nitrogen, must then be used to clear the steam to prevent creation of a vacuum by the condensing steam.

If a vessel is left full of steam with closed valves and is not rated for full vacuum, equipped with vacuum breakers, or otherwise vented, condensation can produce sufficient vacuum to collapse (implode) the vessel. An implosion (also called a *cavitation*) is a sudden collapse in which the vessel's walls burst inward. This is a concern during turnarounds when vessels are steam cleaned. In addition to risk of implosion, the vacuum could draw in air, creating an explosive atmosphere or fire hazard when hydrocarbons are introduced. This is a concern during startup preparations.

Steam heating of isolated equipment that is full of liquid can result in dangerously high pressure. Excessively high pressure first develops within a confined space due to liquid expansion from heating and from phase change after further heating. Blocked heat exchangers or vessels full of water can explode if the water boils. Procedures that typically provide for flare or atmospheric venting should be followed during vessel steaming and drying. Equipment not designed for high temperature and pressure may sustain damage if steaming and drying are not performed properly.

Did You Know?

Did you know that 1 part of liquid water, when boiled, expands to approximately 1600 times that volume as steam?

Steam turbine casing failure normally occurs when casings are exposed to the full pressure of the inlet steam. Therefore, the exhaust valve of a steam turbine must be opened before opening the inlet.

Steam jets, an economical type of vacuum pump, can produce large static electrical charges. This hazard is avoided by grounding equipment properly when using steam or by making sure that in-place grounds are intact.

Steam jets essentially a steam nozzle that discharges a high velocity jet and used to create a particular pattern of spray steam.

During startup or normal operation, **water hammer** can occur when water vapor is present, or when steam condenses and the accumulated condensate is carried along with the steam flowing within the piping. When the condensate is forced to stop or change direction (for example, at a pipe bend), the result is a hydraulic shock due to the momentum change, which sounds like a pipe being hit with a hammer. Water hammer can be extremely dangerous if not controlled. To prevent water hammer, steam is admitted very slowly until equipment warms up, and condensate is removed through drain valves and/or steam traps.

Water hammer hydraulic action associated with a noncompressible fluid in a pipe, so named because it sounds like a pipe being hit with a hammer; the energy developed by the sudden stoppage of fluid in motion.

When **steam tracing** is used to prevent freezing on stainless steel piping or vessels, it is important to use chloride-free steam. Figure 9.3 shows an example of steam tracing. If leaking steam contains chlorides, chloride-induced corrosion of the stainless steel is likely. It is worth noting that 316 stainless steel is more resistant to chloride corrosion than 304 stainless. However, chlorides damage both grades of stainless steel over time. Steam tracing should be drained and purged with an inert gas or air to maintain the integrity of the system when it is not needed. This will eliminate the potential for leaks when the system is returned to service.

Steam tracing tubing that is installed adjacent to a pipeline and is enclosed with the pipeline by insulation. Steam is then passed through the tubing providing heat.

Figure 9.3 Example of steam tracing.

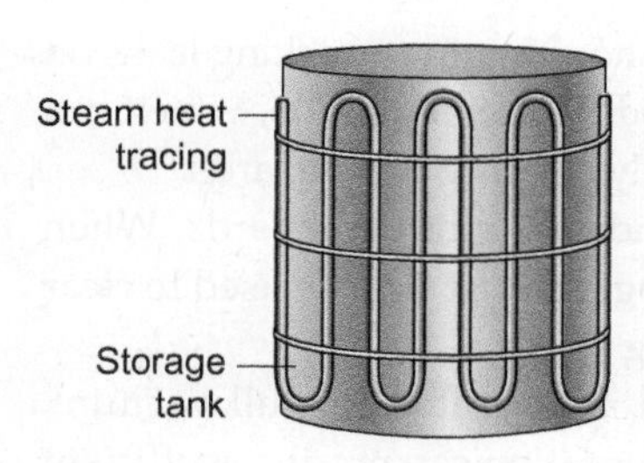

Uninsulated steam piping or tracing can be a source of serious burns. Steam piping should be insulated to prevent personnel from suffering serious burns that may result from unintentional contact. Steam, if breathed, can produce serious lung burns or suffocation.

9.2 Water Systems

Raw water, typically surface water from a channel, river, or canal, supplies the cooling water system, boiler feedwater treatment system, firewater, and utility water stations. Raw water is usually pretreated for sediment removal in sediment clarifiers for all systems except firewater. Potable water is typically supplied from a city water source or a water treatment and chlorination unit within the facility.

Boiler Feedwater (BFW) Treatment System

Water for steam generation is thoroughly treated to remove impurities that would compromise the integrity of process systems and operations. Boiler feedwater (BFW) is treated mechanically and chemically to remove or control contaminants. Poor quality BFW can cause extensive corrosion, pitting, and scale buildup within **steam generators**, steam piping, and exchangers. Steam generators convert BFW to steam and control the stream to a supply header.

Steam generators any shell and tube exchanger or kettle type exchangers using boiler feedwater (BFW) to remove process heat, convert BFW to steam, and then pressure control that steam to a supply header.

When steam is generated, impurities in the water become more concentrated in the boiler, causing a marked reduction in boiler efficiency. Specific chemicals are added to mitigate this problem by keeping these dissolved solids in suspension for blowdown. Often, the chemicals are supplied by the vendor.

Boilers have a continuous blowdown that the process technician adjusts, as needed, based on water sample testing. Boilers also have manual blowdowns that the process technician opens briefly, normally once per shift. When dissolved solids reach an unacceptable concentration, the blowdown process flushes the impurities from the boiler bottom out to a wastewater sewer. Makeup water and additional chemicals replace the water lost through blowdown.

High pressure steam generators require extremely pure water to avoid premature failure. They may use clarified raw water makeup passed through resin bed water softeners, resin bed dealkalizers, and then cation and anion bed demineralizers.

Roughly speaking, water softeners utilize salt-regenerated resin beds to remove heavy minerals, such as iron, by exchanging sodium ions from the resin for iron in the water. Dealkalizers reduce the alkaline mineral compounds, such as calcium and magnesium, which build up in the boiler bottom (Figure 9.4). Blowdown helps reduce or prevent scale buildup. Demineralizers further remove minerals, mineral salts, and ions (positive or negative), including the sodium ions remaining from the softening process.

To mitigate corrosion in a boiler or other steam generator by dissolved gases, especially oxygen, BFW is deaerated. Treated BFW is transferred to a horizontal vessel known as a deaerator. A deaerating heater has two sections; the upper section is the deaerator and the lower section is the storage section. In the deaerator, low pressure steam is mixed with feedwater across plates, causing the stripping of noncondensable gases (mainly oxygen and carbon dioxide) from the water. These gases exit the deaerator via a vent. An oxygen scavenging chemical is also introduced into the storage section, further lowering the oxygen content of the boiler feedwater.

Figure 9.4 Builtup scaling inside a pipe.

CREDIT: zulkamalober/Shutterstock.

To control corrosive minerals, low pressure steam systems may utilize less sophisticated methods, such as introducing lime or soda ash to the BFW to control pH and to aid in dissolved solids suspension for blowdown. The treatment of BFW depends on its application. Boiler feedwater treatment may use some or all of the following methods:

- Clarification
- Sedimentation
- Filtration
- Softening
- Dealkalization
- Demineralization
- Deaeration
- Chemical addition
- Blowdown

The boiler section of the facility produces high pressure steam. Boiler feedwater pumps take suction on the deaerator and transfer the treated BFW to each boiler and steam generator, which heat the water to make steam. The high pressure steam is exported to some users throughout the facility, while for other users the steam flows through a pressure letdown station and a desuperheater. Figure 9.5 illustrates this process.

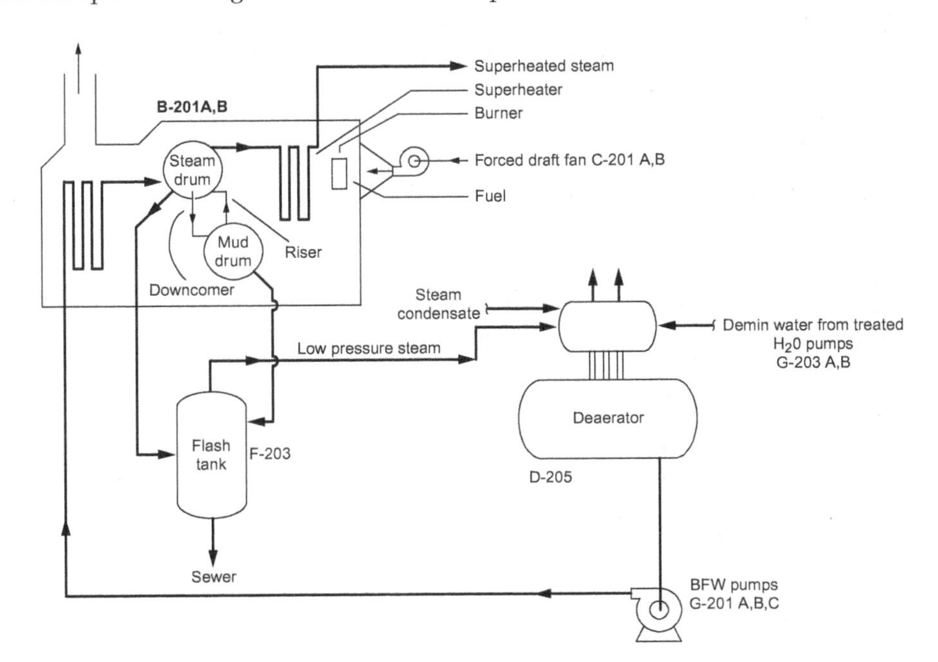

Figure 9.5 Illustration of a boiler feedwater circuit.

Hazards and Mitigation

In addition to being used for process heating, steam is commonly used to purge natural draft gas-fired heaters before igniting the pilots and burners. If a heater or boiler is shut down for any reason, faulty fuel gas isolation valves or a hydrocarbon leak from a tube may create a flammable atmosphere within the firebox. In older heaters, manual block valves are opened to admit steam into the firebox for several minutes to purge any hydrocarbon vapor that might be present. Pilot ignition follows purging.

Today's heaters, boilers, and furnaces are automated, monitored, and managed by what is commonly known as a *burner management system (BMS)*. On heater startup, the BMS automatically takes the heater through the purge cycle. A pilot-ready light will show, and the process technician will press a pilot ignition button. A flame scanner monitors the ignition of pilots and, if all pilot flames are scanned, provides local panel indication for the technician to proceed with burner ignition. The BMS typically shuts the system down if the pilot flames are not seen. Most BMSs do not trip an online heater for loss of a single burner; instead they activate an alarm. Normally, more than one burner must lose flame to shut down a heater. Forced draft and balanced draft heaters use a fan for purging with air.

Snuffing steam is usually lower pressure steam used to extinguish a fire in the heater firebox. It can be applied by opening manual block valves to a heater firebox in the event of tube failure that has resulted in a process fire. The steam displaces oxygen and hydrocarbon vapor to snuff the fire.

The following knowledge and protocols provide for safe, efficient process operation:

- **Safety**—Boiler feedwater and boilers must be kept within prescribed operating parameters to reduce the risk of overheating and failure. Excess water can cause irreparable damage to turbines by infiltrating the steam distribution system. Removing impurities from BFW greatly increases operating efficiency, improves equipment life, and reduces possible upsets. Blowdown systems reduce dissolved solids in steam drums and slow the accumulation of impurities on turbine blades. *Knockout pots*, also known as *knockout drums*, remove liquids from the fuel gas so they do not reach heater or boiler burners.
- **Health**—Safety protocols and proper PPE have been developed to help mitigate the possibility of exposure from dangerous scenarios and environments during normal inspection, routine maintenance, redundant sampling, shutdowns, and turnarounds.
- **Process hazard and boiler firing**—Boiler feedwater is added directly into the boiler's steam drum and therefore must be at a higher pressure than the boiler's output steam pressure. Use caution with high pressure and temperature any time work is being performed on the boiler feedwater. Be especially cautious when opening the block valves on a standby BFW pump. If the discharge check valve fails, the pump will spin backward fast enough to disintegrate if block valves are not closed quickly.

Firewater

Firewater water from plant firewater lines used for emergencies.

The firewater system stores and distributes a supply of **firewater** to all users within the facility in case of an emergency, typically through underground pipelines to prevent damage to the supply lines. Depending on the process facility, the firewater system will generally consist of a firewater storage tank (shown in Figure 9.6), firewater pumps (which may consist of one electric, one diesel, and one backup), control valves (in automated systems), and the piping associated with distribution.

Fire hydrants, monitors, hose reels, and deluge systems are checked routinely for condition, valve operation, valve position, valve handles in place, connections capped for hydrants, and position of monitor nozzles. The firewater system should be tested weekly to ensure that it is functioning properly. Valve operation is checked for free movement opening and closing. Freeze protection requires bleeders to be open. Caps should be on each connection, hand tight only. Monitor nozzles should be in full fog position and aimed toward the equipment they are to protect.

Figure 9.6 Example of a firewater storage tank.

CREDIT: Lev Kropotov/Alamy Stock Photo

Any time a fire or spill alarm is sounded, the firewater pumps are started immediately. The firewater pressure must be held at a minimum set pressure, for example 110 PSIG, during any emergency. If the pressure falls below the set pressure, the spare or diesel pump should start. If either pump were to fail during an emergency, the backup firewater pump is used.

Potable Water

The normal supply of **potable water** comes from an outside source. The potable water system stores and distributes a supply of potable water to all users, which include safety showers, eyewash stations, sinks, restrooms, and water fountains. The potable water system consists of potable water tanks, potable water pumps, and piping associated with distribution. Potable water used in sinks, restrooms, and drinking fountains drains into the facility sanitary sewer. Figure 9.7 illustrates this process.

Potable water water that is safe to drink and use for cooking.

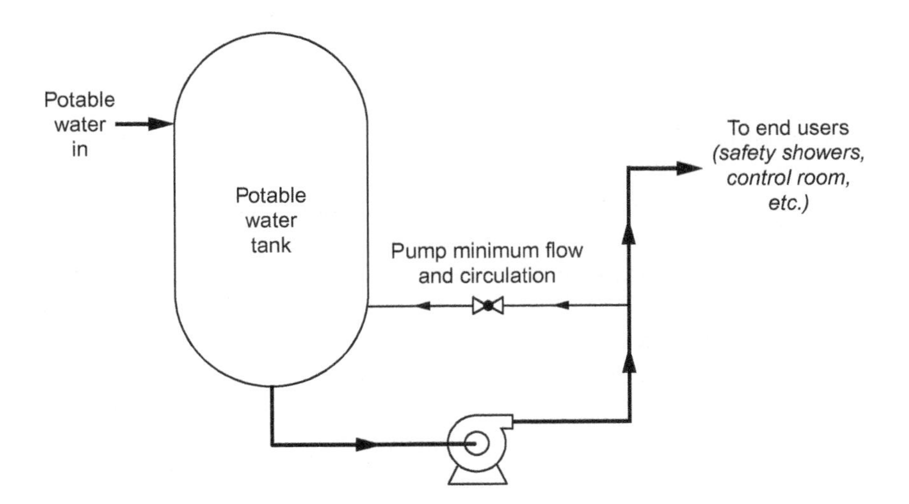

Figure 9.7 Illustration of a potable water system.

The state water commission sets the minimum acceptable operating practices for public potable water systems, which dictate that certain tasks be performed to ensure safe and healthy potable water for use in the facility. The process technician may be required to perform a routine monthly survey to determine if there are any potential sources of contamination in the potable water system. The survey checks for the following:

- Are there any hoses connected to the potable water system that may be inadvertently hooked to an unsafe water system?

- Are there any valves on the potable water system that do not have bull plugs or caps on them to prevent cross-contamination from piping or hoses connecting to unsafe water systems?
- Are the pH, system pressure, and tank level within range?

Potable water flows through filters that remove particulates before distribution to the users. Normally, one filter is in service and another is on standby. The filters are normally switched when the differential pressure across the one in operation increases. Local pressure gauges indicate the differential pressure. A bypass may be temporarily opened if both filters become plugged. Chlorine addition, at prescribed flow rates to meet specific ppm concentration, destroys bacteria and algae growth to meet local drinking standards.

Electric heat tracing series of self-regulating heating cables designed to provide freeze protection and temperature maintenance to metallic and nonmetallic pipes, tanks, and equipment.

The portions of the potable water supply header that are above ground may use steam or **electric heat tracing** (shown in Figure 9.8) and insulation to prevent damage to the piping in freezing weather.

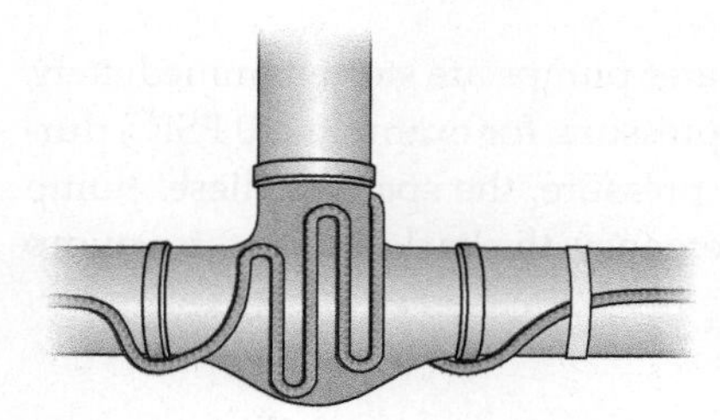

Figure 9.8 Example of electric heat tracing.

A heat trace panel controls electric heat trace circuits. A thermostat mounted inside the panel energizes the circuits when the selector switch is in the AUTO position and the outside temperature decreases to a predetermined temperature as set by the process facility. The heat trace panel may contain many circuits with different temperature settings for different applications.

Wastewater

Most process units collect oil and water from the process and storm drains into an American Petroleum Institute (API) oil-water separator. The API sets operating standards and design. Oil skimmers economically remove hydrocarbons from water to meet water quality objectives. Oil is skimmed from the water and recycled, while the water is pumped into the process wastewater (PWW) header. Water collects in a large sump, also known as an oil guard, from which it is transferred to the wastewater treatment section. In the event of heavy rain exceeding the pump capacity, an outfall occurs. Water will overflow the oil guard into the open ditch. Samples must be collected if an outfall occurs to ensure that it does not contain hydrocarbons. If sampling determines the outfall-contained hydrocarbons, remedial action must be taken for recovery.

Wastewater treatment is a process for cleaning contaminated water before recycling or release. Hydrocarbons and other contaminants, along with solids, are characteristic components of wastewater. Wastewater treatment takes water from processes that do stripping, steam boiler and cooling tower blowdowns, waste neutralization, and other process functions. Figure 9.9 illustrates the process.

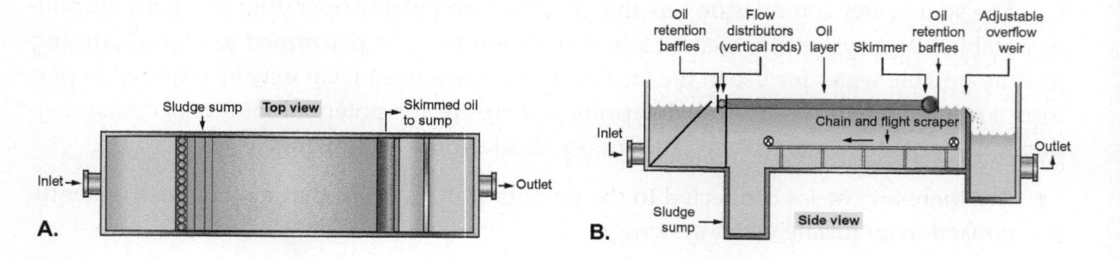

Figure 9.9 Water treatment process.

WASTEWATER TREATMENT OPERATIONS Water treatment operations consist of the following:

- **Pretreatment operations**—The first step in treating wastewater is separating solids and hydrocarbons. API separators, interceptor plates, and settling ponds capture fugitive contaminants, process matter, and sludge by separating, skimming, and filtering. Emulsions of oil and water are typically separated by heating. The difference in the specific gravity of water and oil particles enables fugitive oils to be captured from wastewater surfaces. Wastewater that is acidic ($pH < 7$) is neutralized with ammonia, lime, or soda ash. Wastewater that is alkaline ($pH > 7$) is neutralized with sulfuric acid, hydrochloric acid, sulfur, or carbon dioxide-rich flue gas. Carbon dioxide (CO_2) and water (H_2O) interact to form carbonic acid (H2CO_3), which is a naturally occurring component of acid rain.
- **Secondary treatment operations**—Wastewater containing suspended solids requires a removal process such as sedimentation, air flotation, screening, or filtering. The process of *flocculating* (clumping of particulates in solution) further facilitates separation of particulates that may have escaped the pretreatment and initial secondary removal processes. Flocculating is similar to the initial clarification of raw water in that typically, a polymer is added that brings smaller particles together into larger masses for settling out or removal by filtration. Secondary treatment operations decompose biological matter and oxidize soluble organic matter using biologically activated sludge (anaerobic bacteria added), lagoons, or filters. High adsorption molecules (added polymers) aid fixed bed filters by forming slurries that are removable by sedimentation or filtration. Other methods for removing oils and chemicals from wastewater include stripping and solvent extraction. Stripping is a process for removing sulfides and/or ammonia, and solvent extraction separates phenols.
- **Tertiary treatment operations**—Tertiary treatments including, but not limited to, chlorination, ozonation, ion exchange, reverse osmosis, and activated carbon adsorption capture regulated contaminants to satisfy regulated fugitive emission permit limits. Wastewater must contain a sufficient level of oxygen for some wastewater streams to oxidize specific chemicals and fulfill requirements. Wastewater recycling by cooling or oxidizing alleviates impurities by spraying or air stripping.

HAZARDS AND MITIGATION The following knowledge and protocols provide for safe, efficient process operation:

- **Fire protection and prevention**—During treatment operations, vapor from hydrocarbons in wastewater must be monitored because it can possibly create an explosive atmosphere.
- **Health**—Adhering to safe, responsible work procedures is essential for personal safety, health, and environmental integrity. The proper PPE for process sampling, inspection, maintenance, and turnaround activities must be used to lessen the probability for an accident that may harm people and damage equipment, facilities, or the environment.

9.3 Sanitary Sewer System

The purpose of the sanitary sewer system is to process human sewage waste. Sewage can be treated onsite, such as in a factory-built aeration-type package, or it can be collected and transported via a network of pipes and pump stations to a municipal treatment plant. With onsite treatment, biological matter is progressively converted into a solid mass by microorganisms, then neutralized, and ultimately disposed of or used as landfill in environmentally suitable areas. The treated water may be discharged into a process wastewater system or into a public water system if the discharge meets regulatory guidelines. Figure 9.10 illustrates this process.

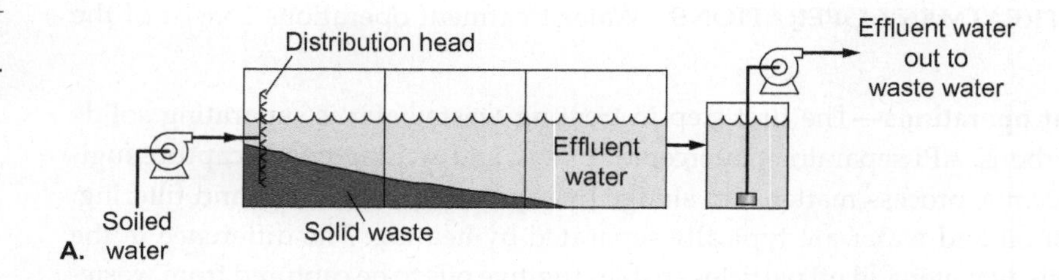

Figure 9.10 **A.** Illustration and **B.** photo of a sanitary sewer system.

CREDIT: B. SKY2020/Shutterstock.

Aeration-Type Package

An aeration-type package is a type of sewage treatment system that consists of lift stations, an aeration section, and a settling section for internal biological treatment of sewage before disposal.

LIFT STATIONS The lift station lifts sewage out of a sump and discharges it into a screen to remove very large solids and debris. Sewage then flows directly into the aeration compartment, which is the heart of the system.

AERATION SECTION Plant air is blown through diffusers that break the air into small bubbles. The turbulence produced by the plant air tends to break up sewage solids and mixes the fresh sewage with the activated sludge in the aeration compartment. The aerobic microorganisms consume and digest the fresh sewage as food. The air also circulates the contents of the compartment to pick up additional oxygen from the atmospheric air above the liquid surface.

Aerobic bacteria and other living microorganisms present in the sewage become active due to the abundance of oxygen, and they digest the organic solids in suspension and solution. The aerobic bacteria divide and multiply. The soluble organic matter metabolized by the bacteria converts to carbon dioxide and bacterial flocculate, which settles from the solution.

SETTLING SECTION The sewage flows from the aerator into the settling basin. Solids are returned from the bottom of the settling basin to the aeration tank for further treatment and for seeding the incoming sewage with living organisms. The settling basin effluent flows over a weir into a trough and through the effluent pipe to the effluent sump. A sanitary sewer pump takes suction on the effluent sump and discharges to the process wastewater system.

A hydraulic skimming system provides continuous automatic removal of floating solids from the settling basin surface (Figure 9.11). The system consists of a skimming trough, which draws floating material from the bottom of the settling basin and discharges it to the aeration compartment through a specially designed eductor. *Eductors* function much like steam ejectors, which pull vacuum, except the eductor usually uses water or air for motive force. Eductors are available in many designs. Figure 9.12 shows one example of an eductor.

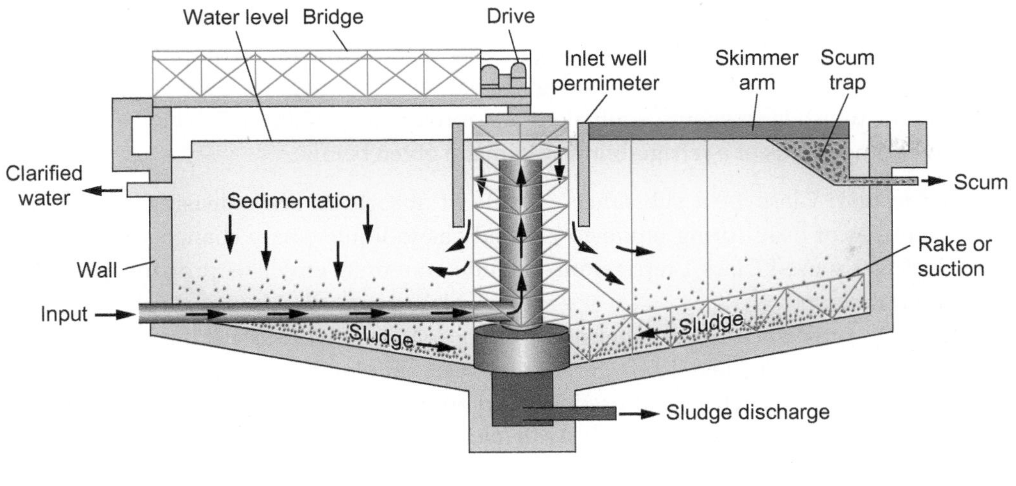

Figure 9.11 Skimming system.

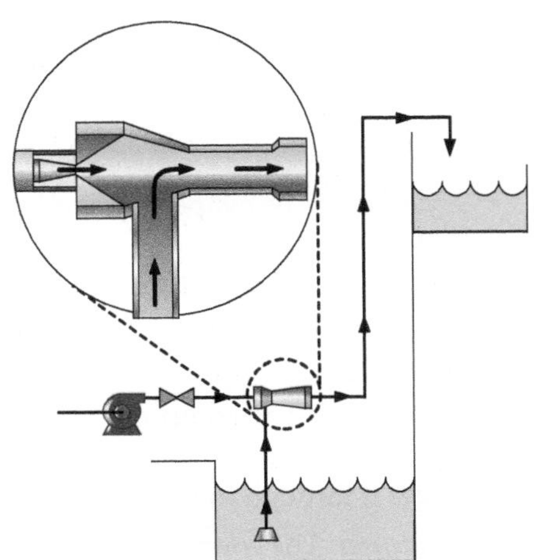

Figure 9.12 Example of an eductor.

Hazards and Mitigation

The following knowledge and protocols provide for safe, efficient process operation of water systems:

- The sanitary sewer may not retain the proper biological activity over long periods and can become septic or poisoned. This can kill the bacteria (sometimes referred to as "bugs") in the sewer system that are used to break down organic material. To provide proper treatment of sewer wastes and maintenance for the sanitary sewer system, it is periodically reseeded with a *bioaugmentation* product, or "bugs." This maintenance addition of bacteria helps prevent the system from going septic and discharging untreated waste to the process wastewater system.
- A tablet chlorination injection system can be used as an added treatment step to improve the quality of the process wastewater. This system should be checked daily to verify that the chlorine tablets are in place to ensure proper water treatment.

9.4 Refrigeration Systems

In 1820, British scientist and inventor Michael Faraday discovered that compressed ammonia chilled the air as it evaporated, and the concept of refrigeration was born. Refrigeration absorbs heat from an enclosed space or a substance and rejects that heat to another location. Some large, commercial refrigeration systems still use ammonia.

Refrigeration is used to remove excess heat from chemical reactions, liquefy process gases, separate gases, and purify products by preferential freeze-out of one component from a liquid mixture. Refrigeration also provides plant air conditioning for comfort and instrument and analyzer cooling. A "ton of refrigeration" is the refrigeration output of melting one short ton of ice at a temperature of 32 degrees Fahrenheit (0 degrees Celsius) over a period of 24 hours, which is an energy equivalent of approximately 12,000 BTU/hr.

Five components of a refrigeration system are given below:

- *Refrigerant:* Class 1—a substance capable of absorbing and releasing considerable quantities of heat during liquid-to-gas and gas-to-liquid phase change, or latent heat. Commonly used Class 1 refrigerants include ammonia, ethylene, propane, propylene, and butane.
  Class 2—typically brine or other nonfreezing solution that is first chilled to the desired temperature by a Class 1 refrigerant and then circulated to remove process heat, or sensible heat. The refrigerant used is based on the needed refrigeration temperature as well as economics, such as readily available light hydrocarbons in the process unit that function well as refrigerants.
- *Expansion valve:* Located immediately before the evaporator, this valve creates a rapid refrigerant pressure drop, which brings about a partial refrigerant phase change from liquid to vapor and auto-refrigeration, which lowers the refrigerant temperature considerably.
- *Evaporator:* The evaporator is a heat exchanger having refrigerant that has undergone a sudden pressure reduction, been partially flashed, and experienced auto-refrigeration flowing through the tubes, absorbing heat from the air or process outside the tubes. The remaining refrigerant liquid flashes from the heat absorption and flows to the compressor suction.
- *Compressor:* The compressor, whether a centrifugal, reciprocal, screw, or rotary type, takes suction on the refrigerant within the evaporator, compresses the refrigerant, and forces it through the condenser.
- *Condenser:* The condenser is a heat exchanger that uses process air or water to cool and condense the hot, high-pressure refrigerant from the compressor discharge.

Figure 9.13 depicts a simple vapor compression refrigeration system, such as household and automotive air conditioning.

Figure 9.13 Vapor compression refrigeration.

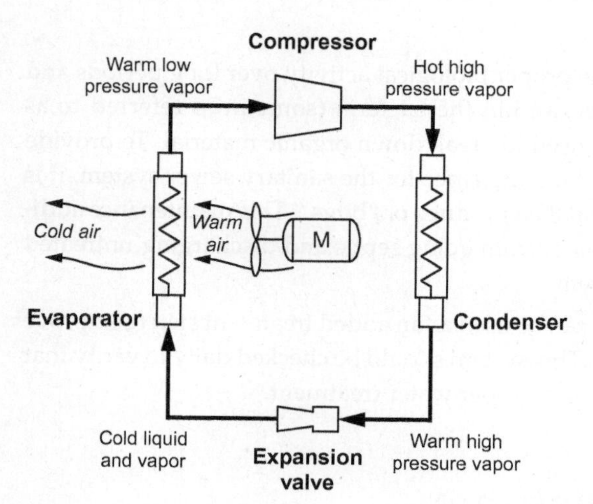

The refrigeration cycle starts with cold liquid and vapor refrigerant in the evaporator that absorbs heat from air or a product and fully flashes to a vapor. The refrigerant leaves the evaporator for the compressor suction as a low pressure, low temperature, or warm vapor that is then compressed. The refrigerant leaves the compressor discharge as hot, high-pressure vapor due to the heat of compression and flows through the condenser. The condenser utilizes either air or water to cool and condense the refrigerant. The cooled liquid refrigerant then flows through the expansion valve, experiences a rapid drop in pressure, and partially flashes. Auto-refrigeration, due to the flashing, lowers the refrigerant temperature considerably, and it returns to the evaporator as a cold mixture of liquid and vapor.

Process Applications

Refrigeration is a widely used process in the industry for cooling and separation purposes. Many large-scale facility applications utilize available ethylene, propane, or propylene for refrigerant cooling. Discharged by a large compressor, the refrigerant flows through a cooling water exchanger and then a control valve provides the pressure drop for flashing and auto-refrigeration. From the control valve, the refrigerant flows to the users.

In a large gas facility, a compressor may take suction on the overhead vapor line of a large tower and compress propylene or propane product for cooling and storage. A refrigeration system may also take a slipstream from the product cooler outlet.

Some processes use a large reciprocating commercial refrigeration compressor that circulates propane in a closed loop with cooling water exchangers serving as condensers and a pressure control valve serving as the expansion valve.

Hazards and Mitigation

The following knowledge and protocols provide for safe, efficient process operation of refrigeration systems:

- **Safety and Health and Operations**—Lines throughout a refrigerant system can be extremely hot or cold; therefore, these lines should be insulated. Extremely high temperatures can lead to thermal burns. Extremely cold temperatures can injure personnel who touch the lines with unprotected skin.

  Compressors generate high noise and vibration levels. Guards should be in place to protect personnel, and hearing protection is required per OSHA and site regulations.
- **Operation**—Lubricating oil tends to accumulate in the cold sections of the refrigeration system. If necessary, a slipstream from the bottom of the evaporator is drained into a reclaimer where oil is removed periodically.

  Lighter constituents in the refrigerant, such as ethane in a propane system or an ammonia system, tend to accumulate in the receiver, causing higher condensing pressure. This can be reduced by periodically purging the vapors from the receiver.

  Moisture, if present, will form ice and plug up the system either at the control valves or in cold spots, like the evaporator. Moisture normally enters the system with the purchased refrigerant charge or due to lack of drying the system before startup. Moisture can be a source of considerable operating problems until it is removed. Some refrigeration systems employ a continuous dryer; some use only a moisture indicator. Moisture must be removed prior to startup by evacuating the system, purging the system with nitrogen or dry gas, injecting methanol, or a combination of these methods.

9.5 Cooling Towers

Cooling towers release waste heat to the atmosphere that was previously absorbed by the circulating cooling water system. Heat exchangers transfer heat to cooling water and the cooling tower rejects that heat to the atmosphere through the latent heat of evaporation, in most cases with the assistance of forced draft or induced draft fans within the cooling tower. Cooling towers are designed as either crossflow or counterflow, where crossflow towers force air to intersect at right angles to water flow, and counterflow towers have process water and air flowing in opposite parallel directions from one another.

In order to maximize contact and heat transfer between the air and cooling water within the cooling tower, spray nozzles are distributed throughout the bottom of the cells that collect returning water on the top of the cooling tower. The nozzles dispense water into the tower, which falls and collects in the tower basin.

Air enters the bottom of the cooling tower and flows upward, counter to the downward falling water. The cooling water pumps take suction on the basin and circulate water through the process and back to the top of the cooling tower.

Two cooling tower types are *forced draft,* where fans are positioned at the air inlet, and *induced draft,* where fans are positioned at the air outlet. Figure 9.14 shows an example of a cooling tower.

Figure 9.14 Cooling tower.

CREDIT: cpaulfell/Shutterstock.

Water Treatment

Water contains oxygen and is naturally corrosive to steel piping and equipment. Corrosion inhibitor is typically added into the cooling water return or the cooling water basin. Vendor supplied totes, such as portable polyethylene chemical tanks equipped with metering pumps, inject the corrosion inhibitor. (Note: A *tote* is a polyethylene container designed for specialized handling of various bulk cargo materials such as hazardous and nonhazardous liquids.)

Vendor supplied biocide, typically injected intermittently from a portable tote, controls algae and microbial growth that could foul piping and exchangers if left unchecked. Sulfuric acid injection controls the cooling water pH and is monitored by analyzer probes. Improper or excessive acid injection is very harmful to piping and equipment.

Hazards and Mitigation

The following knowledge and protocols provide for safe, efficient operation of cooling towers:

- **Fire prevention and protection**—A leaking tube in a heat exchanger that uses cooling water may lead to process chemicals entering the cooling water return piping and pose the risk of ignitable vapors within and above the cooling tower. There are many possible sources of ignition, such as lightning, which can instantaneously ignite lingering flammable vapors and create a fire.
- **Safety**—Power loss to cooling tower fans or water pumps can severely compromise operations in a process facility. Cooling water with contaminants invites corrosion and fouling in pipes and process equipment, accumulation of scale on pipes from impurities, and production of an environment of microorganisms that are detrimental to the structural integrity in wooden cooling towers.
- **Health**—Cooling tower water is prone to contamination from processes that contain substances such as *Legionella*, sulfur dioxide, hydrogen sulfide, and carbon dioxide, as well as other hydrocarbons. *Legionella* is the bacterium that causes Legionnaires' disease, a form of pneumonia. *Legionella* is controlled at acceptable levels by biocide injection or

shock treatment. It is, however, good work practice to utilize proper PPE during process sampling, inspection, maintenance, and turnaround activities to prevent exposure to these bacterium and chemicals. Wastewater systems are also known for the presence of **hydrogen sulfide (H_2S)**. Personal H_2S gas detectors warn of hazardous environment above a preset threshold, measured in parts per million.

Hydrogen sulfide (H_2S) highly toxic, highly flammable, colorless gas with a very distinctive, rotten egg-like odor.

- **Process concerns**—To protect the cooling water piping from freezing during the winter months, the piping is electrically traced. A thermostat controls tracing temperature. It is important to periodically inspect cooling towers, heat exchangers, and pumps. Scheduled preventive maintenance and inspection enhances safety and is the key to trouble free operation. A check of all fans for vibration should be made regularly. Also:
  - Check and replace worn or cracked belts.
  - Inspect fan blades for deflection and for cracks.
  - Grease all bearings and fill oil cups.

9.6 Electrical Systems

Process facilities require electricity to power a variety of equipment and instrumentation within the facility. Facilities either receive electricity from a supplier outside the site, or onsite steam turbines or gas-fired turbines may turn a generator(s) to supply power. Electrical substations transform and distribute power within a facility and are positioned away from vapor laden areas or cooling tower moisture. Transformers, circuit breakers, and feed-circuit switches are devices associated with substations. Substations can be located in hazardous areas, with the appropriate classification requirements.

Even if all power comes from outside the facility, it is important that personnel be aware of the electrical equipment services on the unit that may be managed by the electrical and maintenance departments, as well as the location of all equipment breakers and switchgear for reset and lockout/tagout (LOTO) purposes.

Many facilities may have 230,000 volts (230 KV) entering the facility. Substations and transformers lower this voltage to 34.5 KV, 13.8 KV, 480V, and 230V to supply various equipment and building requirements. Figure 9.15 shows an example of a facility power grid.

Because of the critical nature of power supply to the facility, most plants have uninterruptible power systems (UPSs) that are used to provide power to critical instrumentation, alarms, computers, and control panels during a power outage. This system contains batteries, a battery charger, inverters, transfer switches, and bypass switches. The system ensures critical control equipment has the necessary power to safely bring the unit to a stable mode or shutdown in the case of a power supply interruption.

Hazards and Mitigation

The following knowledge and protocols provide for safe, efficient process operation of electrical systems:

- **Fire protection and prevention**—Generators too close to other process units can increase the possibility of ignition from a process upset if not properly classified.
- **Safety**—Safety policies and practices such as dry footing, high voltage warning signs, and guarding are necessary for prevention against electrocution. Along with other appropriate safety practices, electrical lockout/tagout (LOTO) procedures must be followed for work on high voltage electrical equipment.
- **Health**—Safety practices and utilizing proper PPE prepares the worker for noise and hazardous environments when inspecting, maintaining, and performing other work around transformers and switches. Be cautious of transformer fluid leaks, which may contain hazardous chemicals.

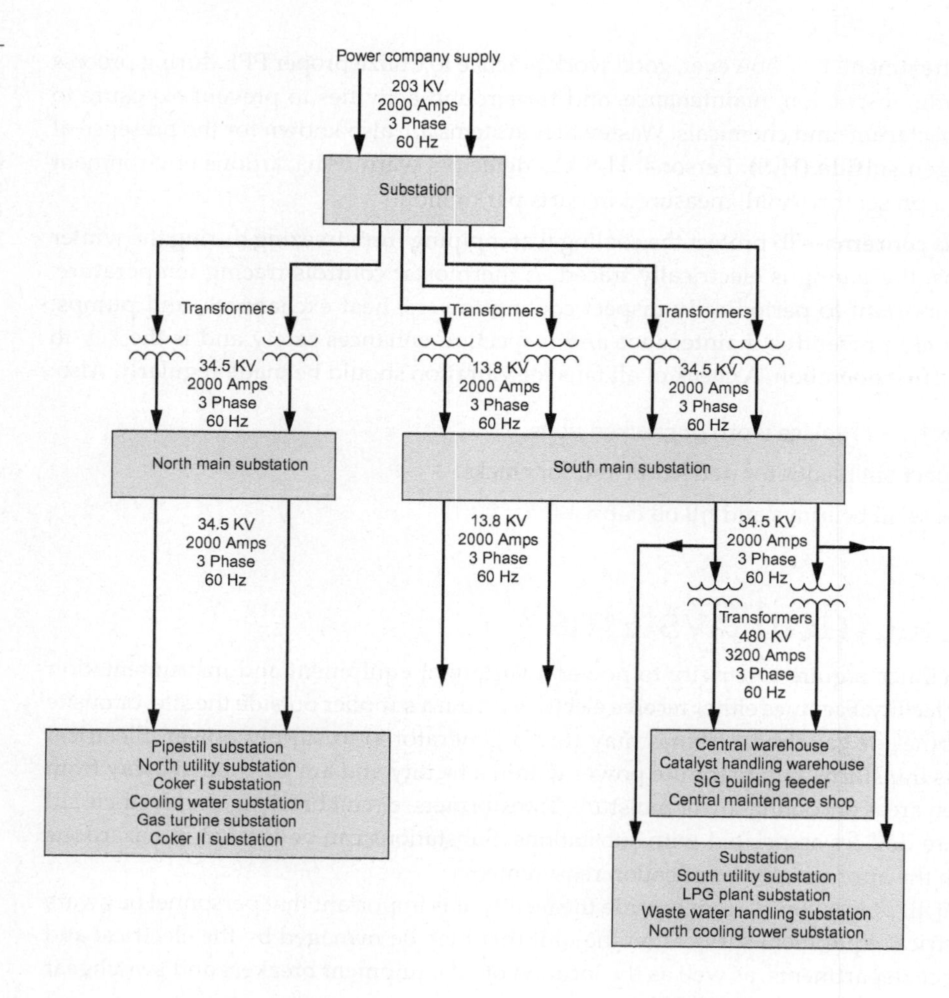

Figure 9.15 Example of a facility power grid.

9.7 Air Systems

The utilities unit provides plant air (sometimes called utility air) and instrument air by using compressors, coolers, receivers, dryers, controls, and piping. Compressed air is discharged from a compressor and through knockout drums where moisture is removed from the air. A portion of this air supplies the plant air header, while the remainder flows to air dryers and provides instrument air.

Plant Air

Plant air is normally used for purging the equipment containing inert gas to allow entry for maintenance and to operate pneumatic tools and pumps. Pneumatic tools and pumps operate at ambient temperatures and have a short residence time for the air—that is, the air passes through very quickly. Therefore, there is little opportunity for condensed moisture to collect. Plant air is normally filtered to prevent debris from entering the tools and is compressed to 90 to 105 PSIG (shown in Figure 9.16).

Instrument Air

Instrument air is supplied throughout the facility to operate pneumatic instrumentation such as control valves, controllers, and indicators. Instrument air is always dried to a dew point that will not allow condensation to take place, typically a dew point of about 20 degrees Fahrenheit (-6.7 degrees Celsius) below the minimum ambient temperature. This reduces the risk of condensation and cold weather freeze up in small bore piping and instruments. Instrument air should be cleansed of any materials that could impede normal operation. The pressure should range between 60 and 105 PSIG (typically 92 PSIG).

Figure 9.16 Plant air schematic.

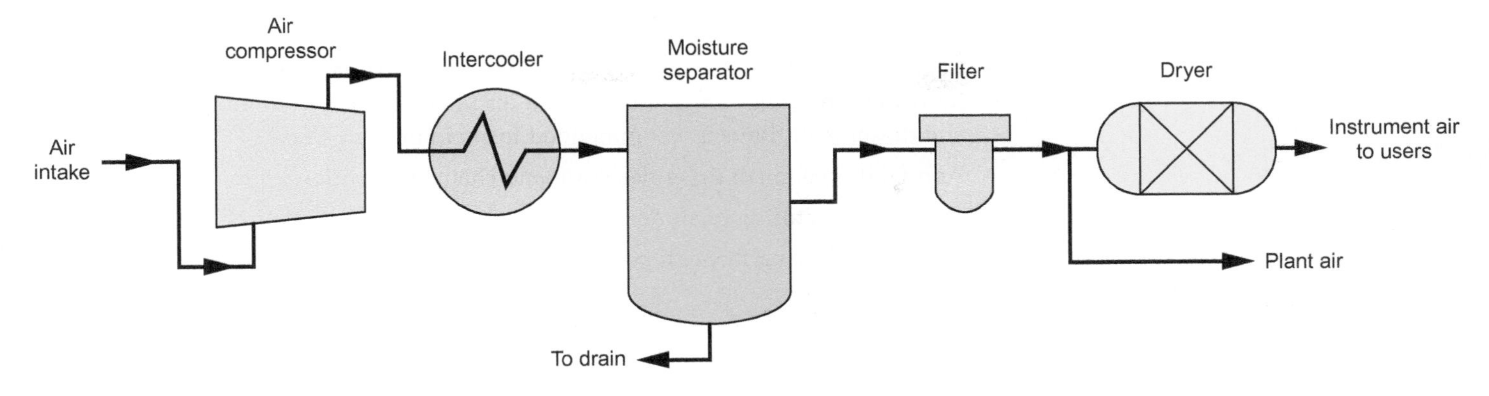

Did You Know?

In some areas of the country, air temperature periodically falls below 32 degrees Fahrenheit (0 degrees Celsius). Moisture may develop in the instrument air system if the dewpoint does not stay between -40 degrees Fahrenheit (4.4 degrees Celsius) and -70 degrees Fahrenheit (-57 degrees Celsius). The lack of dry instrument air can damage instruments and cause corrosion in equipment and lines. In these situations, alternate sources of air can be applied, such as nitrogen and process gas.

Hazards and Mitigation

The following knowledge and protocols provide for safe, efficient process operation of air systems:

- ***Fire protection and prevention***—Air compressors should be located so that the suction inlet is sufficiently distant from flammable vapor and corrosive gas sources in order to reduce the potential for fire and explosion.
- ***Safety***—
  - Knockout drums for removing condensation from instrument air helps keep liquid from entering into the distribution system.
  - Strainers must clean gases containing unwanted materials.
  - Upset of automatic compressor controls can lead to unit upset or shutdown.
  - Air compressors and associated equipment must be capable of handling pressures in the system and if not, pressure relief is essential for preventing equipment damage and process upset.
  - Guarding around exposed moving parts prevents accidental physical encounters with moving or reciprocating equipment.
  - Compressor buildings need correct electrical classification and proper ventilation.
  - Upstream instrument air drying systems help prevent moisture from contaminating instrumentation.
  - In the event of instrument air supply failure due to insufficient power or process equipment failure, nitrogen can be used in place of the instrument air.
- ***Health***—Safe work protocol and/or proper PPE reduces chances of exposure to hazardous environments while inspecting and maintaining processes. Utilizing safety procedures while operating is essential so that plant and instrument air are kept separate from breathing systems and potable water systems.

- *Operation and maintenance*—The maintenance aspects of instrument air systems are relatively straightforward and depend on manufacturer recommendations for compressors and dryers. Equipment typically requires:
  - Regular inspection, usually weekly, of filters downstream of the compressor. If oil lubricated machines are used for plant air systems, then more frequent inspection and draining of filters is recommended to avoid buildup of oil.
  - Annual inspection of dryer desiccant and changeover valves.
  - Monthly monitoring of air dew point.
  - Regularly blowing through take-off points not normally used to check for moisture condensation.

9.8 Pressure Relief and Flare Systems

Pressure Relief Systems

Flare systems serve to contain and control vapor and liquid releases from pressure-relieving devices and blowdowns. Relief valves release pressure automatically when their design release pressure is reached (Figure 9.17). Blowdowns are a purposeful and routine release of material through blowdown valves during process unit startups, furnace blowdowns, shutdowns, and emergencies. Depressurization is a swift disposal of vapor from pressure vessels in the event of a fire and takes place when a rupture disc bursts, usually at a higher pressure than the design release pressure of the relief valve.

Figure 9.17 Relief valves.

CREDIT: eleonimages/Shutterstock.

SAFETY RELIEF VALVE OPERATION Safety relief valves may be used with air, steam, gas, vapor, and liquid. They assist other valves in managing an above normal operating pressure. Safety valves that discharge large volumes of steam tend to open up to their full capacity. The pressure required to open liquid relief valves is typically higher due to higher spring resistance. Pilot-operated safety relief valves have a sixfold greater ability to seal and discharge than do normal relief valves. Nonvolatile liquids are normally sent to oil-water separation and recovery systems, whereas volatile liquids may go to liquid or vapor separation drums located upstream of the flare. Some process relief valves relieve to a lower pressure part of the unit process.

Flare Systems

Closed pressure release and flare systems consist of relief valves and lines from process units to receive discharges, knockout drums, and liquid or vapor seals to prevent air from entering the system. Purge gas is used to prevent **flashback** (also called burnback). Atomizing steam injection is often used at the flare tip to better mix and provide air for complete combustion and reduce detectable smoke. Figure 9.18 shows an example of a flare and vent system.

Flashback situation in which gas vapors ignite and return to the source of the vapors.

Figure 9.18 **A.** Example of a flare and vent system. **B.** Photo of a flare tip.

CREDIT: B. Leonid Eremeychuk/Shutterstock.

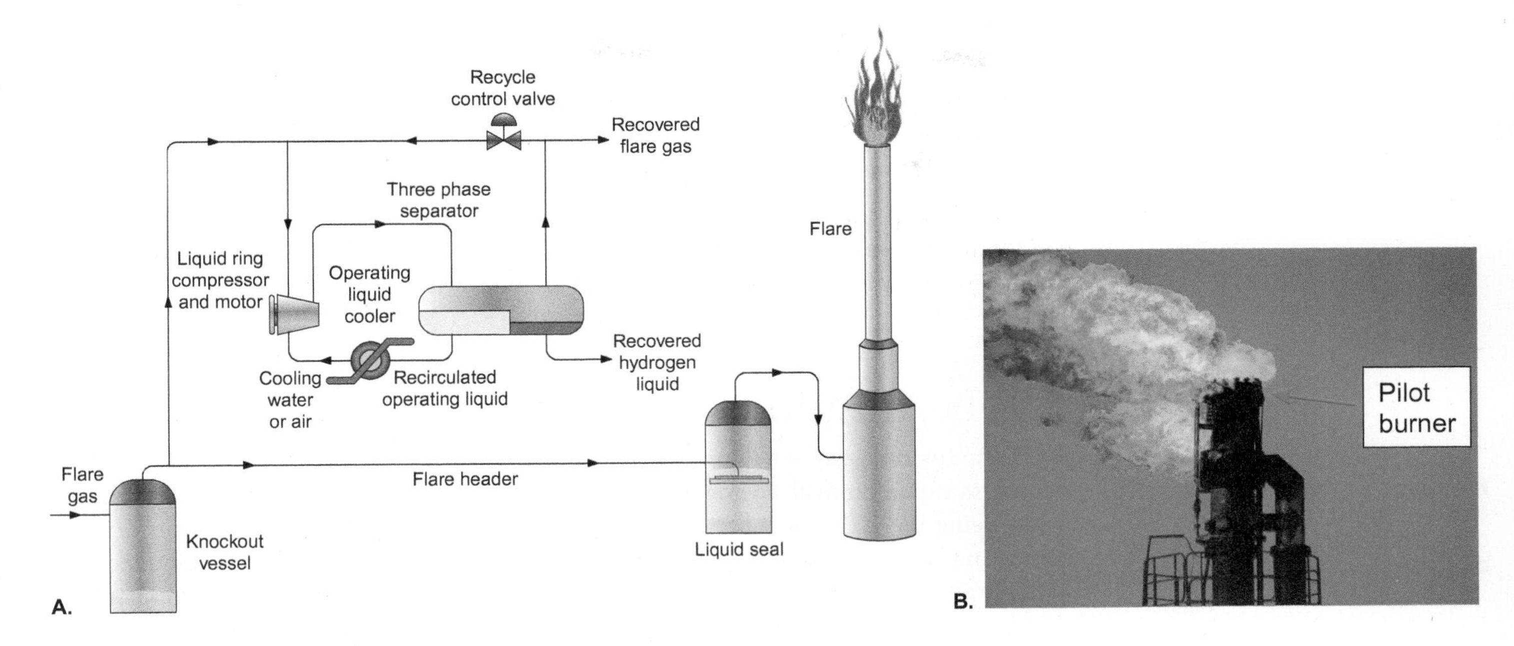

Hazards and Mitigation

The following knowledge and protocols provide for safe, efficient process operation of flare systems:

- **Fire protection and prevention**—Vapor and gas in proximity to ignition sources should be eliminated.
- **Safety**—Liquids should release to liquid–vapor separators upstream of vapor disposal systems. Flare knockout drums and flares should have sufficient capacity to control emergency blowdowns. Drums require relief for overpressurization scenarios.
  Pressure relief valves are necessary for overpressurization in processes because of the following:
  - Loss of cooling water to condensers and coolers can raise the pressure in process units.
  - Loss of reflux can raise the pressure in distillation towers and disrupt the volume of vapor leaving the distillation tower.
  - Sudden vapor and pressure increases occur from the injection of a lower boiling point liquid, such as water, into a process vessel operating at high temperature.
  - Excessive steam pressure may upset heaters or damage equipment and lead to fire, upset automated controls or heat exchangers, and so on.
  - There is a risk of equipment internal explosion, uncontrolled chemical reactions, thermal acceleration, or amassed gases in vessels.

  Proper maintenance of relief valves is imperative, as valves must function as designed. The most prevalent operation problems are listed here:

  - The valve fails to open at threshold pressure due to plugged valve inlets or outlets or corrosion impeding normal operation of disc holders and guides.
  - After opening, reseat fails due to fouling, corrosion, accumulated deposits, or residue from gas streams that deteriorate the valve disc.
  - Operating pressure is too near the valve threshold pressure setting, leading to chattering and early opening.

- **Health**—Safe work protocol and proper PPE are safeguards against hazardous environments while inspecting and normal process operations, as well as during turnaround activities.
- **Pilot and ignition operation**—Dependable flare pilot operation is critical, even during extreme weather. Flaring operations are infrequent and unscheduled. The flare should always be available for emergency releases to mitigate hazards, protect equipment and personnel, and prevent harmful environmental release. Most flares are equipped with multiple pilots to ensure flammable gas is ignited under any circumstance.

Windshields and flame retention devices help keep the pilots lit even in extreme weather conditions. The majority of pilots are intended to function through wind at 100 miles per hour or greater and incorporate remote pilot ignition systems.

9.9 Nitrogen

Nitrogen an odorless, invisible, inert gas, forming approximately 80% of the atmosphere. An important purging and blanketing medium.

Several industrial gases are used as a general utility for *inerting*, a term used to indicate air or process vapor removal. **Nitrogen** is used in manufacturing and air-freeing process equipment. Being inert, or nonreactive, nitrogen is used for blanketing, purging, and drying. Blanketing is done on storage tanks and process equipment. Examples of purging include flare purging after a safety relief valve blows, removing process vapor from equipment on shutdown, and removing air during startup. Hot nitrogen is also used for regenerating dryer desiccant.

The nitrogen header distributes nitrogen to all users within a facility. The nitrogen system consists of a pressure reduction and metering station and the associated distribution piping. Typically, nitrogen is supplied by an industrial gas company and is metered as it enters a facility. Nitrogen pressure is typically lowered by local pressure regulators or control valves to about 115 PSIG for the unit supply header. Users may utilize other local pressure control schemes to maintain lower design operating pressure. The high pressure nitrogen supply may be utilized for special purposes, such as catalyst regeneration.

Did You Know?

Some plants have their own Air Separation Unit, which separates nitrogen from the air and supplies it to the facility's distribution system.

Hazards and Mitigation

The following knowledge and protocols provide for safe, efficient process operation of nitrogen systems:

- *Safety*—Nitrogen is an inherent danger in confined spaces because it is an asphyxiant, which displaces oxygen and causes suffocation. Considerable care must be taken when entering equipment after purging. The equipment must be properly cleared with air and the atmosphere tested prior to entering.
- *Storage*—Compressed nitrogen cylinders must be kept in cool, dry, ventilated areas with proper seals that are clearly marked according to OSHA Hazard Communication Standard 29 CFR 1910.1200. Nitrogen cylinders must be isolated from physical damage, heat, and proximity to ozone.

- *Spills and leaks*—Spills or leaks of gaseous or liquid nitrogen require persons without breathing equipment and protective clothing to remain clear of the hazardous area until removal of the spill or repair of the leak. Liquid nitrogen boils at -320 degrees Fahrenheit (-155.6 degrees Celsius). Such a cold liquid can burn skin similar to a hot liquid. The following steps are helpful after a spill or leak:
  1. Evacuate all personnel to ventilated areas until oxygen level returns to normal in the area of the leak/spill.
  2. Notify proper personnel of the incident.
  3. Isolate the leak, if safe to do so.
  4. Require emergency personnel to use an air supplied breathing device such as a self-contained breathing apparatus (SCBA).
  5. Allow time for fugitive nitrogen to dissipate.

9.10 Natural Gas

The natural gas system distributes natural gas to all users within the facility. **Natural gas** is a combination of light hydrocarbons, predominately methane. Some uses of natural gas include firing furnaces, driving gas turbines, blanketing vessels, and use as a feedstock. The natural gas system consists of a pressure reduction/metering station and the associated distribution piping. Typically, natural gas is supplied by a third party at high pressure—for example, 600 PSIG. Depending on the natural gas use, the pressure is reduced to various pressure levels.

Natural gas a flammable gas associated with gas and oil fields and consisting principally of methane and the lower saturated paraffin hydrocarbons. It may also include impurities such as water vapor, hydrogen sulfide, and carbon dioxide.

Sometimes, natural gas is combined with off-gas (known as fuel gas). Process facilities produce off-gas from various processing units. Process off-gas pressure may be as low as 3 PSIG. It is a colorless gas that is generally high in hydrogen sulfide (H_2S). It is called *sour fuel gas* because hydrogen sulfide (H_2S) smells like rotten eggs. The low pressure off-gas is compressed and sent to scrubbers for H_2S removal and carbon dioxide (CO_2) removal, if present. Without the fuel gas scrubbers, processing of high sulfur crudes would result in high H_2S and sulfur dioxide (SO_2) emissions in the process facility's fuel gas system, which would be environmentally unacceptable. Natural gas is added, as required, for maintaining fuel system pressure and efficiency. The fuel gas system, which contains mostly methane, is the primary source of fuel for firing boilers and furnaces.

Did You Know?

Hydrogen sulfide should be regarded as highly toxic. At concentrations higher than 1000 ppm, it reacts with enzymes in the bloodstream causing pulmonary paralysis, sudden collapse, coma or unconsciousness, and death.

Hazards and Mitigation

Since natural gas is basically odorless, chemicals with specific odors are added to better enable leak detection to reduce the chance of fire or explosions. Methyl mercaptan, t-butyl mercaptan, and thiophane (tetrahydrothiophene) are similar compounds that carry a rotten egg smell and are used as odorants in natural gas.

Summary

Every industrial facility depends on the provision of utilities—steam, water, fuel, compressed air, inert gases, and cooling systems. Utilities play a vital role in industrial operations. Some 70 percent of the energy used on a typical process facility site passes through the utility systems. Besides the cost factor, energy waste contributes to environmental pollution. Therefore, utilities are a serious subject.

Steam drives turbines, operates pumps, provides process heating, warms heat tracing, and provides building heat. Refrigeration and cooling water systems remove waste heat and enable certain process component separations. Wastewater treatment and API separators enable environmental compliance, as do flare systems and relief systems.

Instrument air provides process control. Nitrogen enables purging and drying. Fuel gas use for heaters, boilers, and furnaces improves process facility economics. Proper use of utilities enables the process to function, improves the safety and economics of the process, and helps make process facilities more environmentally friendly.

Checking Your Knowledge

1. Define the following terms:
   a. Boilers
   b. Electric heat tracing
   c. Firewater
   d. Flashback
   e. Hydrogen sulfide (H2S)
   f. Natural gas
   g. Nitrogen
   h. Potable water
   i. Steam
   j. Steam clouds
   k. Steam generators
   l. Steam jets
   m. Steam tracing
   n. Steam turbines
   o. Water hammer
2. Which of the following are common uses of steam in a process facility? (Select all that apply.)
   a. Operating pumps
   b. Operating alarms
   c. Providing process heating
   d. Providing building cooling systems
3. Which of the following statements about "water hammer" in a steam system are correct? (Select all that apply.)
   a. Water hammer can occur when condensate is removed through drain valves and/or steam traps.
   b. Water hammer can occur when steam condenses and the accumulated condensate is carried along with the steam flowing within the piping.
   c. Water hammer is annoying but never dangerous.
   d. Water hammer can occur when condensate is forced to stop or change direction (e.g., at a pipe bend).
   e. Water hammer can occur when steam is admitted very slowly until equipment warms up.
4. Potable water is supplied throughout the facility for use as ___________.
   a. boiler feed water
   b. drinking water
   c. water for firefighting
   d. tower cooling water
5. If volatile mixtures of gas and air accumulate after a burner loses its flame during light off, a(n) ___________ may result.
6. Which of the following does not describe an eductor?
   a. Eductors function much like steam ejectors.
   b. Eductors usually use water or air for motive force.
   c. Eductors are used for processing firewater.
   d. Eductors are available in many designs.
7. Which of the following statements regarding wastewater treatment is true?
   a. Care must be taken to remove as many bacteria ("bugs") from wastewater as possible.
   b. Sanitary sewer systems should be reseeded periodically with a bioaugmentation product.
   c. If a tablet chlorination injection system is used, it is only necessary to verify once a week that chlorination tablets are in place.
   d. The addition of bacteria can cause the system to go septic and discharge untreated waste to the process wastewater system.

8. Which of these are the basic components of a refrigeration system? (Select all that apply.)
   a. Refrigerant
   b. Evaporator
   c. Freezer
   d. Compressor
   e. Expansion valve
   f. Hydrator

9. How can moisture in a refrigeration system be removed prior to startup? (Select all that apply.)
   a. By evacuating the system
   b. Through atomizing steam injection
   c. By purging the system with nitrogen or dry gas
   d. By injecting methanol

10. Which of the following statements about cooling towers is true?
    a. Counterflow towers force air to intersect at right angles to water flow.
    b. An induced draft tower's fans are positioned at the air inlet.
    c. Counterflow towers have process water and air flowing in opposite parallel directions from one another.
    d. A forced draft tower's fans are positioned parallel to the air outlet.

11. What are two problems that must be addressed to maintain a cooling water system?

12. Which of the following measures would *not* help reduce health hazards that cooling towers may pose?
    a. Use of PPE during process sampling, inspection, maintenance, and turnaround activities
    b. Personal H_2S detectors
    c. Shock treatment to help control *Legionella* levels
    d. Withholding chlorination

13. Devices that are associated with substations are: (Select all that apply.)
    a. Transformers
    b. Freezers
    c. Feed-circuit switches
    d. Steam jets

14. Which of the following is *not* a component of uninterruptible power systems (UPSs)?
    a. Batteries
    b. Safety relief valves
    c. Inverters
    d. Bypass switches

15. Plant air is normally used for which of the following? (Select all that apply.)
    a. Operating pneumatic instrumentation such as control valves, controllers, and indicators
    b. Purging equipment containing inert gas to allow entry for maintenance
    c. Helping to prevent moisture from contaminating instrumentation
    d. Operating pneumatic tools and pumps

16. __________ __________ is supplied throughout the facility to operate pneumatic instrumentation such as control valves.

17. Pressure relief valves are necessary in case of overpressurization in processes. Match each of the following overpressurization occurrences with the potential consequences.

Problem	Potential Consequences
I. Loss of cooling water to condensers and coolers	a. Potential equipment internal explosion, uncontrolled chemical reactions, thermal acceleration, or amassed gases in vessels
II. Loss of reflux	b. Pressure in distillation towers increases and disruption of the volume of vapor leaving the distillation tower
III. Sudden vapor and pressure increases	c. Heater upset, damaged equipment, fire, upset automated controls or heat exchangers
IV. Excessive steam pressure	d. Pressure increase in process units

18. Which of the following are common operation problems that are specific to relief valves? (Select all that apply.)
    a. Injuries due to accidental physical encounters with moving equipment
    b. Failure of threshold pressure to open due to corrosion or plugged valve inlets or outlets
    c. Failure to reseat after opening due to fouling, corrosion, accumulated deposits, or residue from gas streams
    d. Chattering and early opening due to operating pressure too near the valve threshold pressure

19. __________ is the term used to indicate air or process vapor removal.
    a. Blanketing
    b. Inerting
    c. Drying
    d. Pumping

20. What is the chemical symbol for *sour fuel gas?*
    a. CO_2
    b. SO_2
    c. H_2S
    d. H_2O

NOTE: Answers to Checking Your Knowledge questions are found in the Appendix.

Student Activities

1. Work with a classmate to develop a simplified block flow diagram of a propane refrigeration system, based on the description of a refrigeration cycle in this chapter. Identify the major components. Use air cooling to condense the propane.
2. Write a paragraph identifying the important factors that require attention when starting up a unit after a turnaround. Refer to the following topics:
   - Steaming of equipment
   - Water hammer in steam piping
   - Heater startup considerations
   - Purging equipment with nitrogen.
3. Given a utility flow diagram (UFD) list the steps required to start up steam systems, including valve alignment:
   - Identify all valves that must be checked for proper alignment.
   - State the proper position for each valve for startup.
   - State whether the valves will be checked via the DCS and/or via the field technician.
   - Position the valves correctly.

Chapter 10
Unit Commissioning

Objectives

After completing this chapter, you will be able to:

10.1 Discuss the term commissioning and the phase of planning in the commissioning process. (NAPTA Operations, Commissioning 1, 2*) p. 142

10.2 Explain the phases of construction, precommissioning, initial startup, and postcommissioning of a new process unit and the role of the process technician in these phases. (NAPTA Operations, Commissioning 2) p. 143

Key Terms

Acceptance documentation—the formal written validation that the unit has achieved its design capacity and specifications, and the facility agrees that the unit will function as engineered, **p. 146**

Commissioning—systematic course of action by which process units are placed into active service, whether it is the initial startup of newly built unit or the recommissioning of a revised process unit, **p. 142**

Commissioning team—group of individuals who play a key role in the planning and implementing of the commissioning or decommissioning of a process unit or facility, **p. 142**

Construction phase—building phase of an initial process unit or facility, **p. 143**

Initial startup—the first commissioning of the unit that involves the introduction of feedstock to produce a defined product at a given purity, **p. 145**

Initial startup procedures—set of guidelines or instructions used to perform the initial startup of a new process unit or facility, **p. 145**

Mechanical completion—documented checking and testing of the construction to confirm the installation is in accordance with construction drawings and specifications, and is ready for commissioning in a safe manner in compliance with project requirements, **p. 144**

*North American Process Technology Alliance (NAPTA) developed curriculum to ensure that Process Technology courses will produce knowledgeable graduates to become entry-level employees in process technology. Objectives from that curriculum are named here in abbreviated form. For example, "(NAPTA Operations, Commissioning 1, 2)" means that this chapter's objective 1 relates to objectives 1 and 2 of NAPTA's curriculum on commissioning.

Nameplate capacity—designed capacity of the unit, **p. 146**
Performance testing—step test of the unit to determine if the process unit is able to achieve its maximum design intent, **p. 146**
Planning phase—phase of the project where justification and plans are developed for the construction of a new process unit, **p. 142**
Postcommissioning—last phase of the commissioning process, which begins after initial startup is completed, **p. 146**
Precommissioning—activities that must be completed prior to moving into the startup phase of a new process unit, **p. 144**
Punchlist—list of uncompleted construction items from contracted design that are not safety critical but must be addressed by the contracted construction firm, **p. 146**
Recommissioning—returning existing process units or equipment to active service after an extended idled period, **p. 146**
Startup—bringing into operation a piece or pieces of equipment or a process facility, **p. 142**

10.1 Introduction

Commissioning systematic course of action by which process units are placed into active service, whether it is the initial startup of newly built unit or the recommissioning of a revised process unit.

Startup bringing into operation a piece or pieces of equipment or a process facility.

Commissioning is the process of starting up a newly constructed process unit or facility. It could also include initial **startup** of an existing unit or facility that has undergone significant process reengineering or repairs. The intent of the commissioning phase is to test the process unit to verify that it functions in accordance with the engineered design and the owner's operational requirements.

Unit Commissioning

The unit commissioning process consists of five phases:

- Planning
- Construction
- Precommissioning
- Initial startup
- Postcommissioning

Each phase of commissioning includes different activities that must be completed prior to proceeding to the next phase. Three key criteria must be met to consider the commissioning process successful:

1. *No lost time accidents or injuries*—commissioning is not a success if the commissioning is not completed safely. Safety is stressed from the beginning of the design, construction, and commissioning phases.
2. *No equipment damage*—if commissioning damages current or new equipment, it is not a success (depends on many disciplines, including design and construction).
3. *On-test product within a reasonable period*—timelines for achieving on-test product vary by process, but typically less than 2 days is considered very good, 7 days is acceptable, and more than 14 days is less than acceptable.

Planning phase phase of the project where justification and plans are developed for the construction of a new process unit.

Commissioning team group of individuals who play a key role in the planning and implementing of the commissioning or decommissioning of a process unit or facility.

PLANNING The **planning phase** begins prior to the construction phase, when the facility owner has determined the need to construct a new process unit or facility or to close an existing unit or facility. During this phase, an engineering firm is selected to aid in the design. Then, together with the engineering firm, a general contractor is selected to provide unit construction. The facility also selects the commissioning team representatives.

A **commissioning team** will be selected to oversee the commissioning or decommissioning of the facility or unit. The lead process technicians serving on the commissioning team

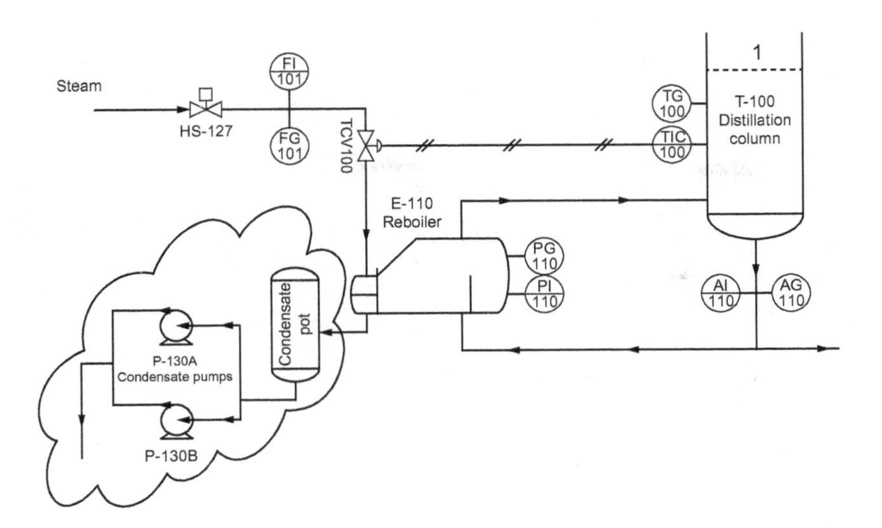

Figure 10.1 P&ID design and equipment selection showing some modification suggestions (see circled section).

CREDIT: Adapted image of Simulation Solutions.

are typically process personnel with more than 10 years of operating experience. These people are able to suggest modifications to proposals based on knowledge, skills, abilities, and all that they have learned in the plant (Figure 10.1).

The commissioning team works jointly with the engineering firm during the design stage. Process technicians provide valuable resources to the engineering firm during the design stage in terms of valve locations, safety-related equipment, equipment location, safety policies, and how they may affect schedules, as well as other valuable information for construction of the unit.

During the planning phase, the following subjects are addressed:

- *Training*—the initial development of training material is likely to begin.
- *Procedures/checklists*—the initial development of operating, emergency, and maintenance procedures and checklists begins.
- *Safety*—process safety management (PSM) items are developed, meetings are held to determine environmental impact of the project, risk assessments begin, and safety strategy documents are developed.
- *Budget*—detailed budget plans are developed.
- *Construction plans*—planning for construction commences and schedules are developed.
- *Staffing plans*—staffing needs are assessed and a timeline for hiring is set.
- *Maintenance items*—a spare parts list is developed for the unit, and the lubrication schedules and routine maintenance programs are developed.
- *Operational planning*—plans are created for staffing needs, run plans, and benchmark targets, as well as for achieving design capacity and performance testing.

Many of these developments continue into the construction phase.

10.2 Role of the Process Technician in Commissioning

Construction

Once the **construction phase** begins (Figure 10.2), the process technician is actively engaged in the field with the construction crew. In this stage, the process technician provides the construction crew with a wide variety of support that includes the following:

Construction phase building phase of an initial process unit or facility.

- Ensuring construction activities are moving along safely and per facility policies
- Issuing work permits, including hot work and confined space entry

Figure 10.2 Construction phase of the initial process facility.

CREDIT: USJ/Shutterstock.

- Performing or providing fire watch duty as requested by the construction crew
- Performing safety audits as required per the project's safety strategy document
- Ensuring that construction of the facility is proceeding per issued and approved isometric drawings, P&IDs, and PFDs
- Assisting new technicians on the project in training and in performing required duties.

During construction and as the unit moves toward mechanical completion, the process technician's job duties will additionally include facilitating or being on hand to assist qualified personnel with these tasks:

- Inspection of vessels and equipment
- Pressure testing of vessels, piping, and other equipment
- Line blows of process piping (for example, nitrogen purging and water washing)
- Flushing and cleaning of vessels and equipment
- Electrical and instrumentation loop checks.

Most of these listed duties are not normally carried out by operating personnel. They should not be expected to have the skills to complete these tasks, but they should probably be present to assist and possibly witness the activities. The process technician gains valuable knowledge of the unit as construction takes place, whether it is a new unit or a revamp of an existing unit.

Mechanical completion documented checking and testing of the construction to confirm the installation is in accordance with construction drawings and specifications, and is ready for commissioning in a safe manner in compliance with project requirements.

Precommissioning activities that must be completed prior to moving into the startup phase of a new process unit.

Construction is considered complete when the **mechanical completion** documentation is signed. Once the unit reaches mechanical completion, the commissioning team moves into the next phase of the process: precommissioning.

Precommissioning

Precommissioning begins after construction and before initial startup of a new unit or facility. During this phase, the process technician will become more familiar with the process piping and equipment. This is generally completed in stages, using either the system approach or the specific equipment approach. During precommissioning, the process technician will gain more hands-on experience (Figure 10.3) and begin to take ownership of the unit. As part of the precommissioning activities, steam and other utilities are commissioned so that necessary utility services are available for other precommissioning activities. The time allotted for completion of the precommissioning activities will vary depending on size of unit and/or extent of construction or repair activities completed. These efforts are typically completed in 2 to 3 months.

Figure 10.3 Preparing equipment after mechanical completion.

CREDIT: Pichit Boonhuad/Alamy Stock Photo.

The following items are critical to startup activities and the ongoing operation of the unit or facility. The process technician may be enlisted to work with construction company personnel to ensure and/or witness that these tasks are completed:

- Hydrostatic leak and pressure testing of piping and equipment
- Equipment inspection (for example, towers and reactors)
- Flushing, as well as chemical and mechanical cleaning
- Installation of temporary screens, strainers, and blinds
- Purging and removing air from equipment
- Drying out equipment
- Verification of instrumentation.

Normal operating unit lineups would be verified using the following:

- Red lining—This identifies a lineup that is not ready for precommissioning.
- Green lining—This identifies a lineup that is ready for precommissioning.
- Yellow lining—This identifies a lineup that might be ready for precommissioning but that needs to be verified.

As the unit nears completion of the precommissioning activities, a process safety review takes place as part of the process safety management (PSM) requirements for the facility. The process technician is a valuable part of this process.

Initial Startup

The **initial startup** is the most critical time to focus on safety. For this reason, process technicians should pay attention to all vessels and equipment during startup. This is the phase of commissioning of the unit when feedstock is first used to produce product. The initial startup consists of many unknown factors and is potentially hazardous because unit operability has not been established.

Initial startup the first commissioning of the unit that involves the introduction of feedstock to produce a defined product at a given purity.

The process technician gains valuable experience during the unit startup. The unit is prepared for startup using **initial startup procedures**. These procedures are different from the normal startup procedures that will be followed for subsequent startups. Utilizing the initial startup procedures, the process technician:

Initial startup procedures set of guidelines or instructions used to perform the initial startup of a new process unit or facility.

- Prepares the unit piping and instrumentation for startup
- Slowly brings raw materials into the unit
- Fills vessels, process lines, pump cases, and compressors per procedure
- Starts pumps, agitators, mixers, and compressors, as required
- Takes readings, as required, for startup purposes
- Assists unit maintenance personnel, as required.

Initial startup procedures are extremely important and should be followed in detail. They are done slowly and deliberately to ensure safety of personnel, equipment, and the environment. If deviations are required, the process technician should follow the facility procedure deviation policy. The process technician is required to identify, troubleshoot, and correct problems as they occur during the initial startup period.

The operation run plan (a temporary operating procedure) defines how the unit process technicians initially run the unit. The unit typically runs at a reduced rate until the on-test product specifications are met. On-test specifications should be met within a reasonable period, normally from two to seven days.

Acceptance documentation the formal written validation that the unit has achieved its design capacity and specifications, and the facility agrees that the unit will function as engineered.

Performance testing step test of the unit to determine if the process unit is able to achieve its maximum design intent.

Nameplate capacity designed capacity of the unit.

After achieving on-test production, the unit is slowly brought to full operation, or 100 percent capacity. This begins the performance trial of the unit. The unit runs at 100 percent capacity for a fixed period, and then the **acceptance documentation** is signed.

As a planned event at some time after acceptance documentation is signed, the unit is pushed to maximum capacity to see if it is capable of maximum production. This push is known as **performance testing** (or step testing) of the unit to determine if the process unit can meet its intended specifications. The unit will not always run at maximum capacity, but time and changing market conditions may require an increase in the unit's maximum capacity requirement, so it is good to know how it will run at current maximum capacity. Once the unit successfully runs at maximum capacity, plans are made and the unit eventually debottlenecks to increase its **nameplate capacity**.

Postcommissioning last phase of the commissioning process, which begins after initial startup is completed.

Punchlist list of uncompleted construction items from contracted design that are not safety critical but must be addressed by the contracted construction firm.

Postcommissioning

Postcommissioning is the final phase of commissioning. By this time, the unit is on-stream and in normal production. The process technician continues to make adjustments, troubleshoot, and solve problems as they occur. During this phase, any outstanding issues must be addressed by the contracted construction firm. These items may be compiled on a **punchlist**, which may include paint, bracing, insulation, or other items (such as hot line protection) that may have an impact on the health and safety of operating personnel (Figure 10.4).

Figure 10.4 Example of a punchlist.

Project: T150 Gas Scrubber Tower		Job Note: 20-000-5555	Initiator: R. Culbreath	Date: 7.17.22		
		Engineer: R. Gamble				
Item.	**Contractor**	**Punchlist Discrepancy**	**Priority**	**Punchlist Closure**		
				Project Engineer	Field Coordinator	Complete
1.	Gamble Enterprises	Paint nitrogen line orange from tie point 400 to tie point 401A	2			
2.	Gamble Enterprises	Remove scaffolding around F150 Gas Scrubber	3		R. Culbreath	Yes
3.	Gamble Enterprises	Remove roll-off boxes near P150A pump	3			
4.	Louis Tools	Install isolation valve on discharge line on P150B pump	4		R. Culbreath	Yes
5.	Gamble Enterprises	Install bracing on 50# steam line at OSBL	2		R. Culbreath	Yes
6.	Gamble Enterprises	Reinstall coupling guard on P250A pump	3			
7.	Gamble Enterprises	Install chain wheel on T150 vent valve	3		R. Culbreath	Yes
8.						

Recommissioning returning existing process units or equipment to active service after an extended idled period.

Recommissioning

Recommissioning is performed on a unit or equipment that has been out of service. The steps in recommissioning are the same as those for an initial commissioning project. A team is assembled (one that includes several process technicians) to oversee the recommissioning.

Summary

Process technicians play a key role during the design, construction, and initial startup of a new process unit. The unit commissioning process consists of five phases:

- Planning
- Construction
- Precommissioning
- Initial startup
- Postcommissioning.

Each phase of commissioning involves activities that must be completed prior to proceeding to the next phase. Unit commissioning has three key criteria that must be met to consider the commissioning process successful:

- No lost time accidents or injuries
- No equipment damage
- On-test product within a reasonable period.

A commissioning team is a group of individuals who play a key role in the planning and implementing of the commissioning or decommissioning of a process unit or facility. The commissioning team works jointly with the engineering firm during the design phase and is actively engaged in the field with the construction crew. During precommissioning, the process technician will gain hands-on experience and begin to take ownership of the unit.

One of the most critical periods for safety is the initial startup. This startup includes the introduction of feedstock to produce a defined product at a given purity. The startup process consists of many unknown factors and is potentially hazardous because unit operability has not been established.

The process technician gains valuable experience during the unit's initial startup. The unit is prepared for startup using initial startup procedures (a set of guidelines or instructions to perform the initial startup of a process unit or facility).

During postcommissioning, the unit is on stream and in normal production. The process technician makes adjustments, troubleshoots, and solves problems as they occur. Any outstanding punchlist items (the list of uncompleted items from construction that are not safety critical) are completed. Recommissioning follows a similar process to commissioning, but it is performed on a unit or on equipment that has previously been in service.

Checking Your Knowledge

1. Define the following terms:
   a. Acceptance
   b. Commissioning
   c. Commissioning team
   d. Construction phase
   e. Initial startup
   f. Initial startup procedures
   g. Mechanical completion
   h. Nameplate capacity
   i. Performance testing
   j. Planning phase
   k. Postcommissioning
   l. Precommissioning
   m. Punchlist
   n. Recommissioning
   o. Startup

2. What is the first phase of the commissioning process?
   a. Construction
   b. Initial startup
   c. Planning
   d. Precommissioning

3. A commissioning process would be considered less than successful if the timeline to achieving on-test product ___________.
   a. took more than 6 hours
   b. took more than 2 days
   c. took more than 7 days
   d. took more than 14 days

4. Which of the following are responsibilities of the process technician during the *construction* phase? (Select all that apply.)
   a. Installation of temporary screens, strainers, and blinds
   b. Line blows of process piping
   c. Pressure testing of vessels, piping, and other equipment
   d. Purging and removing air from equipment

5. Which of the following might a process technician do in conjunction with construction personnel during the *precommissioning* phase?
   a. Drying out equipment
   b. Equipment inspection (i.e., towers and reactors)
   c. Hydrostatic leak and pressure testing of piping and equipment
   d. Learning the procedures for each instrument
   e. Verification of instrumentation
6. Punchlist items _________. (Select all that apply.)
   a. are post-startup items.
   b. always include safety-critical issues.
   c. may not have been included in the contracted design.
   d. must be addressed by the contracted construction firm.
7. Which of the following is a responsibility of the process technician while the unit is in the *initial startup* phase?
   a. Drying out equipment
   b. Taking readings, as required, for startup purposes
   c. Installation of temporary screens, strainers, and blinds
   d. Monitoring vessels and equipment
   e. Starting pumps, agitators, mixers, and compressors, as required
8. What term describes the documented checking and testing of the construction to confirm the installation is in accordance with construction drawings and specifications and is ready for commissioning in a safe manner in compliance with project requirements?
   a. Acceptance documentation
   b. Mechanical completion
   c. Performance testing
   d. Postcommissioning
9. Which of the following are process technician responsibilities during the *construction* phase? (Select all that apply.)
   a. Electrical instrumentation loop checks
   b. Flushing and cleaning of vessels and equipment
   c. Budgeting for personnel
   d. Issuing work permits
   e. Assessing staffing needs and setting the timeline for hiring
   f. Performing or providing fire watch duty as requested by the construction crew
10. What is the level of participation by process technicians on a *recommissioning* team?
    a. No process technicians are included.
    b. Only one process technician is allowed on the team.
    c. Several process technicians are included.
    d. The team is entirely made up of process technicians.

NOTE: Answers to Checking Your Knowledge questions are found in the Appendix.

Student Activities

1. Perform research on companies that offer commissioning and precommissioning services. Select one and complete a one-page report on its processes. Explain why you chose that particular company.
2. Work with a classmate to simulate the process of commissioning a new unit. Develop a timeline from idea creation to production and complete a two-page report that explains the timeline details.
3. Perform research on safety-related incidents that have occurred when commissioning a new unit or recommissioning an existing facility. Write a two-page report covering the research, and explain how the incidents could have been avoided.

Chapter 11
Unit Startup

Objectives

After completing this chapter, you will be able to:

11.1 Differentiate between the different types of startups:
- Normal/routine
- After an emergency shutdown
- Equipment startup after maintenance
- After a turnaround. (NAPTA Operations, Normal Startup—Overview and Communication 1; Normal Startup—Process Unit 3) p. 150

11.2 Describe the process technician's role in the execution of unit startup. (NAPTA Operations, Normal Startup—Process Unit 4*) p. 154

11.3 Describe the risks and hazards associated with unit startup. (NAPTA Operations, Normal Startup Overview and Communication 3, 4; Normal Startup—Process Unit 1, 2) p. 157

11.4 Describe the safety, health, and environmental (SHE) activities associated with a unit startup, and how these activities are covered by OSHA's Process Safety Management (PSM) of Highly Hazardous Materials standard. (NAPTA Operations, Normal Startup—Overview and Communication 2, and Normal Startup—Process Unit 2, 5) p. 158

Key Terms

Air free—purged of any oxygen from process piping and equipment prior to the introduction of process chemicals, **p. 154**

Feed forward—the process of feeding a process stream to the next processing area, **p. 152**

*North American Process Technology Alliance (NAPTA) developed curriculum to ensure that Process Technology courses will produce knowledgeable graduates to become entry-level employees in process technology. Objectives from that curriculum are named here in abbreviated form. For example, "(NAPTA Operations, Normal Startup—Process Unit 4) means that this chapter's objective 2 relates to objective 4 of NAPTA's curriculum about the normal startup of the process unit.

Hydro test—strength and integrity test, using water, for process piping and equipment, **p. 155**

Startup—bringing into operation a piece or pieces of equipment or a process facility, **p. 150**

Tightness test—pressurization test, typically using nitrogen or other inert gas, for process piping and equipment to ensure that equipment is leak free prior to the introduction of hydrocarbons; also known as the *leak test*, **p. 154**

11.1 Introduction

Startup bringing into operation a piece or pieces of equipment or a process facility.

This chapter provides an overview of various types of unit startups in a process facility. A unit **startup** is the introduction of raw materials (feedstock) to manufacture a defined product at a given purity. This entails systematically putting process equipment into service in order to start the process. Unit startups are an important part of process operations that are executed when a process unit is ready to be brought into service for production.

Commissioning (see Chapter 10, *Unit Commissioning*) is the systematic process by which a process unit is *initially* placed into active service and can refer to either the initial startup of a newly built unit or facility *or* the recommissioning of a revised process unit or facility. Equipment that has been replaced should go through a commissioning phase to ensure that the equipment or system is performing as designed. The commissioning phase (planning, construction, precommissioning, startup, and postcommissioning) is performed to test the process unit to verify that it functions in accordance with the engineered design's intent and the operational requirements.

There are several types of unit startups. Each startup brings both common and unique hazards if not managed correctly. One type of startup is the planned sequenced event, such as an initial commissioning startup after a new process has been constructed, as discussed in Chapter 10, *Unit Commissioning*. Another is a startup after a turnaround (TAR), a planned, scheduled process unit or facility shutdown for maintenance and repair, discussed in Chapter 14, *Maintenance*. There are also startups that involve individual systems and equipment, and there are startups that take place after an emergency shutdown. The hazards associated with unit startups are unit specific.

Unit startups and the basic planning requirements are further defined later in this chapter. Unit startups are a diversion from normal operations and carry increased levels of risk. If not managed properly, they can cause injury to personnel or damage to equipment, the environment, and surrounding communities. Their relation to OSHA's Process Safety Management (PSM) of Highly Hazardous Materials is covered later in this chapter.

Normal/Routine Startup

A normal or routine startup takes place after a planned outage, shutdown, or turnaround (TAR) has been completed. The operations team follows unit specific standard operating procedures (SOPs) for equipment and system startup that ultimately lead to the start of production.

Unit startup procedures are executed by the process technicians in a sequenced order during a normal or routine startup. With the assistance of the necessary personnel, material sampling and specification verifications are made while process systems are brought online. Maintenance people and contractors are available to troubleshoot problems associated with equipment startup.

Even during normal, routine startups there can be electrical, mechanical, and instrumentation problems that must be solved before the startup can continue. For this reason, engineers, craftspeople, and maintenance technicians are often staffed around the clock to provide the necessary support for startup activities (Figure 11.1).

Figure 11.1 Mechanical engineer inspects gas booster compressor engine before startup.

CREDIT: Oil and Gas Photographer/ Shutterstock.

Startup Execution Plan

One of the key activities in preparation for a normal, routine startup is development and communication of a startup execution plan. This strategic document typically includes consideration for, or makes reference to, instructions for the following:

- Required staffing
- Coordination with other units
- Utility and auxiliary systems commissioning
- Hazardous chemical inventory procedures
- Detailed equipment and unit startup procedures
- Notification of the EPA or other regulatory agencies in advance of the scheduled startup.

Part of the startup planning process includes establishing communication with raw material suppliers and other vendors to ensure raw materials, tools, and equipment are available and delivered on time for startup. There must also be communication with adjacent and connected process units so that they can provide raw material for the equipment or unit that is in the startup mode or receive waste or off-spec materials once the unit is making product.

The sequence of a unit startup is determined by the systems that are required to be placed into service first. Safety systems like firewater, deluge, and chemical detection systems are given priority at the start of a unit startup. Auxiliary systems such as hot oil, seal oil, dry gas seal, water, steam, condensate, plant air, instrument air, nitrogen, and other utilities must also be started in the proper sequence, based on the needs of the process. These utility systems are usually started up first followed by the process equipment and systems.

Systems for hazard mitigation, preventing chemical exposure, and protecting personnel and the environment are usually the next to start up. Flare and vent systems, API separators, and closed drain systems are examples of systems placed in service early during startup activities to manage chemicals safely, prevent exposure to personnel, and prevent potential harm to the environment or surrounding communities. Raw materials from tank farms, pipelines, marine vessels, and specialty materials from tank cars, tank trucks, and other vendors can be safely brought into the process unit once these types of systems are in service.

Major pieces of equipment and systems, which are considered the heart of a process unit, are only started up once the required utilities and safety systems are in operation. Key pieces of process equipment, like centrifugal compressors in process gas or refrigeration service, can be brought online for circulation purposes prior to the actual start of production. Operation of large and more expensive pieces of rotating equipment requires maintenance people to monitor and evaluate their performance prior to start of production.

Likewise, after fuel gas systems are in service, large cylindrical or cabin-type furnaces and hot oil systems can be started up to provide heat for fractionation and dehydration towers and reactor systems.

Refrigeration systems can be placed in service for raw material and product cooling. Placing seal oil and dry gas seal systems in service allows operation of pumping equipment for circulation and transfer of material.

Once all major equipment and systems of the process unit are either in operation or on standby, raw material can be introduced on a continuous basis. If the unit being started takes its feed from a separate processing area or unit, coordination between the units is critical to start the flow from one unit to the next. Feeding a process stream to the next processing area is called **feed forward** flow. This effectively begins the processing of the finished or intermediate product.

Feed forward the process of feeding a process stream to the next processing area.

After startup, when production has been successfully established, the site staff transition to routine operations and equipment monitoring. Process technicians use methods such as *audio visual olfactory* (AVO), a method to monitor the sounds, sights, and smells of a process unit during unit walkthrough inspections.

Startup after an Emergency Shutdown

Safe management of an emergency shutdown and, ultimately, the unit restart depend heavily on the knowledge, skills, and abilities of the process technician. Due to the large quantity of equipment and instrumentation on a process unit, many variables and scenarios can cause an emergency shutdown or interruption of production. Emergency operating procedures should be in place to minimize the effect of an emergency shutdown and to place the unit in a safe condition quickly.

The emergency operating procedures should include step-by-step instructions for securing a process unit for each possible type of emergency. They should also include the effect on personnel, the process, the environment, and surrounding communities. Emergency shutdowns can cause, or be caused by, the sudden failure of major equipment, such as compressors and furnaces, or by failure of utilities such as instrument air, steam, or electricity. Faulty trip or shutdown instrumentation can also cause emergency shutdowns.

Emergency shutdowns caused by equipment or instrument failure require complete and careful evaluation of the unit equipment, instrumentation, and safety systems prior to restart activities. The evaluations often dictate postemergency startup activities. Process technicians must exercise caution when starting up after an emergency shutdown and must be aware of the hazards that may be present due to process interruption. Once the unit is deemed safe, procedures for restarting the unit after emergency shutdown should be followed.

Failure or shutdown of minor process equipment and subsequent operation of spare process equipment may have only a minimal effect on the unit and may not require removal of the unit's raw material or feed. Spare equipment and instrumentation can effectively minimize process interruptions and eliminate the hazards that can result from upset conditions.

Many times, minor equipment failure can be evaluated, repaired, and restarted in a short period, so that the unit is returned to normal process conditions relatively quickly. In these cases, once the hazards have been fully evaluated and spare or failed equipment is ready for service, personnel on shift can develop a logical startup plan. However, even during minor equipment outages, the applicable unit specific SOPs must be used for equipment and system startup.

Failure or shutdown of major process equipment has the greatest effect on unit production and usually requires temporary removal or diversion of the unit raw material or feed. When major equipment failure and long-term process interruptions occur, operations, engineering, and maintenance staff need to work together to evaluate equipment and process conditions in order to determine a logical startup plan. After the unit condition is evaluated and deemed safe to restart, the applicable unit specific SOPs should be followed in order to bring the unit back to normal operations.

Equipment Startup after Maintenance Activities

Occasionally, auxiliary and individual pieces of equipment and some process systems are removed from service to conduct maintenance activities without an entire unit shutdown. There are precautions that must be taken in order to mitigate the hazards associated with such equipment shutdown, repair, and restart. Precautions and hazards should be identified and managed with the use of shutdown, isolation, repair, and startup procedures. These procedural tools are critical to managing the hazards safely.

The complexity of the equipment being removed from service predicts the associated hazards. When not shutting down an entire process unit, these hazards can range from limited to severe. Proper communication and coordination between operations and maintenance personnel is critical to the safe removal, efficient repair, and startup of process equipment. Planning and execution of the repair, evaluation of the hazards, and the effect on the rest of the unit can normally be completed with the process technicians and maintenance technicians who are on shift.

Safely managing equipment startup after completion of maintenance activities can be as simple as placing a repaired pump back in service. It could have a minimum effect on the process, or it might be as complex as placing an entire system back in service, with major impact on all process operations.

In industries like oil and gas, the hazards associated with even minor repairs to equipment and process service can be severe. Process technicians must follow these guidelines at all times:

- Operating procedures must be used to shut down safely and remove equipment from service and to minimize the effect of the shutdown on the rest of the unit.
- Control of work (COW) procedures must be used to identify, isolate, and prepare the equipment for repair activities.
- Maintenance procedures must be used to ensure a quality repair and to maximize equipment integrity.
- Operating procedures must be followed to start up safely and place the equipment back in service.
- Proper personal protective equipment (PPE) must be used to protect individuals from exposure and associated hazards. PPE includes hard hats, safety glasses, goggles, and protective clothing, as shown in Figure 11.2.

Adhering to site policies and procedures, as required by the OSHA 1910 PSM standard, ensures that the proper shutdown procedures, COW procedures, and startup procedures are used to place equipment back in service safely once maintenance people have completed their work.

Figure 11.2 Personal protective equipment. Hard hat and safety goggles with side shields, respirator, and gloves.

CREDIT: Bannasak Krodkeaw/Shutterstock.

Initial Unit Startup and Startup After Turnaround

An initial commissioning startup and a startup after a turnaround (TAR) are the most involved startup procedures and require the same amount of effort and focus. This is because, in both cases, there are many pieces of process equipment which must be evaluated and made safe for the introduction of process chemicals, steam, high pressure, high temperature, corrosive service, and many other potentially hazardous services.

These startups are planned events that require the maximum level of coordination and communication among site personnel and usually require an elevated level of effort for a prolonged period. Operations and other personnel involved in major unit startups include:

- Process technicians and operations management team
- Maintenance planners, craftspeople, and inspection staff
- Process, control, mechanical, and electrical engineers
- Warehouse and procurement staff
- Safety, health, and environmental (SHE) and project safety management (PSM) staff
- Contractor managers, contractor engineers, contractor craftspeople, and specialty contractor staff.

All of the staff work together to plan and execute unit commissioning startups and post-TAR startups. Planning for a new unit commissioning startup may begin six months or longer before the actual startup date. See Chapter 10, *Unit Commissioning*, for more on unit commissioning.

Maintenance and operating procedures play a critical role in the tracking and completion of startup activities. These procedures define the necessary steps required to prepare process equipment and systems for service and include activities such as given below:

- Line and equipment flushing
- Hydrostatic testing
- Line and equipment dry out
- **Tightness test**
- Making piping and equipment **air free** prior to the introduction of process chemicals
- Verifications of alignment and rotational direction for rotating equipment
- Performance tests for specialty equipment.

Tightness test pressurization test, typically using nitrogen or other inert gas, for process piping and equipment to ensure that equipment is leak free prior to the introduction of hydrocarbons; also known as the *leak test.*

Air free purged of any oxygen from process piping and equipment prior to the introduction of process chemicals.

11.2 Process Technician's Role in Planning and Executing Startups

Process technicians play a key role in the planning and safe execution of unit startups. A startup for a large process unit with many systems, and sometimes hundreds of pieces of equipment, can take several days to complete. Process technicians are one of the most important personnel groups on the unit during startups because of their knowledge of the unit's process technology, design, equipment, safety and control systems, and hazards. Process technicians have the following responsibilities before and during startup:

- Assist in the prestartup safety review (PSSR).
- Execute unit startup and inventory procedures to facilitate safe, efficient, controlled startup.
- Line up process equipment, piping, and control valves.
- Start up, monitor, and control rotating equipment.

- Prepare and start special equipment utilizing any normal or special control of work (COW) procedures.
- Establish and control of process conditions within operating limits.
- Coordinate all work activities while the startup is in progress.
- Monitor all site and contractor technicians to ensure compliance with safe work practices and health, safety, and environmental (SHE) policies, including those for PPE.
- Ensure that hazards to personnel, the environment, and equipment are managed correctly, and report any deviations from the site safety, health, and environmental policies.
- Maintain all unit safety equipment in good order so that it is available to mitigate emergency situations, alleviate environmental hazards, and provide personnel protection.
- Participate in employee health monitoring programs when the potential for unique exposure hazards is present.
- Complete COW procedures and permit processes for managing work activities surrounding process equipment.
- Complete lockout/tagout (LOTO) procedures for equipment preparation and energy isolation when there is a need to inspect, repair, or replace process equipment.
- Complete **hydro test** for piping and equipment.
- Complete strength and integrity tightness tests for process piping and equipment, and ensure that the equipment is leak free prior to the introduction of process chemicals.
- Maintain the facility in a clean, orderly, safe condition. This is difficult due to the vast amount of equipment preparation and related activities that take place. Some hazard mitigation and other benefits of maintaining good housekeeping during a startup are as follows:
  - Eliminating slipping and tripping hazards
  - Eliminating potential for chemical exposure
  - Eliminating environmental hazards
  - Improving operating and maintenance efficiency
  - Ensuring that tools and equipment are in the proper place and available for use when needed
  - Helping maintain a higher level of equipment integrity and reliability.

Hydro test strength and integrity test, using water, for process piping and equipment.

During a unit startup, coordination is critical between the process technicians assigned to field duties and those assigned to control room duties. The field technician is engaged in proper lineup of process equipment and piping and control valves, as well as the startup of rotating equipment. The control room technician is engaged in establishing and maintaining the operating conditions and process variables as systems and equipment are brought online. (Figure 11.3)

The startup of a process unit can require manipulation and control of hundreds or even thousands of process variables. The variables temperature, pressure, level, and flow, along with unit specific systems and equipment, are unique to the products made in each process facility.

Newly constructed process units are usually equipped with a fully instrumented distributed control system (DCS) for controlling the process. A DCS is an automated control system consisting of field instruments and field controllers, usually connected by wiring, that carries a signal from the controller transmitter to a central control monitoring screen. Networks for communication and monitoring connect the entire system of controllers.

Older process units may have a combination of previous generation electronic and pneumatic instrumentation and control systems. Regardless of the type or sophistication of control technology available to the process technician, it is the process control and operating conditions that, once established within the correct control limits, will fulfill the process unit's purpose by delivering on-specification products.

Figure 11.3 A control room technician monitoring operating conditions and process variables.

CREDIT: Sahan Nuhoglu/Shutterstock.

Unit startups serve an added function. They can provide a unique learning experience for new or inexperienced technicians. The diversion from normal operations enables personnel to execute operating procedures, SHE policies, and work practices that are seldom encountered during routine or normal operations. This is especially true of a new unit commissioning startup or post-TAR startup due to the high volume of work and the introduction of new work processes and procedures. Such activities give process technicians an opportunity to increase their knowledge and to get hands-on experience with new or repaired equipment. Unit specific, as well as site specific, knowledge can be gained about how unit startup activities can affect an entire process facility or community.

During unit startups, process technicians may need to revise operating procedures where corrections to or deviations from previously established work practices are necessary. Each facility should have guidelines in its SHE policy that define the steps required for deviations from and corrections to operating procedures. Certain corrections or deviations fall within the OSHA PSM guidelines for management of change (MOC). These could require a hazard analysis of the deviation prior to executing the deviation.

Many of the risks and hazards associated with the process industry can be eliminated with the proper development and use of safe operating procedures. It is critical for the process safety information contained in the procedures to be accurate. The practice of revising process safety information provides process technicians with the opportunity to make a job or work practice safer for themselves and for other site personnel.

The process technician's familiarity with the site SHE policies is also critical to the safe execution of unit startups. SHE policies are implemented by process facilities to minimize the risks and hazards associated with the process industry and to ensure that the facility is operated within the guidelines provided by applicable regulatory agencies.

These policies should be readily available through a site computer system, through local area network (LAN), or in hard copy manual form. (**Note**: Most procedures are now time and date stamped and are set to expire after a given time period. Many industries destroy printed copies after completion of a procedural task. These measures help prevent use of obsolete procedures.)

It should be a common practice for process technicians to use these SHE policies as a reference to understand and implement the established safe work practices for a given activity. Typical SHE policies that are utilized during unit startups include:

- *Employee health monitoring*—defines the need for employee health monitoring while activities are conducted in hazardous areas, in hazardous chemical sampling, or where extended exposure to hazardous chemicals may occur during TARs

- *Environmental reporting*—defines the requirements and reportable quantities of hydrocarbons or other hazardous substances that, when released to the environment, require reporting to the proper regulatory authorities
- *Housekeeping*—defines activities that must be completed in order to maintain the facility in a clean, orderly, safe condition
- *Management of change (MOC)*—defines the need and application for managing changes associated with an industrial process in support of the OSHA PSM regulation
- *Material release reporting*—defines reporting requirements of regulatory authorities such as the EPA when venting, purging, or draining equipment or in the event of a material release
- *Operating procedures*—defines unit specific procedures used for the purpose of equipment and system startup, shutdown, normal operation, as well as emergency situations
- *Personal protective equipment (PPE)*—defines equipment that must be worn by personnel when working in process areas or conducting specific activities such as sampling of hazardous materials, entering hazardous areas, or opening process equipment (Figure 11.4)

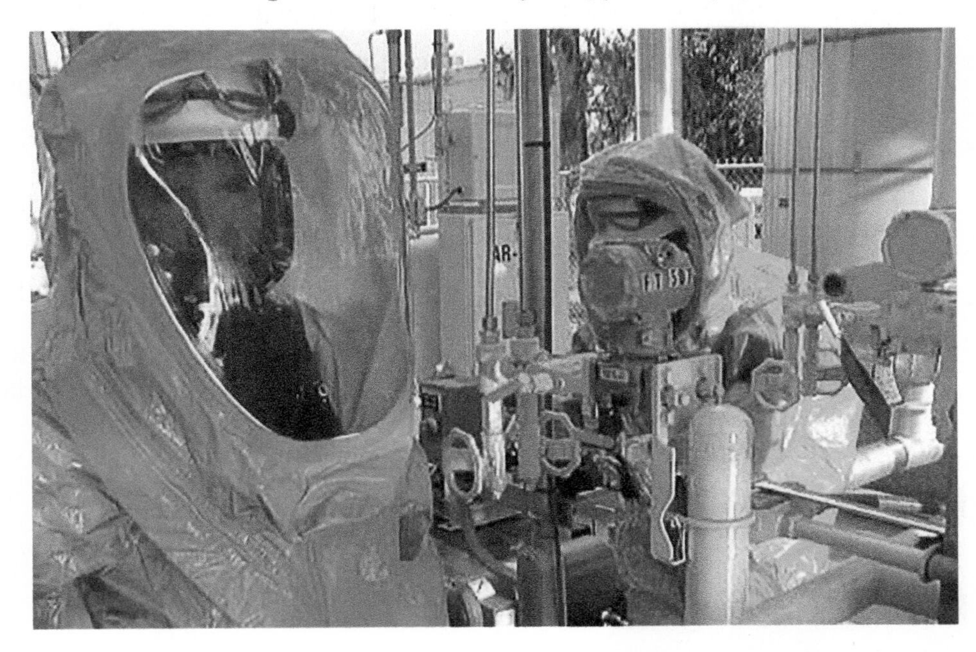

Figure 11.4 Process technicians must wear the right PPE, such as these chemical-resistant suits, depending on the hazards in the area.

CREDIT: Courtesy of Brazosport College.

- *Process hazard analysis*—defines the need and application for the systematic assessment of the potential hazards associated with an industrial process in support of the OSHA PSM regulation
- *Process safety information*—defines the type of documentation that is considered process safety information in support of the OSHA PSM regulation, including but not limited to operating procedures, inspection and maintenance procedures, operating manuals and training material, process and instrument drawings (P&IDs), electrical one-line diagrams, instrument loop drawings, and electrical classification drawings
- *Process safety management*—defines the 14 elements of the OSHA 1910 regulation; includes requirements related to management of change (MOC), process safety management (PSM), incident investigations, employee participation, HAZOP, operations procedures, mechanical integrity, inspection, training, trade secrets, contractors, prestartup safety review, compliance audits, and emergency planning and response.

11.3 Risks and Hazards

Hazards on a process unit are present during unit startups, shutdowns, emergencies, and even during normal operations. Much like unit shutdowns, unit startups occur infrequently and are considered a nonroutine activity. The human element associated with performing unfamiliar activities raises the hazard level. Demand for product usually dictates the startup timing, duration, and schedule.

Process units can be systematically started up to manufacture the intended product while consciously managing the hazards. Hazardous conditions can develop quickly, and emergencies can occur during the startup of a process unit, which can lead to personnel injury, irreparable damage to equipment, material release, fire, or explosion. Every process facility has the responsibility to identify potential emergency scenarios and to have documented mitigation plans (emergency procedures) in place to manage such scenarios safely.

Detailed startup planning and the strict use of SOPs and safe work practices should help eliminate hazardous incidents. The ability of process technicians to maintain control of the process throughout startup is critical to eliminating these hazards, which can also affect an entire process facility and surrounding communities. Startup hazards include, but are not limited to, the following:

- Atmospheric release or a material release due to poor equipment, valve, or flange tightness testing (Figure 11.5)
- Uncontrolled release of fuel gas to flare and vent systems prior to establishing process control
- Pipe and equipment damage due to thermal expansion or contraction
- Failure of new, repaired, or replaced rotating equipment
- Hazards from slips, trips, or falls due to poor housekeeping
- Personal injury due to the failure to wear proper PPE once fuel is introduced and equipment is placed in service
- Hazards associated with continuation of work activities paralleling startup activities (for example, hot work; vehicle entry; inspection and x-ray; excavations; heavy lifting; installation of new piping, equipment, and support structures).

Figure 11.5 Equipment hazards.

CREDIT: Serg_Kr/Shutterstock.

11.4 Unit Startup and OSHA's PSM Standard

The Occupational Safety and Health Administration (OSHA) is a U.S. government agency created to establish and enforce workplace safety and health standards, conduct workplace inspections, propose penalties for noncompliance, and investigate serious workplace incidents. Many of the activities that take place during a unit startup are covered by OSHA's

PSM (Process Safety Management) of Highly Hazardous Materials standard (1910.119), including the following:

- Process safety information must be in place for use as reference material and must reflect the details of the process; this includes P&IDs. These detailed drawings graphically represent the equipment, piping, and instrumentation within a process facility. They show the interconnection of process equipment and the instrumentation used to control the process. They are the primary schematic drawings used for laying out a process control installation in a facility. Additional process safety information includes instrument and control loop diagrams, plot plans, electrical one-line diagrams, operating procedures, training material, and operating manuals. See Chapter 3, *Reading Process Drawings* and Chapter 8, *Shift Change/Relief*, for specifics of documentation.
- Hazard analysis of the process must be complete, and all safety and environmental action items resulting from the analysis must have been completed prior to startup.
- Process technicians must be trained for startup with sufficient knowledge and understanding of the process equipment and associated hazards.
- Activities related to mechanical integrity and inspection of process equipment must be complete, including maintenance procedures and work practices, equipment files, inspection files, inspection frequencies, and equipment health monitoring programs that are intended to prolong or improve equipment integrity.
- Guidelines for emergency planning and response, prestartup safety review (PSSR), incident investigations, and contractor management are also defined in the OSHA 1910 PSM requirements that must be in place for startup.

Prior to every startup, a prestartup safety review should be completed in an effort to verify that the unit is ready for safe startup. The review team consists of site operations; maintenance; SHE staff; technical and contractor managers; as well as process technicians, maintenance staff, and contractor technicians. The PSSR activities include verification that:

- Rotating equipment is installed per manufacturer specifications and is ready for startup.
- Electrical equipment and control systems are installed per manufacturer specifications and are ready for startup.
- Piping and structural equipment are installed per manufacturer specifications and are ready for startup.
- Construction and mechanical completion checklists defining equipment turnover requirements are completed.
- Process safety information, operating procedures, and training materials are in place prior to startup.
- Operations staff and maintenance craftspeople are trained on the process and are ready for startup.
- Hazard and operability (HAZOP) study has been done to determine potential hazards associated with process systems, equipment, process materials, and work processes, and all of the safety and environmental action items are closed.

Unit startups require a vast amount of planning that incorporates the skills and abilities of the operations staff. Unit process technicians, process engineers, maintenance technicians, and contractor technicians are engaged in startup activities. The operations team will have unit specific SOPs for equipment and system startup or shutdown in normal operation, as well as in emergency situations. The procedures developed for startup are used for equipment and system startup that ultimately leads to product manufacturing.

Summary

Unit startups are an important part of process operations and provide valuable experiences for process facilities, site personnel, and process technicians. Startups require all of the skills and abilities of process technicians and most site personnel. The deviation from normal, routine duties provides learning experiences for the site staff that occur only during startups. It is also a time of increased risk because of this deviation from the norm.

The safe startup of process equipment after maintenance, repair, or replacement has taken place can provide opportunities to improve equipment performance, longevity, and integrity. When implemented during a startup, new technology can result in a process that is safer to operate, with reduced or eliminated risks to personnel, the facility, and surrounding community. In many cases, the installation and startup of new technology can often reduce operating costs, leading to higher profitability for a process facility.

A unit startup brings personnel together from many areas of a facility to work more closely than during normal operations. This provides opportunities to establish working relationships that otherwise might not have occurred and enables sharing of the unique skills of each individual. New career choices may be discovered when opportunities are experienced by personnel related to startup planning; process hazard analysis team participation; writing and updating process safety information, procedures and training material; and maintenance activities not experienced during normal operations.

Startup planning and execution, when managed and completed safely, can constitute one of the most gratifying experiences in the process industry. It offers new experiences, the chance to add to the combined effort of many work teams, and the opportunity to work with new site personnel and contract personnel for the common benefit of a process facility.

Checking Your Knowledge

1. Define the following terms:
   a. Air free
   b. Feed forward
   c. Hydro test
   d. Startup
   e. Tightness test
2. The raw materials used to manufacture a defined product are referred to as ________
3. Which of the following statements are true? (Select all that apply.)
   a. Unit startups must be executed long before a process unit is ready to be brought into service for production.
   b. Unit startups always carry identical risks regardless of the unit involved.
   c. Unit startups entail systematically putting process equipment into service in order to start the process.
   d. Unit startups are a diversion from normal operations and carry increased levels of risk
4. Safely managing an emergency shutdown and unit restart depends heavily on which of these factors? (Select all that apply.)
   a. Abilities of process technicians
   b. Number of years of service
   c. Knowledge of process technicians
   d. Size of the unit
5. Failure or shutdown of ________ ________ ________ has the greatest effect on unit production.
6. The OSHA 1910 Process Safety Management standard covers: (Select all that apply.)
   a. contractors
   b. process safety information (PSM)
   c. management by objectives
   d. business etiquette
   e. trade secrets
   f. prestartup safety review
   g. compliance audits
   h. training
   i. customer satisfaction
   j. overtime compensation

7. Identify two of the primary responsibilities of the process technicians during startups. (Select all that apply.)
   a. Execution of unit startup and inventory procedures to facilitate a safe, efficient, and controlled startup
   b. Establishment and maintenance of the operating conditions and process variables as systems and equipment are brought online
   c. Proper line-up of process equipment, piping, and control valves as well as the startup of rotating equipment
   d. Establishing and maintaining control of process conditions within operating limits
8. Process technicians are important personnel on the unit during startups because they __________. (Select all that apply.)
   a. know the equipment and interconnecting valves and piping.
   b. know the operating limits and design criteria
   c. understand the potential hazards.
   d. know how to run the entire facility.
9. Match the following potential startup hazards one of the potential causes:

Startup Hazard	Potential cause:
I. Atmospheric release or a material release	a. failure to wear proper PPE
II. Pipe and equipment damage	b. poor housekeeping
III. Hazards from slips, trips, or falls	c. poor testing of equipment, valve, or flange tightness
IV. Personal injury	d. thermal expansion or contraction

10. Prestartup safety review (PSSR) should be completed in an effort to verify that the unit is ready for startup. PSSR activities include verification that: (Select all that apply.)
    a. Rotating equipment is installed per manufacturer's specifications and ready for startup.
    b. Vending machines and other worker amenities are fully stocked.
    c. Electrical equipment and control systems are installed per manufacturer specifications and are ready for startup.
    d. All painting and other aesthetic considerations are completed.
    e. Piping and structural equipment are installed per manufacturer specifications and are ready for startup.

NOTE: Answers to Checking Your Knowledge appear in the Appendix.

Student Activities

1. Select one of the unit startup procedures, and perform procedure review together with shift personnel. Using information learned to date, review for accuracy to determine the following:
   - Is the procedure format correct, based on established guidelines for operating procedures?
   - Are prerequisites identified that must be completed prior to starting the procedure, and if so are they correct?
   - Are hazards identified and caution statements included to mitigate each hazard?
   - Are upper and lower control limits included in the procedure steps to help establish process control?
2. Select a piece of equipment. Describe a startup for that equipment, utilizing any normal or special control of work procedures.

Chapter 12
Process Technician Routine Duties: Normal Operations

Objectives

After completing this chapter, you will be able to:

12.1 Compare the routine duties of a field technician with those of a control room technician. (NAPTA Operations, Normal Operations; Field Technician 2–8; Normal Operation; Control Room Technician 1–9; Introduction to Operations 5) p. 163

12.2 Describe the tools commonly used in performing routine tasks in the field. (NAPTA Operations, Normal Operations; Field Technician 6*) p. 165

12.3 Describe how to monitor the following unit equipment:

- Compressors
- Exchangers
- Motors
- Cooling towers
- Safety equipment
- Valves
- Drums/Vessels
- Pumps
- Control valves
- Instrumentation (NAPTA Operations, Normal Operations; Field Technician 1, 2; Normal Operations; Control Room Technician 1–9) p. 166

*North American Process Technology Alliance (NAPTA) developed curriculum to ensure that Process Technology courses will produce knowledgeable graduates to become entry-level employees in process technology. Objectives from that curriculum are named here in abbreviated form. For example, "(NAPTA Operations, Normal Operations; Field Technician 6)" means that this chapter's objective 2 relates to objective 6 of NAPTA's curriculum about the field technician's role in normal operations.

12.4 Describe equipment health checks, and how to check for equipment leaks. (NAPTA Operations, Normal Operations; Field Technician 3, 4, 5) p. 169

12.5 Explain the duties of a process technician in preparing equipment for maintenance and returning equipment to service. (NAPTA Operations, Equipment Maintenance; Overview and Communications 1, 2) p. 169

12.6 Describe the personal protective equipment (PPE) required for performing routine field tasks in special operating environments. (NAPTA Operations, Normal Operations; Field Technician 6, 7; Normal Operations; Control Room 8; Normal Operations—Other Duties 9) p. 170

12.7 Explain the methods used to document the technicians work in the field. (NAPTA Operations, Normal Operations; Field Technician 8) p. 171

Key Terms

Control board technician—process technician whose primary job function is to remotely monitor and control the process unit within normal operating parameters; also called *console operator*, *board operator*, or *inside operator*, **p. 164**

Control room—room from which operators and technical personnel control the unit; it houses the facility's distributed control system (DCS), which may incorporate all of the facility's operating control boards, **p. 164**

Equipment health monitoring (EHM)—efficient system for protecting rotating equipment and facility operations from unscheduled downtime that provides personnel with the equipment knowledge they need to schedule maintenance, manage inventories, and support efficient workflow scheduling, **p. 169**

Field technician—process technician whose primary job is to monitor the fixed and rotating field equipment, perform sampling, and ensure that the unit operates within normal operating parameters; also called *field operator* or *outside operator*, **p. 165**

Monitor—to observe or watch; observing and listening to the equipment routinely to prevent process upsets, **p. 166**

Normal operations—actions performed or procedures followed when a process unit is operating within design parameters, **p. 164**

Rounds—routine walkthrough of the unit, monitoring the fixed and rotating equipment, and performing other routine tasks, **p. 164**

Route—sequential path followed in order to perform equipment health monitoring (EHM), **p. 169**

Routine duties—duties performed that are rigidly prescribed by control over the work or by written or verbal procedures, or well-defined, constant, and repetitively performed duties that preclude the need for procedures or substantial controls, **p. 165**

Vibration readings—measurement and documentation of rotating equipment to ensure vibrations are within an acceptable range, **p. 167**

12.1 Introduction

Process technicians play key roles in the operational success of the unit. The primary goal of the process technician is to ensure safe and reliable operations during the shift. Technicians should know and understand every process parameter of their unit. Process technicians must be familiar with how each piece of equipment operates, how the equipment sounds, the normal operating temperatures and pressures of the equipment, and how the instrumentation

Normal operations actions performed or procedures followed when a process unit is operating within design parameters.

Rounds routine walkthrough of the unit, monitoring the fixed and rotating equipment, and performing other routine tasks.

Control board technician process technician whose primary job function is to remotely monitor and control the process unit within normal operating parameters; also called *console operator*, *board operator*, or *inside operator*.

Control room room from which operators and technical personnel control the unit; it houses the facility's distributed control system (DCS), which may incorporate all of the facility's operating control boards.

on the unit controls the process. These parameters determine when **normal operations** will be followed.

Process technicians not only read the documentation for the unit but also make **rounds**. Some facilities require rounds every four hours, some every six hours, and others at the beginning and end of a shift. Each technician should be trained in both field and control board duties to ensure greater understanding of the process.

Routine Duties

The **control board technician** (also called *console operator, board operator*, or *inside operator)* works within the central **control room**. The primary job function of the control board technician is to run the distributed control system (DCS). A DCS is a subsystem of a supervisory control system used to control a process unit. It consists of field instruments and field controllers connected by wiring that carries a signal from the controller transmitter to a central control monitoring screen. Process diagrams, controllers, valves, and operating variables are displayed, and the DCS operator can manipulate the parameter set points or valve positions. Automated systems such as the DCS allow greater control and optimization of one or many processes simultaneously, ease communication between field and control room, and provide easy transmission of large amounts of data to and from a central location.

The process facility develops guidelines and checklists specifically for the operating parameters. Routine duties of both inside and outside technician may include the following:

- Having a thorough exchange of information during shift change. The exchange of information may include the following:
  - Safety and environmental issues that exist or were corrected
  - Alarms and their current status
  - Equipment conditions/problems, including corrective actions taken
  - Procedures in progress, such as dryer or reactor regeneration
  - Process status, including problems and corrective actions taken
  - Process trends
  - Maintenance activities: completed, in progress, and planned
  - Contractor activities: completed, in progress and planned
  - Presence of nonoperating personnel, or personnel other than process technicians, such as engineers, members of the management team, maintenance personnel, and contractors who work on the unit
  - Status of permits in force
  - Current coordination or cooperative efforts with other process units
  - Special operating instructions
  - Product quality issues (off test or specification).

In addition, a control room operator's duties include the following:

- Monitoring all process parameters and equipment on the unit
- Thoroughly inspecting the unit utilizing the control system at the beginning and throughout the shift at regular intervals
- Monitoring alarm screens and taking corrective action as required
- Coordinating process activities with the field technician
- Coordinating maintenance, contractor, and technical department activities with the field technician
- Recording all lab data

- Recording shift activities in the unit logbook or eLog, including the recording of all personnel on the unit
- Performing other duties as directed by the facility.

The role of the **field technician** (also called a *field operator* or an *outside operator*), differs from the control board technician in that the field technician is hands-on in the field with the equipment. Field technicians have a somewhat wider range of **routine duties**. Field technicians follow the guidelines and checklists described above for both control room and field technician routine duties. In addition, normal routine duties of the field technician may include the following:

Field technician process technician whose primary job is to monitor the fixed and rotating field equipment, perform sampling, and ensure that the unit operates within normal operating parameters; also called *field operator* or *outside operator*.

Routine duties duties performed that are rigidly prescribed by control over the work or by written or verbal procedures, or well-defined, constant, and repetitively performed duties that preclude the need for procedures or substantial controls.

- Making a thorough inspection of the unit at the beginning of the shift and at regular intervals throughout the shift
- Overseeing and assisting maintenance personnel, contractors, and technical personnel working on the unit
- Performing safety checks as required by the facility
- Performing equipment inspections or surveys as directed by the facility
- Checking the technician's area of responsibility for leaks
- Checking rotating equipment for proper lubrication
- Checking for proper operation of the cooling tower
- Preparing equipment for maintenance using accepted facility practices and guidelines
- Catching routine samples and special samples as directed
- Receiving supplies for the unit, such as lube oil, soap, chemicals, and other supplies as required
- Notifying the control board technician of any process or equipment problems and suggesting corrective actions
- Performing equipment health monitoring as directed by site policies
- Performing housekeeping as required
- Recording normal duties performed in the logbook or eLog.

Effective communication between the field and control board technician is necessary in running the unit safely, efficiently, and reliably. Both roles are critical to the sustainability of the process facility. They are a team working together for a common objective.

12.2 Tools

Field technicians are required to use a variety of tools while performing routine duties. Tools that are commonly used by the field technicians include valve wrenches, pliers, and oil cans. The facility may provide specialty tools for performing certain tasks.

Valve wrenches of different sizes are available for opening and closing valves. The valve wrench in Figure 12.1A is a type used in opening and closing smaller-sized valves ranging from $^1/_2$ inch to 2 inches. The facility may equip each technician with a small valve wrench, which should be carried at all times while on the process unit.

The valve wheel wrench (Figure 12.1B), comes in various sizes, is generally used in the opening and closing of valves ranging from 3 inches to 48 inches and is manufactured out of carbon steel or aluminum. This type of valve wrench is generally placed at specific equipment locations throughout the process unit to allow easy access when needed.

The operator wrench (see Figure 12.1C) is a double-ended wrench, fitted with a valve wrench on one end, which is used to open and close valves ranging from $^1/_2$ inch to 2 inches. The adjustable end of the wrench is used to install and pull bull plugs and is not used for opening and closing valves because it can cause burring of the valve wheel rim, which could lead to hand injuries. Process technicians often carry this style of wrench during routine rounds.

Figure 12.1 **A.** Small double-ended valve wrench. **B.** Large aluminum valve wheel wrench. **C.** Double-ended wrench.

CREDIT: **A.** and **C.** Courtesy of Steve Ames.

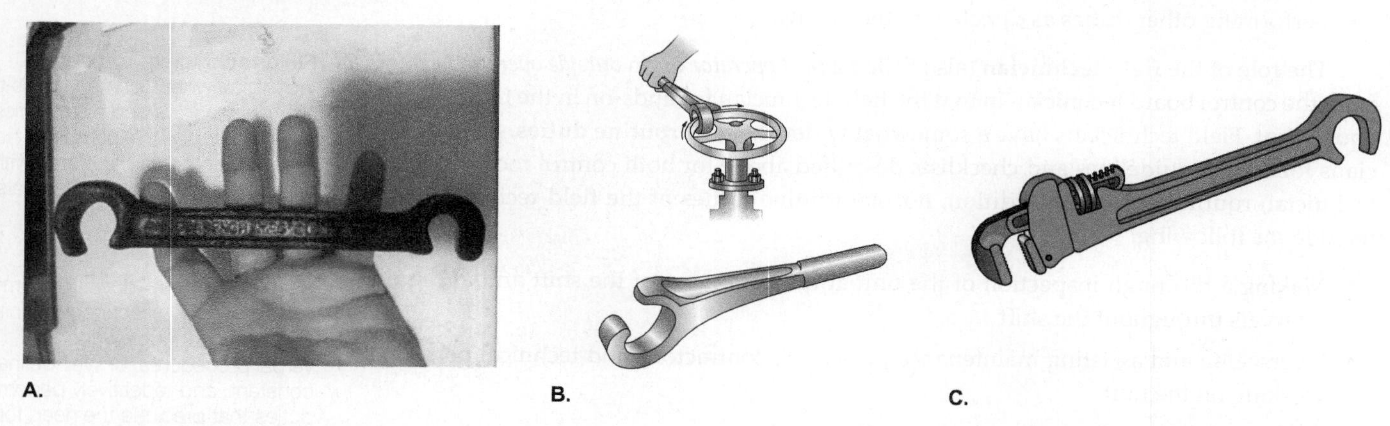

Technicians should also carry a set of pliers when making routine rounds (Figure 12.2). Adjustable pliers may also be known as tongue and groove pliers. They are used for tightening packing nuts and installing or removing bull plugs. Cutting (lineman's) pliers are useful tools for removing tie wraps that attach lockout/tagout tags to equipment.

Figure 12.2 Pliers are essential tools for operators in the field.

CREDIT: (left to right) DruZhi Art/Shutterstock; Chawalit Chanpaiboon/Shutterstock; Yellow Cat/Shutterstock; exopixel/Shutterstock.

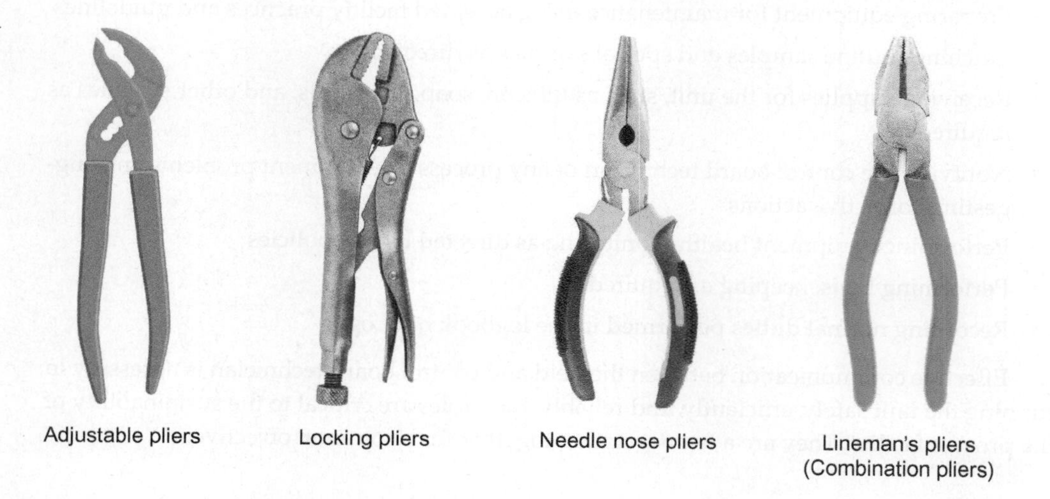

Field technicians are also responsible for maintaining adequate oil levels in rotating and reciprocating equipment. Oil levels should be checked and the correct type of oil added as needed, using oil cans with flexible spouts for small quantities or cans with goose-neck spouts for larger quantities. Figure 12.3A depicts a typical oil can that keeps oil clean for storage. Oil cans come in various sizes and can be closed or open top. On some models, the lid is removed and a flexible spout is added prior to use. Such closed top containers prevent water from entering and contaminating the oil, which could ultimately cause equipment damage.

A grease gun (see Figure 12.3B) is used to lubricate valves, usually on a monthly basis, to ensure that the valves operate easily.

12.3 Equipment Monitoring

Monitor to observe or watch; observing and listening to the equipment routinely to prevent process upsets.

The unit must always be monitored. To **monitor**, the field technician makes routine rounds at regular intervals to ensure that the equipment and process are running optimally (shown in Table 12.1). The technician is also aware of normal sounds and smells in the unit and takes note of and investigates anything abnormal. Control board technicians routinely monitor

Figure 12.3 **A.** Oil can. **B.** Grease gun.

CREDIT: A. Africa Studio/Shutterstock. **B.** Evgenii Kurbanov/Shutterstock.

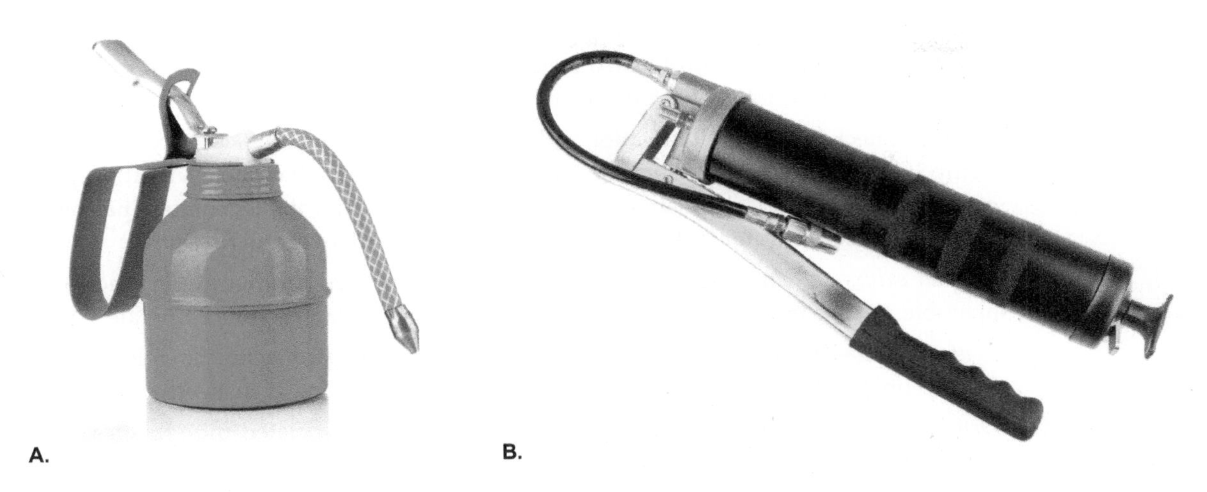

A. B.

Table 12.1 Monitoring Key Inspection Points

Equipment	Key Inspection Points
Pipes, tubing, hoses, and fittings	Monitor for leaks, external corrosion, temperature gauges, temperature differences, and vibration in piping.
Valves	Monitor for leaks, external corrosion, excessive wear, valve stem lubrication; ensure hand wheels are in their proper position.
Process vessels	Monitor proper operating temperature, pressures, flows, and levels; look for hot spots on vessel walls; if thermal heating exists; listen for abnormal noises; check for excessive vibration of equipment.
Pumps	Monitor oil levels; examine seals for leaks; inspect suction and discharge pressures; monitor DP screens; listen for abnormal noises, vibration, and excessive heat.
Compressors	Monitor oil levels and suction, discharge pressures, suction drum levels; look for leaks on flanges; listen for abnormal noises and abnormal heat.
Turbines	Look at oil levels and color; monitor cooling water flows and temperatures; look at seals and flanges for leaks; observe governor for operations; check for excessive vibration, heat, or noises.
Motors	Look for loose covers, shrouds, leaks, and excessive vibrations; listen for abnormal noises; check for excessive heat and vibrations, oil level, and seals.
Engines	Look for loose covers, shrouds, and leaks; listen for abnormal noises; check for excessive heat and vibrations, and monitor oil level.
Heat exchangers	Inspect for external leaks; look for abnormal pressures and temperatures; inspect insulation; listen for abnormal noises; check for excessive vibration.
Cooling towers	Observe for leaks, water levels; check filter screens for plugging, temperature differentials; look for broken fill materials, uneven water distribution; listen for abnormal noise; check vibrations.
Furnaces	Look for flames impinging on tubes, poor flame distribution, hot spots on structure, smoke from stack; check burner spider for proper flame and air flow (draft).
Boilers	Look for flames impinging on tubes, poor flame distribution, hot spots on structure; check water level, air flow (draft), operation of mud drum, and smoke from stack; listen for abnormal sounds; check burner spider for proper flame
Reactors	Monitor operating temperatures, pressures, flows, and levels; look for hot spots on vessel walls; listen for abnormal noises; check for excessive vibration of equipment.

the process and equipment remotely. A proactive monitoring approach by a control board technician helps avoid process alarms during the shift.

When monitoring equipment, the process technicians may consider the following:

- *Compressors*—Check oil flow; verify pressure and temperature instrumentation; verify control valve positions; check motor amperage; and take **vibration readings**.
- *Exchangers*—Perform periodic exchanger surveys to verify that the exchangers do not experience fouling. Local instrumentation allows checking the temperature and/or pressure differential across exchangers. Monthly surveys are typically for records and reviews by engineering. The field technicians should be familiar with the normal

Vibration readings measurement and documentation of rotating equipment to ensure vibrations are within an acceptable range.

differentials from making routine rounds and should always bring attention to a high differential on any exchanger. Surveys may be performed utilizing control loop data for some equipment.

- *Motors*—Regularly check vibration using a handheld vibration pen on pumps, mixers, and compressors. The control board technician may check motor amps and vibration remotely if the technology has been installed.
- *Pumps*—Check for proper lubrication and vibration and add correct viscosity oil per the pump lubrication manual. Vibration can be checked using a handheld vibration pen, or the control board technician may check pump vibration if the technology has been installed. Check seals for leakage and seal pots for normal level and pressure.
- *Manual valves*—Visually inspect valves to ensure that valve handles are in place and operable. Grease the valves on a monthly basis to ensure that the valves operate easily.
- *Control valves*—Inspect visually once a month. The control board technician relies on the field technician to check these valves if sluggishness or sticking is suspected.
- *Drums and vessels*—Inspect gaskets for leaks on vessels, associated pipe flanges, and manways several times per shift.
- *Cooling towers*—Check for proper levels, adequate chemical composition, temperature, and proper fan and pump operation.
- *Safety equipment*—Visually inspect weekly. Remove defective equipment from service and replace immediately, including any air bottles for escape packs not at full pressure.
- *Instrumentation*—Check periodically for leaks. The control board technician should have the field technician check and verify instrument readings.

Checking for Leaks

Field technicians check for small leaks in the process during routine rounds using the senses of sight and smell. If a leak is found, the technician must first notify the control board technician, who then reports the leak to the next level of supervision. Additional notifications are made as necessary.

The field technician may attempt to stop the leak if it is small, but proper personal protective equipment (PPE), such as hard hat, safety glasses, goggles, and protective clothing, is required when attempting to stop a leak. Large leaks should never be handled by one field technician alone due to the potential danger to the technician and to the equipment.

The leaks a field technician may encounter and the actions to mitigate them include the following:

- *Valve packing leaks*—Alternately tighten each packing nut equally until the leak is stopped.
- *Analyzer tubing leak*—Tighten fittings up to one-half turn. If the leak does not stop, notify the control board technician to request the appropriate maintenance technician to repair the equipment.
- *Control valve packing leaks*—(**Note**: A field technician must not adjust control valve packing until after contacting and coordinating with the control board operator to ensure the process is not negatively impacted.) Alternately tighten packing nuts one revolution and then request the control board technician to stroke the valve to ensure the valve is not sticking. Repeat if needed. Confirm continued functionality with control board technician.
- *Flange leaks*—Tighten opposite bolts equally until the leak is stopped.
- *Vapor leaks*—Use proper PPE for the situation. If the chemical component is ethylene or propane, then water or steam may have to be used to de-ice the suspected area.
- *Leaks under insulation*—Strip the insulation away slowly to assess the leak. Report condition to supervisor or appropriate maintenance personnel for insulation repairs.

- *Exchanger leaks*—Whether internal or external, could results in the lack of heat transfer, off-specification material, or water or other components in the outlet stream. If safe to do so, wear appropriate PPE and verify an exchange tube leak by following these steps:
  - Block in the shell side.
  - Open a low point bleeder on the shell side of the exchanger slightly.
  - Check for material leaking out of a tube and into the shell side.

  Note: Coordinate any exchanger checks with control board technician to prevent process interruption or upset.

The primary responsibility of process technicians regarding any process leak is to (1) protect themselves, (2) notify the control board technician, and (3) assist in securing the leaking area.

12.4 Equipment Health Monitoring (EHM)

Many facilities today have an established **equipment health monitoring (EHM)** program. EHM is designed to increase uptime, reduce maintenance costs, and monitor equipment after startup or repair. Field technicians are responsible for performing daily checks and recording the data for specific pieces of rotating equipment. The technicians may be required to record the vibration levels on pumps and motors, oil levels in pumps and other pieces of equipment, temperature and pressure variables at various locations, storage container levels (such as bulk oil or chemical drums), and other readings as determined by the process facility.

Equipment health monitoring (EHM) efficient system for protecting rotating equipment and facility operations from unscheduled downtime that provides personnel with the equipment knowledge they need to schedule maintenance, manage inventories, and support efficient workflow scheduling.

Vibration readings are generally taken with a vibration pen, and temperature readings are taken as recorded from local gauges and/or taken with a temperature gun (shown in Figure 12.4). The readings are taken on a technician's designated **route**. The readings may be recorded on a route log sheet in the event a computerized handheld device is not available. The EHM program can also produce data that can be utilized for better planning and root cause failure analysis.

Route sequential path followed in order to perform equipment health monitoring (EHM).

Figure 12.4 **A.** Vibration pen. **B.** Temperature gun.

CREDIT: A. Tum ZzzzZ/Shutterstock. **B.** Mr.PK/Shutterstock.

A.

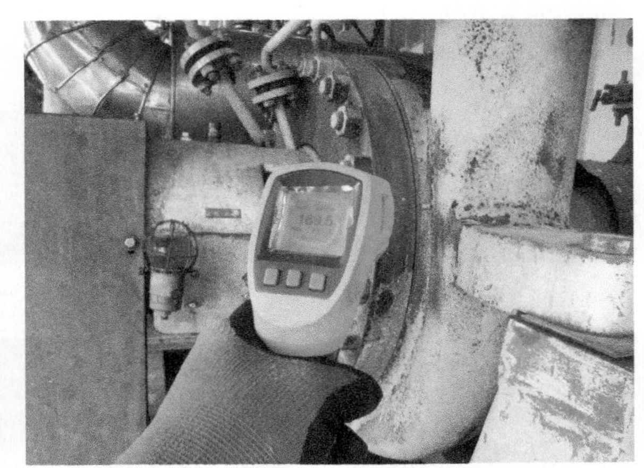

B.

12.5 Starting/Stopping Equipment

Field technicians often prepare equipment for maintenance and return it to service after maintenance. Process technicians are required to start and stop rotating equipment when necessary during their shift. During shift change, the incoming field technician should learn about any maintenance preparations that must be made during the shift. During the first round of the shift, a process technician should become knowledgeable about equipment in service and spare equipment available.

When starting pumps, the technician should follow normal operating procedures. Fundamental considerations for starting up spare pumps include the following:

- Verify that the spare pump has proper oil levels.
- Ensure that seal oil, external or internal, is lined up properly.
- Ensure that the spare pump suction and discharge valve are in the open position.
- Notify the control board technician, and ensure that the unit is prepared for pump swap.
- Start the spare pump.
- Allow time for flow conditions to stabilize, then shut down and secure the pump to be prepared for maintenance or service.

Field technicians must ensure that all pumps in standby mode with auto-start capabilities have fully open suction and discharge valves. He or she must notify the control board technician prior to the startup or shutdown of any equipment. The use of normal startup procedures for returning equipment to service is equally important to prevent an accident or incident.

12.6 Personal Protective Equipment (PPE)

Field technicians use personal protective equipment for a variety of routine duties during a shift (Figure 12.5). Personal protective equipment (PPE) is designed to protect employees from workplace injuries or illnesses from contact with chemicals, radiological materials,

Figure 12.5 Technicians wear PPE for a variety of duties. **A.** Hard hats, safety glasses, and hearing protection. **B.** Hard hats, safety glasses, flame retardant clothing, and gloves. **C.** Steel-toed footwear. **D,** Face shield. **E.** Flash suit being used while working on wind farm transmission system. **F.** SCBA (self-contained breathing apparatus).

CREDIT: A. LuYago/Shutterstock. **B.** ZoranOrcik/Shutterstock. **C.** BushAlex/Shutterstock. **D.** Reggie Lavoie/Shutterstock. **E.** Cl2/Cavan Images/Alamy Stock Photo. **F.** Zulkamalober/Shutterstock.

A.

B.

C.

D.

E.

F.

electrical power, mechanical equipment, falling objects, and other workplace hazards. Some types of PPE a technician may use during a normal shift include the following:

- Hard hats
- Goggles
- Safety glasses with side shields
- Hearing protection
- Gloves
- Steel-toed footwear
- Respirators
- Face shields
- Flame-retardant clothing
- Chemical-resistant suits
- Flash suits
- SCBA (self-contained breathing apparatus).

Many facilities in the industry have adopted minimum PPE requirements, such as a hard hat, safety glasses, steel-toed boots, flame-retardant clothing, and hearing protection whenever anyone (even a plant visitor) enters a process area.

The Occupational Safety and Health Administration's (OSHA) Title 29 of the Code Federal Regulations (CFR) Part 1910 Subpart I regulates the PPE standard. OSHA's general PPE standard mandates that employers conduct a hazard assessment of their workplaces to determine the hazards present that require the use of PPE and provide workers with the appropriate PPE. It also requires employers to train employees to use and maintain PPE in a sanitary and reliable condition. The employer must also provide training for employees in the following:

- Proper PPE use
- Necessary PPE use
- Proper PPE needed
- PPE limitations
- Correct wear and adjustment of PPE
- Proper PPE maintenance.

Personal protective equipment is often essential to performing tasks. When working with materials such as lube oil, cooling tower chemicals such as sodium hypochlorite or sulfuric acid, catalysts for certain processes, or other specialty chemicals, consult the safety data sheets (SDSs) for the proper PPE requirements prior to handling.

12.7 Procedures and Documentation

Process technicians commonly use standard operating procedures (SOPs) during the performance of routine duties. Operating procedures are very important documents in the industry for preparing equipment for maintenance, returning equipment to service after repair, loading or unloading bulk materials, changing out pressure gauges, and other tasks that have been identified by a facility.

Procedures are critical for performing certain tasks. Poorly written or improperly executed procedures lead to accidents, injuries, and deaths. Technicians should follow all procedures, initialing and providing a time of completion for each step, and signing the procedure when the tasks are completed. Technicians may deviate from a procedure if required, but must follow the deviation policy of the facility.

Process technicians may also be asked to create or review existing procedures for accuracy or to update an existing procedure due to a process change. When a procedure is being reviewed for clarity and or accuracy, any misalignment should be noted on the procedure and turned in to the responsible procedure writer for updating. Management of Change (MOC) policy must be followed.

Documenting Routine Duties

Field technicians are required to document work activities in the unit logbook or unit eLog. Technician should provide as much detail as possible for his or her shift relief regarding what took place during the shift. Some items that should be documented include the following:

- Process alarms that activated during the shift and the corrective actions taken to help the relieving shift troubleshoot in the event of a similar alarm, or to eliminate the alarm
- Vibration and temperature alarms experienced on equipment, especially those critical for the continued safe operation of pumps and compressors. Documenting these types of alarms helps to identify vibration and temperature trends, which may prevent catastrophic equipment failure
- Equipment oil levels and the amount of oil added, to establish rotating equipment oil usage trends for planning pump seal and bearing maintenance
- Chemicals used in the process, such as cooling tower chemicals, to help identify trends that may prevent cooling water exchanger fouling
- Maintenance activities performed and planned
- Unusual events, such as leaks in the process, process upsets, and equipment malfunctions.

Summary

The routine duties of the process technician are important for the safe, efficient, and reliable operation of a process unit. A field technician should be very observant during rounds, investigating any abnormal conditions, taking appropriate corrective action, communicating to the control board technician as well as to those in the oncoming shift, and documenting all conditions properly. The field technicians must monitor fixed and rotating equipment on site, perform the equipment health monitoring route, and report any unusual observations. The control board technician works remotely, using the DCS. Unit equipment must be monitored to ensure the process will continue to run safely and smoothly.

Process technicians should be alert and aware of their surroundings at all times. Proper PPE must be utilized, as needed, and technicians should check the PPE prior to use. Wearing PPE prevents exposure to hazardous chemicals, excessive noise, falling objects, and other hazards.

It is imperative that the field and control board technicians have effective communication during both normal operations and critical events. Communication between the field and control board technicians ensures that all relevant information is gathered, documented, and passed on to the next shift. Technicians should always follow procedures and be thorough in documenting all steps in order to prevent future accidents and injuries. Procedures must be followed to prevent accidents and injuries. Any deviation from procedure should follow the facility deviation policy.

Checking Your Knowledge

1. Define the following terms:
   a. Control room
   b. Equipment health monitoring (EHM)
   c. Field technician
   d. Monitor
   e. Normal operations
   f. Rounds
   g. Route
   h. Routine duties
   i. Vibration readings
2. Which of following duties do field technicians perform on a routine basis? (Select all that apply.)
   a. Overseeing and assisting maintenance personnel, contractors, and technical personnel working on the unit
   b. Notifying the control board technician of any process or equipment problems and suggesting corrective actions
   c. Monitoring all process parameters and equipment on the unit
   d. Monitoring alarm screens and taking corrective action as required
   e. Making a thorough inspection of the unit at the beginning of the shift and at regular intervals throughout the shift
   f. Coordinating process activities with the field technician
3. Which of following duties do control room operators perform on a routine basis? (Select all that apply.)
   a. Recording normal duties performed in the logbook or eLog
   b. Preparing equipment for maintenance using accepted facility practices and guidelines
   c. Thoroughly inspecting the unit utilizing the control system at the beginning and throughout the shift at regular intervals
   d. Receiving supplies for the unit, such as lube oil, soap, chemicals, and other supplies as required
   e. Recording shift activities in the unit logbook or eLog, including the recording of all personnel on the unit
   f. Recording all lab data
4. Which of the following is useful to have when removing LOTO tags?
   a. Tongue and groove pliers
   b. Valve wrench
   c. Cutting pliers
   d. Phillips head screwdriver.
5. Which of the following are routine monitoring checks or actions that a field technician performs on a PUMP? (Select all that apply.)
   a. Add correct viscosity oil per the pump lubrication manual.
   b. Check seal pots for normal level and pressure.
   c. Verify pressure and temperature instrumentation.
   d. Check motor amperage.
   e. Check oil flow.
   f. Check for proper lubrication.
6. Which of the following are routine monitoring checks or actions that a field technician performs on a COMPRESSOR? (Select all that apply.)
   a. Verify pressure and temperature instrumentation.
   b. Check motor amperage.
   c. Check oil flow.
   d. Perform periodic surveys to verify that the item does not experience fouling.
   e. Verify control valve positions.
   f. Check for proper lubrication.
7. Which of the following are routine monitoring checks or actions that a field technician performs on a HEAT EXCHANGER? (Select all that apply.)
   a. Verify control valve positions.
   b. Check seals for leakage.
   c. Check the temperature and/or pressure differential across exchangers.
   d. Verify pressure and temperature instrumentation.
   e. Check seal pots for normal level and pressure.
   f. Perform periodic surveys to verify that they do not experience fouling.
8. What is the purpose of an equipment health monitoring (EHM) program?
   a. Increases downtime, decreases cost, and monitors equipment after repair or startup.
   b. Increases the field technician workload.
   c. Increases production time, decreases cost, and monitors equipment after repair or startup.
   d. None of the above.
9. Indicate the correct order of steps to follow when starting spare pumps:
   a. Ensure that the spare pump suction and discharge valve are in the open position.
   b. Allow time for flow conditions to stabilize, then shut down and secure the main pump to be prepared for maintenance or service.

c. Ensure that any seal oil, external or internal, is lined up properly.

d. Verify that the spare pump has proper oil levels.

e. Start the spare pump.

f. Notify the control board technician and ensure that the unit is prepared for pump swap.

10. The primary purpose of ________ ________ ________ is to protect the employee from workplace injuries or illnesses.

11. Which of these is not a PPE training point that OSHA requires employers to provide for employees?

a. Proper PPE use

b. Proper PPE needed

c. PPE costs

d. Necessary PPE use

12. Name four routine work activities that process technicians are required to document in the unit logbook (or unit eLog).

NOTE: Answers to Checking Your Knowledge appear in the Appendix.

Student Activities

1. Perform research on OSHA 29.CFR.1910 Subpart I, select one of the PPE topics listed in the subpart, and provide a brief report to the class on your selected topic.

2. Using information learned in Chapter 8, *Shift Change/Relief* and Chapter 12, *Process Technician Routine Duties: Normal Operations*, work with three other classmates to exchange information between two field technicians and two control board technicians. Using the scenario provided below, and working as a group, develop a working solution for the problem. Complete a written report and simulate a shift change.

   Scenario: You are working in a small chemical facility when a hydrocarbon leak develops in a pump discharge flange. The pump can be isolated, and you will require no outside assistance. The unit has been running smoothly, and you would like to avoid an upset condition that may result from the leak. The field technician and the control board technician are the only two personnel that are immediately available to handle this situation.

Chapter 13

Sampling

Objectives

After completing this chapter, you will be able to:

13.1 Explain the process technician's role in the sample analysis. (NAPTA Operations, Normal Operations; Other Duties 3*) p. 176

13.2 Explain the importance of following the sample procedure precisely. (NAPTA Operations, Normal Operations; Other Duties 2) p. 177

13.3 Describe the sampling procedures and equipment used for different sampling events. (NAPTA Operations, Normal Operations; Other Duties 1) p. 178

13.4 Describe the personal protective equipment that must be used while performing different sampling activities. (NAPTA Operations, Normal Operations; Other Duties 1) p. 180

13.5 Explain the importance of the sample analysis in relation to the proper operation of the unit. (NAPTA Operations, Normal Operations; Other Duties 5) p. 181

13.6 Explain the various types of analysis (methods and equipment) conducted on process samples. (NAPTA Operations, Normal Operations; Other Duties 4) p. 183

Key Terms

Analyzer—a device used to measure physical and/or chemical compositions of materials, **p. 177**

Chromatography—an analytical technique used by the laboratory and unit analyzers for determining the individual components in a sample of liquid or gas, **p. 183**

*North American Process Technology Alliance (NAPTA) developed curriculum to ensure that Process Technology courses will produce knowledgeable graduates to become entry-level employees in process technology. Objectives from that curriculum are named here in abbreviated form. For example, "(NAPTA Operations, Normal Operations; Other Duties 3)" means that this chapter's objective 1 relates to objective 3 of NAPTA's curriculum about the process technician's other duties during normal operations.

Color—visual comparison scale used in the process industry to determine product color purity; generally refers to the color of liquid chemicals (for example, the clarity and hue of yellow represents a poor quality; a high quality is water-white or clear); also known as the *ASTM International* or the *Saybolt color scale*, **p. 183**

Flame resistant—characteristic of a fabric to resist ignition and to self-extinguish if ignited, **p. 181**

Flame retardant—characteristic of a fabric that has had a chemical substance to impart flame resistance, **p. 181**

Gas chromatography (GC)—a system of identifying and quantifying a gas sample's compounds by their boiling points. It consists of an oven, column, autosampler, and detector. The autosampler injects the sample into the column, which is packed with special material to help separate the components of the sample. A carrier gas sweeps the sample through the column while the oven adds temperature to speed the sample along. The mixture separates in the column and elutes as individual compounds into the detector where each is individually quantified. As a result, each component leaves the column separately and in a predictable sequence and rate, **p. 183**

Karl Fischer (KF) water method—a process that uses a chemical solution to determine the quantity of water in refined chemicals; also known as *Karl Fischer titration*, **p. 183**

Repeatable—able to be done in exactly the same manner each time so that results can be scientifically compared, **p. 181**

Sample loop—a continuous circulation of process liquid or gas from a higher-pressure source to a lower-pressure return, to ensure capture of a representative sample, **p. 179**

Sample point—section of small diameter-valved tubing that extends from the main process piping system for collecting samples, **p. 178**

Sampling—process of collecting a representative portion of a material for analysis, **p. 176**

Specifications—the quality limitations for a product; product purity parameters that have been agreed on by the company and the customers or regulated by governmental agencies; commonly called *specs*, **p. 177**

13.1 Introduction

Process technicians are responsible for keeping their facilities operating safely and efficiently to ensure the production of a high-quality product. They control and monitor process systems, inspect equipment, and conduct routine system operations. Process technicians also routinely sample and test process fluids and solids at various stages of the production process (Figure 13.1). These tasks are necessary to ensure the reliability of continuous stream analyzers, maintain correct process operating parameters, and ensure product stream specification.

Samples are taken from process systems for early problem detection to prevent equipment damage and product waste. If testing of a sample indicates that a material is unacceptable, actions are taken to correct the problem before larger problems and lost revenue occur. Samples are also taken to verify that waste products discharged into the environment comply with company and government regulations.

The Importance of Sampling

Sampling process of collecting a representative portion of a material for analysis.

Catching a sample is a common and very important duty that process technicians perform. **Sampling** is the series of steps for collecting and preserving a liquid, gas, or solid for analysis.

Figure 13.1 Process technicians routinely sample process materials.

CREDIT: Rawi Rochanavipart/Shutterstock.

Sampling takes place throughout the process and serves many purposes. Sampling during a product transition helps to minimize off-specification production. An off-specification product has the same production costs as a marketable product, but sells at a lower rate than prime product, sometimes at a loss. Some off-spec products have no market and are a total loss if the process cannot recycle the material.

Sampling is the primary means for analyzer verification. An **analyzer** is an instrument or device that performs continuous sample stream analysis. Analyzers provide real-time composition and quality data for the control board process technician and are integral to safe and effective production. However, analyzers are not foolproof. If an analyzer is plugged, damaged, or goes out of calibration, it provides false data, and other process controllers downstream may be incorrectly controlled and magnify the problem. Independent sampling verifies the accuracy of the online analyzers. The more often a sample is caught and tested, the sooner problems are identified and the sooner corrections can be made to return to prime production. It is critical that samples be caught correctly and promptly.

Analyzer a device used to measure physical and/or chemical compositions of materials.

Samples are also used for customer quality assurance. In some facilities, product runs for different customers may have to meet different physical or chemical requirements or **specifications**. These specifications are product purity parameters that have been agreed on by the company and the customers or that are regulated by governmental agencies. The customer pays the negotiated price for the product that meets the agreed upon specifications. If the product fails to meet these specifications, the customer pays a lower price or rejects the product. The result is lost revenue for the facility.

Specifications the quality limitations for a product; product purity parameters that have been agreed on by the company and the customers or regulated by governmental agencies; commonly called *specs*.

Sampling also determines the start and stop points for certain tasks. For example, most process units use heat exchangers and condensers to control temperature and pressure in equipment and to control reactions. After a period, the exchanger components may become fouled and lose efficiency. Often, the exchangers may be cleaned chemically to restore efficiency and extend the useful life of the tube bundle. During the cleaning process, samples are caught and tested to determine when each stage of the cleaning process is complete.

Samples also protect personnel from mistakes. For example, if a vendor filled a self-contained breathing apparatus (SCBA) air bottle with nitrogen, it would result in a fatality. A lab test to verify the contents of the bottle could save a life.

13.2 Following Proper Sampling Procedure

Not only is the sampling process essential to efficient operation, but the way the sample is collected is also critical to achieving reliable results and to maintaining personnel safety. Sampling procedures are intended to preserve the integrity of the sample and provide safety measures to protect the process technician while collecting the sample. Deviating from procedure can result in bad data, serious injury to the technician, and/or equipment damage.

There are hazards associated with sampling. Most samples are collected from in-service equipment, and the associated piping and equipment may be operating at high temperature and pressure. The sample may be dangerous, or it may contain a flammable or explosive compound, and it may be necessary to keep the sample container electrically grounded to eliminate the potential of a static spark while collecting the sample from a process line. The sample may also be toxic to humans and require specific PPE or handling instructions to reduce exposure.

All process facilities have procedures on sampling techniques for the various samples. The facility's written sampling procedure should specify the type of container to be used to collect the sample. Using the appropriate sample container is important for personal protection, as well as for correct collection and containment. For example, a sample with harmful vapor would be collected in an airtight container, and a corrosive product would be collected in a container resistant to corrosion.

13.3 Sample Points, Sample Loops, and Sample Containers

Sample point section of small diameter-valved tubing that extends from the main process piping system for collecting samples.

A **sample point** generally consists of a short section of small diameter-valved tubing that extends from the main process piping system to allow collection of samples. The valve in the sample line can be operated to start and stop the flow of the sample to the sample container. If the setup is not a continuous flow sample loop, the sample may need to purge or drain into a closed drain system for a period in order to retrieve a representative sample. The locations of sampling points can be found on process facility diagrams or by consulting facility procedures. The central lab should have a master printout of the location of all sample points in the process facility.

The container often consists of a bottle, sealed with a cap and septum. The resealable membrane on top of a bottle or vial is a septum. (The vial that the phlebotomist puts your blood samples in has a septum.) When sampling with a bottle, a sample is drawn from the process and arrives at atmospheric pressure in the sample container. This is accomplished by inserting the sampling bottle into the sleeve until the septum is pierced by the needles extending from the needle assembly (Figure 13.2). Once in position, the product can flow into the sample bottle via the process needle while a vent needle vents air and gases. When the required amount has been collected, the process technician stops the product sample flow (or it is stopped automatically), and the bottle is pulled out of the sleeve. The septum reseals automatically.

Figure 13.2 Example of a septum.

CREDIT: Courtesy of Ray Player.

A **sample loop**, shown in Figure 13.3, provides a continuous circulation of the process liquid or gas from a higher-pressure source to a lower-pressure return, such as a pump or compressor discharge back to the suction. Circulation ensures capture of a representative sample. A loop is typically equipped with a sample station appropriate for the sampling required, including a sample cooler if needed. Sample loops are installed with online analyzers and where sample line purging of the process is not convenient or safe to perform. Continuous circulation of the process stream is necessary to provide the online analyzer with a current sample of the process.

Sample loop a continuous circulation of process liquid or gas from a higher-pressure source to a lower-pressure return, to ensure capture of a representative sample.

Figure 13.3 Sample point and sample loop.

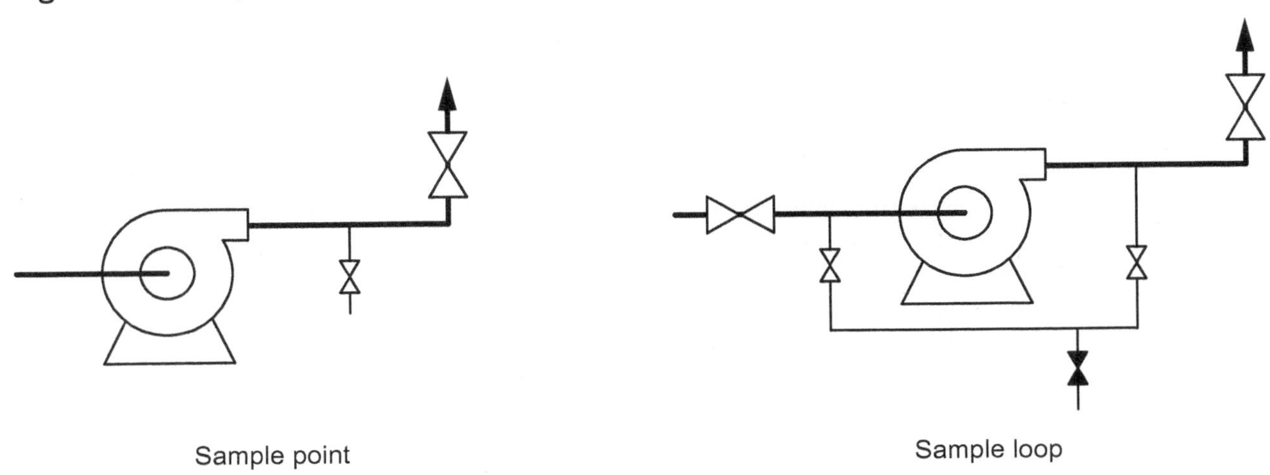

Sample containers are vessels used to collect and contain a sample for analysis. The containers may be glass bottles or vials of various sizes, plastic bottles, plastic bags, metal tins or cans, metal sample cylinders (also called *bombs*), or others depending on the type and quantity of sample needed. A pressurized gas sample or very hazardous liquid requires a sample bomb. Figure 13.4 shows examples of sample containers. Some samples may react

Figure 13.4 **A.** Sample bomb for collection of high pressure fluid samples. **B.** Sample jars holding a variety of solids samples. **C.** Inline automatic sample system (autosampler).

CREDIT: A. lalalala.dragonfly/Shutterstock. **B.** Photographic Services, Shell International Limited. **C.** INEOS.

A.

B.

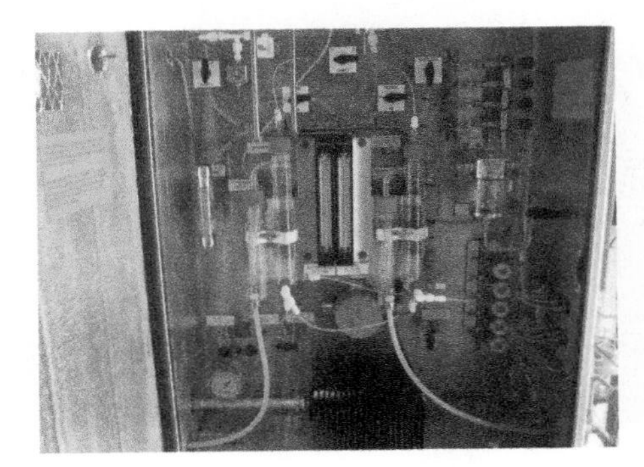

C.

with certain materials, such as aluminum, and require a sample container of a specific construction for collecting samples. Following correct sampling procedures is important not only for personnel safety but also for reliability of data.

13.4 Wearing Proper PPE

Proper personal protective equipment (PPE) must be used at all times during the sampling process to protect the process technician. The materials and work areas determine the hazards associated with taking samples. Process technicians should become familiar with the materials used in their process systems to determine the hazards that exist. For instance, several hazards are associated with acidic and caustic materials, and if these types of materials encounter skin or eyes, they could cause serious burns or blindness. Toxic fumes are another hazard that may be present when some materials are sampled. If harmful fumes are inhaled, they could cause serious respiratory injury or death.

Some materials may be under high temperature or pressure, and when these materials are sampled, care should be taken to avoid being sprayed. There are even materials at a very low temperature which could cause a cryogenic burn if contact is made with the skin. The materials in some process systems are flammable or explosive and could be ignited by the smallest spark, flame, or static electricity in the atmosphere. To avoid serious injuries in these locations, smoking is prohibited, and open flames are not allowed around these materials.

The type of sample being collected dictates the type of PPE required. Figure 13.5 shows some types of PPE that are used when sampling. The PPE may include:

- Safety glasses with side shields
- Goggles
- Face shield
- Slicker top
- Protective gloves (heat resistant, chemical resistant, and so on)
- Rubber boots
- Fire resistant clothing (FRC) or chemical resistant apron or clothing
- Respiratory protection.

Wearing the required PPE provides the process technician protection from any potential exposure in the event of unforeseen incidents. Protective eyewear, such as goggles or a face

Figure 13.5 Personal protective equipment. **A.** Technician with hard hat, safety glasses, face shield, and protective gloves. **B.** Fire retardant clothing. **C.** Chemical suit.

CREDIT: A. Copyright © BASF SE 2021. **B.** Tawansak/Shutterstock. **C.** andrey polivanov/Shutterstock.

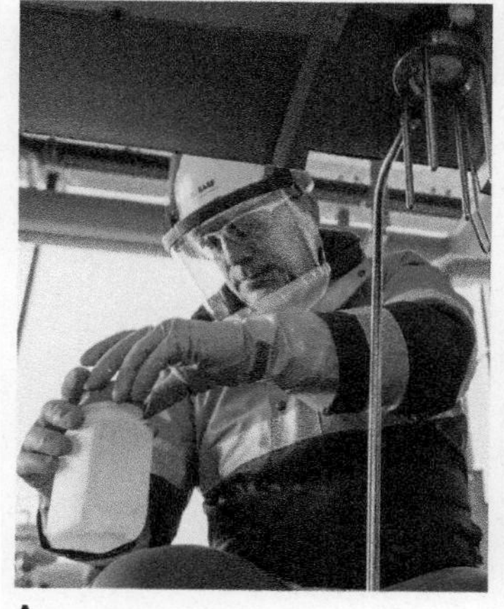

A.

B.

C.

shield, should be worn if there is a possibility of overspray, misting, or splash. Protective specialty gloves should also be worn when sampling from hot or cold lines and when collecting samples of corrosive chemicals that could cause chemical burns if contact is made with exposed skin.

The process technician's uniform also provides a small degree of protection for skin. Most uniforms worn in process facilities are made of fire-resistant or fire-retardant fabric. Fire resistant/retardant clothing, which is used in situations where there is risk of electrical arc or flash or thermal burns, is regulated by NFPA-70E, American Society for Testing and Materials (ASTM International), and OSHA standards. **Flame resistant** fabrics, such as Nomex®, are slow to ignite and self-extinguish if they are ignited. **Flame retardant** fabrics normally have an added chemical substance that provides flame resistance; they are generally made of cotton or cotton blends.

Flame resistant characteristic of a fabric to resist ignition and to self-extinguish if ignited.

Flame retardant characteristic of a fabric that has had a chemical substance to impart flame resistance.

Fire resistant/retardant clothing, shown in Figure 13.5B, should be worn while in a process unit. This type of apparel is designed to provide very brief protection in the event of a flash fire. These materials, however, are not resistant to acids and other corrosives. When sampling corrosive substances, it may also be necessary to wear a protective rubber apron, slicker suit, or chemical suit, like the one shown in Figure 13.5C.

Respiratory protection equipment helps guard the process technician from inhalation hazards that may be encountered in certain sampling activities. The type of respirator used depends on the hazards that are likely to be encountered. A half-face respirator employs filter cartridges to clean incoming air of specific substances (user must select the proper cartridge). The full-face respirator uses filter cartridges as well, but also covers the eyes and can achieve a superior mask to face seal due to increased seal area.

Some substances or environments are designated as Immediately Dangerous to Life or Health (IDLH). These conditions normally require supplied-air respirators. Supplied-air equipment provides a complete source of contaminant-free breathing air to the user's face mask. One type of supplied-air equipment is the self-contained breathing apparatus (SCBA). The breathing air for this equipment is supplied by user-worn tanks. Another type of supplied-air equipment is the airline respirator system. In this equipment, the air to the user's face mask is supplied through a hose that is connected remotely to breathing air cylinders located out of the hazard zone. Typically, the SCBA is selected whenever short term use is anticipated (30–60 min. tank capacity) or when obstructions to movement of a hose may be encountered. In addition to proper PPE, special care should be taken to prevent spills when materials are sampled because a spill could cause a fall and injury. If a spill occurs, it should be reported immediately to the appropriate first-line supervisor and cleaned up promptly and properly to prevent an accident.

13.5 Contamination, Consistency, and Reliability

Samples must be collected and maintained free of outside contamination while being transported to the lab and during analysis. Failure to do so will render samples unusable. For example, if a collected gas sample were exposed to the outside atmosphere prior to testing, it would become useless if testing was for oxygen content.

Consistent sampling and record keeping are important troubleshooting tools. Samples gathered and tested over a long period can give insight and help reveal or prevent equipment failures or changes in the process. A single off-specification sample result may not be significant by itself, but when a pattern can be detected over time, it is easier to determine the root cause of a problem. The earlier a problem is detected, the easier it is to correct, and it often saves labor and money.

Repeatable able to be done in exactly the same manner each time so that results can be scientifically compared.

Reliability is paramount for usable results. When collecting samples, it is essential that the test be **repeatable**, regardless of who collects the sample. The samples need to be caught in exactly the same way each time they are collected so that results can be scientifically compared. For example, suppose polymer powder samples are collected to determine the amount of volatile gas present in the powder at a given point in the process. A large difference in results could be seen based on the length of time taken to place an airtight cap on the powder sample vial. The longer the vial remains open to atmosphere, the more gas will escape, yielding an inaccurate result. If one process technician catches the sample and caps it immediately, but another process technician caps it several minutes after it has been caught, the results will be inconsistent. When the samples are analyzed at the lab, the lab technician may not be aware of the variances in sampling techniques and would present the results based on the information found in these inconsistently obtained samples. This could lead to an error in process adjustments.

As another example, variations may occur if samples are typically collected automatically and the sampling system breaks down. A grab sample (manually collected sample) may vary slightly from automatically collected samples. In such a situation, it would be important for the technician collecting the sample to label it as a grab sample.

Proper Labeling and Quantity

The laboratory at a refinery or a process facility is a busy place. Lab technicians receive dozens of samples each day and must test each according to procedure, then record and report the results. The lab technicians must pay very close attention to ensure that human error does not compromise results. For example, a decimal point in the wrong place can be the difference between large amounts of off-specification product versus highly profitable prime product. The process technician needs to make sure both the sample quantity and the sample labeling are correct to help ensure the tests are free from human error.

The quantity of the sample also needs to be consistent. The sample size needs to be sufficient in quantity to be a representative sample of the product at the time of sampling. Small sample amounts are often collected over a period to make a composite sample to test a product in a time-lapse manner. This is especially useful in blend or batch operations where a product is sold in specific quantities. A composite sample is more representative of the batch than a single sample, which is a snapshot at a specific time.

Labeling may seem insignificant, but it is as important as any other part of the sampling and testing procedure. The sample can be correctly gathered following a detailed procedure and tested by qualified lab technicians on precise testing equipment, but if the product is incorrectly labeled, the sample may be worthless. The sample label, shown in Figure 13.6, should contain relevant information that may include:

- Process unit name
- Date and time
- Vessel of origin, or sample point ID
- Quantity
- Current analyzer result
- Variables to be tested.

It is important to remember that excess or unused portions of samples should be disposed of properly, according to company procedures.

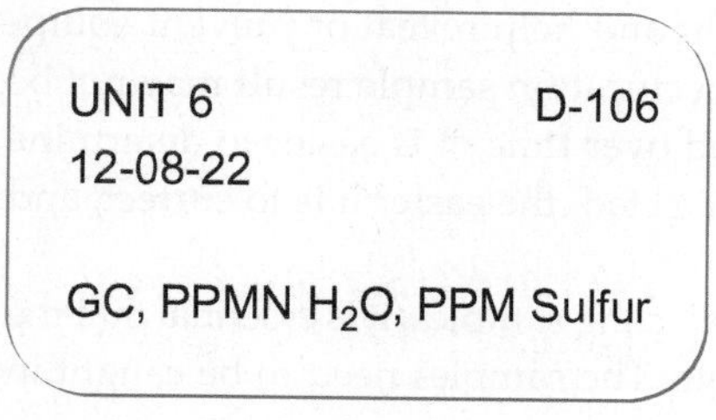

Figure 13.6 Example of a sample label.

UNIT 6 D-106
12-08-22

GC, PPMN H_2O, PPM Sulfur

13.6 Sample Analysis

Sample testing may be performed within the process unit. Testing cooling water for pH and chlorine levels can easily be completed in a small water lab within the unit. Other samples require more time, equipment, and expertise and can be better performed in a dedicated lab. The type of facility determines the materials and tests performed. Sampling and testing methods may be developed by the specific process facility to suit its particular needs. Some sample analyses that the process technician could perform include the following:

- **Color**—a visual comparison scale helps to determine product color purity. A common scale used in the process industry is the ASTM International color scale, which is also called the *Saybolt scale.* The Saybolt scale determines the color of refined oils, such as undyed motor and aviation gasoline, jet propulsion fuels, naphtha, and kerosene, as well as petroleum waxes and pharmaceutical white oils.
- Lead acetate test—a simple means of identifying the presence of sulfur or sulfur-based compounds.
- pH—measure of the acidity or basicity of a solution. The process technician may be required to measure the pH of cooling tower and condensate samples.
- **Gas chromatography (GC)**—a laboratory technique for the separation of mixtures (Figure 13.7). Gas chromatography is a common type of **chromatography** used in organic chemistry for separating and analyzing compounds that can be vaporized without decomposition. It can be completed online or offline. The gas chromatograph consists of an oven, column, autosampler, and detector. The autosampler injects the sample into the column that is packed with special material to help separate the components of the sample. The mixture separates in the column and exits as individual compounds. They are quantified in the detector. Components leave the column separately and in a predictable sequence and rate. The amount of time it takes for each compound to separate out is called *retention time* (RT). The RT provides information to determine the exact makeup of the gas sample. GC analysis can be utilized in a variety of processes when the composition of a gas is important to know.

Color visual comparison scale used in the process industry to determine product color purity; generally refers to the color of liquid chemicals (for example, the clarity and hue of yellow represents a poor quality; a high quality is water-white or clear); also known as the *ASTM International* or the *Saybolt color scale.*

Gas chromatography (GC) a method of identifying and quantifying a gas sample's compounds by their boiling points. The sample mixture separates in the column and elutes (exits) as individual compounds into the detector, where each is individually quantified.

Chromatography an analytical technique used by the laboratory and unit analyzers for determining the individual components in a sample of liquid or gas.

Figure 13.7 Gas chromatograph.

CREDIT: LnP images/Shutterstock.

- **Karl Fischer (KF) water method**—analytical method for quantifying water content in a variety of products, also known as *Karl Fischer titration.*
- Basic sediment and water (BS&W)—(also called *bottom solids* or *sediments*)—analytical term for emulsions of oil, water, and mud which settle out of crude oil during storage. The BS&W is used to see if the crude oil meets the technical specification limits set by the buyer. (When extracted from an oil reservoir, the crude oil will contain some amount of saltwater and particulate matter from the reservoir formation. To perform a BS&W

Karl Fischer (KF) water method a process that uses a chemical solution to determine the quantity of water in refined chemicals; also known as *Karl Fischer titration.*

test, a sample is taken in a test tube and placed in a centrifuge. The sample is spun for a specified time. This causes the heavier sediment and water to separate from the oil. It then can be measured to find the ratio of BS&W to oil.)

- Instruments for moisture, hydrogen, and oxygen analysis—instruments used in most facilities to measure gas flow to the flare, determining the mass flow rate and molecular weight. Panametric® is General Electric's trade name for this type of instrument.
- Gas detector tubes—glass vials filled with chemical reagents that react to specific chemicals. A sample of air is drawn through the tube with a bellows pump. If the targeted chemical is present, the reagent in the tube changes color and the length or depth of the color change indicates measured concentration.
- Multigas LEL meter—portable instrument for detecting the presence and concentration of O_2, H_2S, CO, and combustible gases in manholes, in confined spaces, and around tank sample-pulls.

Process technicians must know the operating parameters of their unit and make adjustments accordingly after receiving the sample analysis to maintain product specification and operating efficiency.

Summary

Sampling is a crucial part of process operations and something that every process technician should master and perform exactly as per procedure. Proper sampling and testing techniques diagnose many operating problems, verify analyzers, and allow the process technicians to react more quickly when a problem arises.

Sampling saves the facility money, increases customer satisfaction, and reduces off-specification production. Sampling helps to minimize downtime and excessive labor by preventing problems before they arise. In addition, sampling can help identify and prevent hazards that could be harmful to process technicians and product end users.

The purpose of sampling is to capture a small representative portion of the process so that analysis may determine if it meets the desired specifications. If other materials are allowed to contaminate the sample, the test results could be inaccurate. The process technician must ensure that the sample container is clean and free from contaminants. When obtaining samples, it is important that the amount of material drawn be representative of the actual material in the process system (purge sufficiently if not a continuous-flow sample loop).

Regardless of who collects the samples, it is essential that the test be repeatable. The samples need to be caught in exactly the same way each time they are collected so that results can be compared scientifically. Using correct sampling procedures will preserve the integrity of the sample and provide safety measures to protect the process technician while collecting the sample. Any deviation from procedure can result in not only bad data but also serious injury to the technician and/or equipment damage.

While sampling or handling materials, process technicians must take all applicable precautions to prevent exposure to hazards. Wearing proper PPE, using correct sampling containers, and correct labeling of samples is vital for the safety of the process technician and the laboratory staff as well. Excess or unused portions of samples should be disposed of properly, according to company procedures.

Checking Your Knowledge

1. Define the following terms:
   a. Analyzer
   b. Chromatography
   c. Color
   d. Flame resistant
   e. Flame retardant
   f. Gas chromatography (GC)
   g. Karl Fischer (KF) water method

h. Repeatable
i. Sample loop
j. Sample point
k. Sampling
l. Specifications

2. Process technicians routinely sample and test process fluids and solids at various stages of the production process __________. (Select all that apply.)
   a. to ensure the reliability of continuous stream analyzers
   b. to maintain correct process operating parameters
   c. to be supercooled and stored for subsequent comparative analyses
   d. to ensure product stream specifications are met

3. Which of the following is an important part of catching a sample?
   a. Knowing the locations of the sample points
   b. Understanding the hazards of the material to be sampled
   c. Wearing the proper PPE
   d. All of the above
   e. Only a and c

4. When a sample contains a flammable or explosive compound, what special preventive measure may be necessary?
   a. Using a special chilled canister to collect the sample
   b. Electrically grounding the sample container to prevent a spark
   c. Wearing static-free clothing
   d. Using a nonmetallic sample container

5. A resealable membrane top on a bottle or vial used to collect samples is called a __________.

6. Metal containers for high pressure samples are sometimes called __________.
   a. vials
   b. cans
   c. bombs
   d. flasks

7. Sample __________ provide continuous circulation of the process stream to be collected in the sample.

8. If there is a chance of overspray when catching samples, which of the following PPE should be worn?
   a. A hard hat
   b. Protective eyewear, such as goggles or a face shield
   c. A pair of regular sun shades (not wraparound)
   d. Flame resistant fabrics

9. Which of the following is NOT a PPE item that the process technician may need to wear for catching samples?
   a. Rubber boots
   b. Respiratory protection
   c. Fire-resistant clothing (FRC)
   d. Steel-toed shoes

10. Which of the following are items that will be listed on a sample label? (Select all that apply.)
   a. Process unit name
   b. Date and time
   c. Company's current fiscal year
   d. Vessel of origin, or sample point ID
   e. Type of PPE worn by technician
   f. Quantity of the sample
   g. Name of supervisor
   h. Current analyzer result
   i. Variables to be tested

11. Match the tests or equipment used by process technicians with their primary purpose.

Tests and Equipment	Description
I. Basic sediment and water (BS&W)	a. A laboratory technique to separate mixtures to determine the exact makeup of the gas sample
II. Color	b. A simple means of identifying the presence of sulfur or sulfur-based compounds
III. Gas chromatography (GC)	c. A visual comparison scale to determine product color purity
IV. Gas detector tubes	d. Analytical method for quantifying water content in a variety of products
V. Karl Fischer (KF) titration method	e. Glass vials filled with chemical reagents that react to specific chemicals in order to determine if the targeted chemical is present
VI. Lead acetate test	f. Measurement of acidity or basicity of a solution
VII. pH	g. Test to see if the crude oil meets the technical specification limits set by the buyer

NOTE: Answers to Checking Your Knowledge questions are in the Appendix.

Student Activities

1. Conduct online research on sample containers known as "sample bombs" used in the refining and process industries. Write a two-page report on the types and sizes available.

2. Perform online research on the following chemicals and write a two-page report on their hazards and the type of PPE that must be worn when catching such samples.
   a. Naphtha
   b. Crude oil
   c. Paraxylene
   d. Benzene
   e. Gasoline

Chapter 14

Maintenance

Objectives

After completing this chapter, you will be able to:

14.1 Describe the characteristics of predictive, preventive, and reactive maintenance and the advantages and disadvantages of each. (NAPTA Operations, Equipment Maintenance; Economic Impact 1*) p. 188

14.2 Explain the process technician's role in the performance of various preventive maintenance activities. (NAPTA Operations, Equipment Maintenance; Economic Impact 3; Maintenance; Documentation and Permits 1–5) p. 192

14.3 Discuss tasks involved in lubrication and the process technician's role in lubrication. (NAPTA Operations, Equipment Maintenance; Economic Impact 3; Maintenance; Documentation and Permits 1–5) p. 197

14.4 Describe what a turnaround is and the tasks to be completed in order to prepare adequately for a turnaround. (NAPTA Process Operations, Turnarounds 3, 9) p. 200

14.5 Explain the role of the process technician in turnarounds. (NAPTA Operations, Turnarounds 5*) p. 202

Key Terms

Friction—force resisting the relative lateral or tangential motion of solid surfaces, fluid layers, or material elements in contact, **p. 197**

Deinventory—reduction or emptying of contents or residuals in a tank to various levels, based on activities and possible exposure of personnel to contents, **p. 206**

Inspection—examination of a part or piece of equipment to determine if it conforms to specifications, **p. 202**

*North American Process Technology Alliance (NAPTA) developed curriculum to ensure that Process Technology courses will produce knowledgeable graduates to become entry-level employees in process technology. Objectives from that curriculum are named here in abbreviated form. For example, "(NAPTA Operations, Equipment Maintenance; Economic Impact 1) means that this chapter's objective 1 relates to objective 1 of NAPTA's curriculum about the economic impact of equipment maintenance.

Isolation—separation; requirements for preventing stored energy (electrical, compressed air, kinetic, pressurized liquid, and so on) from entering or leaving a piece of equipment; may also be referred to as *lockout/tagout (LOTO)* or *lock/tag/try*, **p. 192**

Lubricants—any substances interposed between two surfaces in motion for the purpose of reducing the friction and/or the wear between them, **p. 197**

Lubrication—the process or technique employed to reduce friction and remove heat for reducing equipment wear and increasing longevity and safety, **p. 197**

Mechanical integrity—the state of being whole, sound, and undamaged; capable of functioning at design specification, **p. 188**

Predictive maintenance (PM)—maintenance strategy that helps determine the condition of in-service equipment to predict when maintenance should be performed, **p. 189**

Preventive maintenance—equipment maintenance strategy based on replacing, overhauling, or remanufacturing an item at a fixed interval, regardless of its condition at the time, **p. 191**

Reactive maintenance—equipment maintenance strategy in which equipment and facilities are repaired only in response to a breakdown or a fault, **p. 190**

Routine maintenance—work routinely performed to maintain equipment in its original manufactured condition and maintain operability, **p. 189**

Statistical process control (SPC)—statistical procedures that keep track of a process in order to reduce variation and improve quality, **p. 189**

Thermal expansion—tendency of matter to increase in volume in response to an increase in temperature, **p. 206**

Turnaround maintenance—work done during the shutdown period for an operation unit, usually for mechanical reconditioning, **p. 189**

14.1 Introduction

Maintenance is a key activity in the process industry, and it is done on a day-to-day, month-to-month, and year-to-year basis. It is important to maintain the equipment in a safe and reliable condition to prevent unplanned incidents or accidents. The maintenance frequency of a unit or piece of equipment is determined by the facility strategies for maintaining the **mechanical integrity** of equipment. Some periodic maintenance is also necessary to fulfill licensing requirements for equipment such as fired boilers. Maintaining mechanical integrity ensures that equipment functions at design specifications and prevents the types of failure that leads to a chemical release or other hazards (Figure 14.1).

Mechanical integrity the state of being whole, sound, and undamaged; capable of functioning at design specification.

Figure 14.1 Process technician during inspection tour. In this plant, maintenance reports created on the tablet are recorded and processed centrally at the same time.

CREDIT: Copyright © BASF SE 2021.

Several maintenance strategies are used in the process industry today. Each strategy has advantages and disadvantages. The role of the process technician is similar for each maintenance strategy.

Turnaround (TAR) is a planned, scheduled shutdown of a process unit or facility for maintenance and repair. Turnarounds involve extensive preparation, and many precautions are taken to reduce hazardous operations during the shutdown and subsequent startup. **Turnaround maintenance** is required maintenance performed on pieces of equipment that cannot be performed unless the unit has been shut down and deinventoried. Each facility has developed its own maintenance strategy with a common goal of having a high-performing organization and a safely operating plant.

Turnaround maintenance work done during the shutdown period for an operation unit, usually for mechanical reconditioning.

The process technicians may be required to perform minor maintenance at their facility, such as lubricating equipment, installing blinds, changing out filters, changing out pressure gauges, checking certain types of temperature indicators, and cleaning pump strainers. When performing these tasks, they should always use a standard operating procedure (SOP) or a unit specific procedure for equipment and system startup or shutdown in normal operation. This also applies to tasks performed in emergency operations.

Routine Maintenance

Routine maintenance is regularly done to make sure that equipment is kept close to its original manufactured condition and remains operable. There are several different strategies for providing routine equipment maintenance in a process facility. These strategies include predictive, reactive, and preventive maintenance. Turnarounds, discussed later in the chapter, are considered a long-term maintenance strategy.

Routine maintenance work routinely performed to maintain equipment in its original manufactured condition and maintain operability.

Predictive Maintenance

The **predictive maintenance** strategy seeks to use data to determine the condition of in-service equipment and therefore to be able to plan when maintenance should be performed. This approach offers cost savings over time-based strategies because tasks are performed only when warranted. Predictive maintenance requires that the process technician or maintenance technician attempt to utilize either periodic or continuous health monitoring of equipment as an evaluation tool.

Predictive maintenance (PM) maintenance strategy that helps determine the condition of in-service equipment to predict when maintenance should be performed.

Predictive maintenance determines when best to schedule equipment maintenance and when the maintenance activity is most cost effective (typically prior to the equipment losing its optimal performance). This approach uses principles of **statistical process control (SPC)**. The four basic steps of statistical process control include:

Statistical process control (SPC) statistical procedures that keep track of a process in order to reduce variation and improve quality.

- Measuring the process
- Eliminating variances in the process to make it consistent
- Monitoring the process
- Improving the process to its best target value.

A predictive maintenance strategy provides the following advantages:

- Increases component operational life
- Allows for preemptive corrective actions
- Reduces equipment and/or process downtime
- Lowers costs for parts and labor
- Provides better product quality
- Improves worker and environmental safety
- Raises worker morale

- Increases energy savings
- Results in an estimated 8 to 12 percent cost savings over savings that result from a reactive maintenance program.

There are a few disadvantages when using a predictive maintenance strategy. Those disadvantages include:

- Increased investment in diagnostic equipment
- Increased investment in staff training
- A savings potential that is not readily apparent to management.

The role of the process technician is to monitor equipment daily and report to a designated person in the organization. The process technician also prepares the equipment for maintenance, as necessary.

The American Petroleum Institute (API) has developed a standard (579) entitled Fitness-for-Service/Remaining Life that uses analysis and calculation to predict the remaining life of a piece of equipment. For example, it is possible that only some of the process tubes in a heater will need replacement based on thickness measurements. Prior to these advanced predictive techniques, all of the tubes would probably have been replaced at the same time, even if only some showed thinning tube walls.

Reactive Maintenance

Reactive maintenance equipment maintenance strategy in which equipment and facilities are repaired only in response to a breakdown or a fault.

Reactive maintenance is called a maintenance strategy. It is actually more like an absence of strategy because repairs to equipment and facilities are only done when there is a breakdown or a fault. Reactive maintenance allows continuous operation of equipment until it becomes inoperable. No actions are taken or efforts made to maintain equipment at design specification, to prevent failure, or to ensure that the designed life of the equipment is reached. Reactive maintenance is much more costly than preventive maintenance because it results in unplanned equipment downtime when equipment is out of service for repair or replacement (Figure 14.2).

Figure 14.2 A leaking pump.

CREDIT: PCPartStudio/Shutterstock.

The advantages of a reactive maintenance strategy follow:

- Lower initial costs
- Fewer personnel required
- Work performed only on out-of-service equipment.

The disadvantages of a reactive maintenance strategy are:

- Increased costs due to unplanned downtime of equipment
- Increased costs associated with repair or replacement of equipment (items purchased under time pressure, overtime labor)
- Possible secondary equipment or process damage from equipment failures
- Increased risk of accident, injury, or fatality and environmental contamination due to process upsets caused by equipment failure.

In a reactive maintenance strategy, the role of the process technician is to prepare the equipment for repair by the maintenance technician. Reactive maintenance leads to longer equipment downtime and greater loss of production. It carries a higher risk of incidents that result in uncontrolled product releases, explosions, or fires. Process technicians should always report suspicious equipment conditions to supervisors. If management does not respond by organizing repairs in a timely way, it can negatively affect morale, and personnel safety might be compromised.

Preventive Maintenance

A **preventive maintenance** strategy attempts to prevent any unplanned equipment shutdowns by performing scheduled maintenance, no matter what the current condition of the equipment might be. This type of maintenance strategy requires planned maintenance activities designed to prevent breakdowns and failures while the equipment is in service. The primary goal of preventive maintenance is to replace equipment before it actually fails, in order to preserve and enhance process reliability.

Preventive maintenance equipment maintenance strategy based on replacing, overhauling, or remanufacturing an item at a fixed interval, regardless of its condition at the time.

The strategy involves routine checks for wear, partial or complete overhauls at regular intervals, oil changes, lubrication, bolt tightening, checking pump couplings, and other checks deemed necessary by the process facility or equipment supplier. These routine checks are scheduled on a monthly, bimonthly, quarterly, semiannual, or annual basis. Discovery of deteriorating parts during these checks allows them to be replaced prior to equipment failure. Most of the elements of preventive maintenance can be incorporated without major interruptions to the process or production.

There are numerous advantages of a preventive maintenance strategy:

- Improved cost-effectiveness for many capital-intensive processes and equipment
- Flexibility for maintenance frequency
- Increased equipment life
- Energy savings
- Reduced equipment and/or process failure
- An estimated 12 percent to 18 percent cost savings more than savings found in a reactive maintenance program.

A few disadvantages also exist. The disadvantages of a preventive maintenance strategy include the following:

- Catastrophic failures are not eliminated.
- It is more labor intensive.
- It requires performing noncritical maintenance that increases opportunities for damage to other components.

The role of the process technician in a preventive maintenance program is primarily to prepare the equipment for maintenance and to ensure the equipment is in a safe energy state (lockout/tagout) for the maintenance technician to perform the work. It involves daily and weekly checks for wear and oil condition, seal monitoring, or lubrication addition.

14.2 Process Technician's Role in Maintenance Preparation

Isolation separation; requirements for preventing stored energy (electrical, compressed air, kinetic, pressurized liquid, and so on) from entering or leaving a piece of equipment; may also be referred to as *lockout/tagout (LOTO)* or *lock/tag/try.*

The role of process technicians is similar in each of the maintenance strategies. Their primary goal is to prepare the equipment and ensure that the maintenance technician is working on a piece of equipment that is safe and energy free. **Isolation** of the equipment is paramount in preparing the equipment for a safe energy state during the maintenance work. It is important to ensure the isolation of any potential machinery or equipment from energy sources during repair, service, or maintenance work in accordance with OSHA regulations. The process technician should fully utilize the standard operating procedures (SOPs) for lockout/tagout (LOTO) when preparing equipment for maintenance. Figure 14.3 shows a schematic for isolation of a pump.

Figure 14.3 Schematic for isolation of a pump.

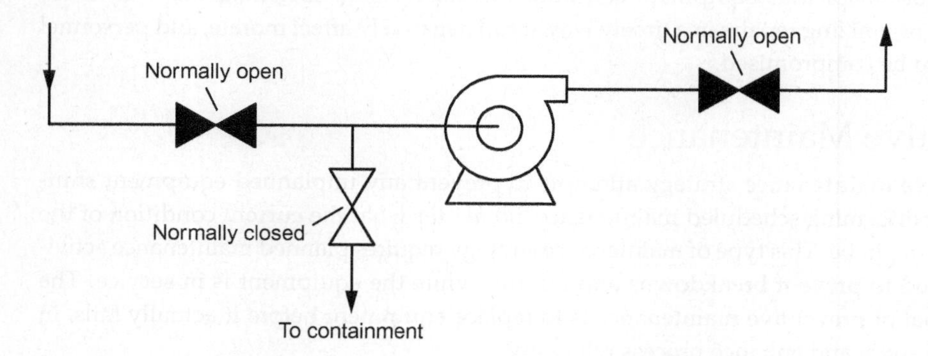

The process technician must execute the following in order to prepare equipment for maintenance and to ensure it is and will remain in a safe energy state throughout the maintenance work:

- Isolate the equipment from energy sources.
- Drain the equipment to a safe location.
- Purge the equipment to a safe location.
- Tag the equipment properly.
- Lock out the equipment properly.
- Verify the equipment is ready and safe to be worked on.

Documentation

In order to document and properly communicate control of workplace hazards during non-routine work, there are many lists, forms, and permits that must be used for guidance. The process technician uses the following:

- Safe work permits to document the description of the work to be performed, the hazards involved, precautions to take, authorization to perform the work, and acknowledgement of the individual performing the task. The Safe Work Permit is the overarching document that should incorporate all aspects of workplace hazard controls.
- Hot work permits associated with heat or spark generating work such as grinding or welding; these might involve electrical and instrument technician or power distribution verification.
- Confined space permits used when entering any space, vessel, pit, tower, and so on, with limited entry or exit.
- Specialty work permits associated with excavation, radioactivity, or any hazards designated for special consideration. Excavation permits, for example, may require involvement from civil engineering.

In the procedure shown in Figure 14.4, the process technician must prepare pump P-101A for routine annual preventive maintenance. Documentation of the work is needed. The technician must keep an accurate time for each event and initial each step as it is completed.

P-101A Pump PM	Time	Tech
1. Start up P-101B.		
2. Allow flow to stabilize; then shutdown P-101A.		
3. **Close** and **Lock** P-101A 4" discharge valve; tag do not operate.		
4. **Close** and **Lock** P-101A 4" suction; tag do not operate.		
5. **Close** and **Lock** ¾" seal oil supply valve; tag do not operate.		
6. **Close** and **Lock** ¾" seal oil return valve; tag do not operate.		
7. **De-energize** and **Lock** P-101A 480V breaker.		
8. Depress the start button on the start/stop button on P-101A and check to see if pump starts; tag start button do not operate.		
9. **Open** the ¾" case drain valve on P-101A to the closed drain system.		
10. **Open** the ¾" checkvalve bypass valve on P-101 discharge; tag do not operate.		
11. Install a nitrogen fitting and hose on the high point bleeder on P-101A discharge line.		
12. **Open** the ¾" high point bleeder valve and start a nitrogen purge to the closed drain system.		
13. Purge P-101A to the closed drain system for 20 min.		
14. **Close** the ¾" high point bleeder valve; close and depressurize the nitrogen hose.		
15. **Close** the ¾" case drain valve on P-101A to the closed drain system.		
16. **Open** the ¾" suction bleeder valve and check for liquid, if pump and lines are dry, allow maintenance to perform PM, otherwise repeat steps 7-13 until pump and lines are dry. Tag bleeder valve do not operate.		
Process technician verifying pump is energy free.		
Maintenance technician verifying pump is energy free.		

Figure 14.4 Example of a preventive maintenance checklist.

Once the process technician and maintenance technician have checked to ensure that P-101A is energy free, properly locked, tagged out, and the permit signed, the process technician allows the maintenance technician to begin the pump preventive maintenance (PM). The short procedure in Figure 14.4 described the items that the process technician must execute, including isolating, draining, purging, tagging, locking, and verifying a safe energy state for the pump throughout the work period.

The maintenance technician is given a list of activities that must be completed for the annual pump preventive maintenance. These are generally computer generated and given to the maintenance technician at the beginning of the day. An example of a typical pump PM is shown in Figure 14.5. The process technician verifies that the work has been completed and signs the PM order.

The process technician is required to document the status of equipment undergoing maintenance on the equipment status sheet and must share the information at shift change. An example of an Equipment Status Sheet is shown in Figure 14.6. The process technician should include the following information on the Equipment Status Sheet:

- Equipment number
- Procedure number
- Equipment LOTO status
- Verification of LOTO
- Drain status

- Purge status
- Time purged
- Tags hung
- Current equipment status.

Figure 14.5 Example of a typical pump PM.

P-101A Annual Pump PM Note any unusual items in comments section	Machinist Technician
Verify P-101A is in a safe energy state.	
Remove P-101A coupling guard; remove and inspect Rexnord coupling; replace if needed.	
Rotate P-101A by hand, checking for proper ease of rotation; if dragging is noticed, report to planner to schedule an internal overhaul of pump.	
Drain and replace oil (Note: this pump uses 68WT. oil).	
Clean and replace oil bubbler.	
Replace Rexnord coupling.	
Replace coupling guard.	
Notify operations technician that the pump has been released to operations.	
Comments: Verified complete by:____________	

Figure 14.6 Example of an Equipment Status Sheet.

Equipment Status Sheet
Daily Record for Maintenance Work

Equipment number:____________ Procedure number:____________	Quantity	Yes	No
1. Is the equipment locked out tagged out per procedure?			
2. Has the lock out tag out been verified?			
3. Has the equipment been drained?			
4. Has the equipment been purged?			
• With nitrogen			
• With steam			
• With water			
• With other - please list ____________			
5. Purge time: Start:____________ Stop:____________			
6. Have tags been hung?			
• Do Not Operate			
• Do Not Close			
• Do Not Open			
• Do Not Enter			
7. What is the current equipment status?			
• Equipment ready for work			
• Maintenance currently working on			
• Maintenance work completed			
• Equipment ready for service			
Process technician reporting equipment status:____________			
Process technician receiving equipment status:____________			

Estimating Costs

Equipment maintenance can be costly. The process technician can act to manage costs by efficiently running equipment in a safe and reliable manner. Table 14.1 shows an example of a cost estimate of the PM that was performed on pump P-101A earlier in the chapter.

The estimated rate for the facility labor is $70.00 per hour contracted service and as much as $100.00 per hour for full time maintenance technicians, including benefit makeup.

If the facility has a PM scheduled monthly for the pump in our example, what is the estimate for the monthly maintenance cost on a process unit with over 100 pumps? Extend

Table 14.1 P-101A PM Cost estimate (assumes 2 hours to complete the PM).

Task	Craft	Number of Technicians	Cost
PM P-101A	Machinist Technician	2	$280.00
Replace Rexnord coupling (if needed) --->			$105.58
Replace oil bubbler (if needed) -->			$86.38
Estimated total maintenance cost for this PM -->			$471.96

that to quarterly, semiannual, and annual PMs, and it is clear that the cost of a preventive maintenance strategy is high. However, preventive maintenance is the most cost-effective of the maintenance strategies.

Cost estimating may be part of the process technician's duties when generating maintenance orders. In estimating cost, the process technician must list the tasks required and identify a cost for each task. The employer provides the cost of labor and equipment to the process technician. The cost estimate in Table 14.2 is for a seal repair job for P-101A.

The cost estimate assumes the wage scale for company employees is $100.00 per hour, including benefits, and $70.00 per hour for contractor employees, including markups. Notice that the cost estimate is broken out by task, craft, quantity of staff required, estimated hours to complete the repair, and labor or equipment cost.

Table 14.2 Cost estimate for P-101A seal repair.

Task	Craft	Quantity	Time	Cost
Erect scaffold to install discharge blind	Contractor	3	1.0 hr	$210.00
Blind pump	Pipefitter	2	1.0 hr	$200.00
Remove pump from case and take to shop	Machinist	2	0.5 hr	$100.00
Order seal from warehouse and deliver to shop	Planner	Cost of seal ---------->		$5,700.00
Replace seal on P-101	Machinist	2	4.5 hrs	$900.00
Reinstall pump	Machinist	2	0.5	$100.00
Remove blinds	Pipefitter	2	1.0	$200.00
Remove scaffold	Contractor	3	1.0	$210.00
+10% contingency				$760.00
Total estimated cost to replace P-101 seal				$8,380.00

Avoiding Hazards

Hazards may be present when preparing equipment for maintenance. Process technicians must remember to wear proper personal protective equipment (PPE) when preparing equipment for maintenance. The P-101A pump PM procedure used earlier in this chapter is a good example of the potential hazards when preparing a piece of equipment for maintenance. Hazards might include:

- Chemical exposure from draining and purging
- Slips, trips, and falls when closing or opening, locking, or tagging valves
- Strains or sprains when bending or stooping to open or close bleeders, pulling or moving hoses for purging, and closing or opening valves
- Cuts, scrapes, or bruises when closing or opening valves, tagging out equipment, removing bull plugs, and installing purge hoses (Figure 14.7)
- Electrical shock when de-energizing or re-energizing breakers.

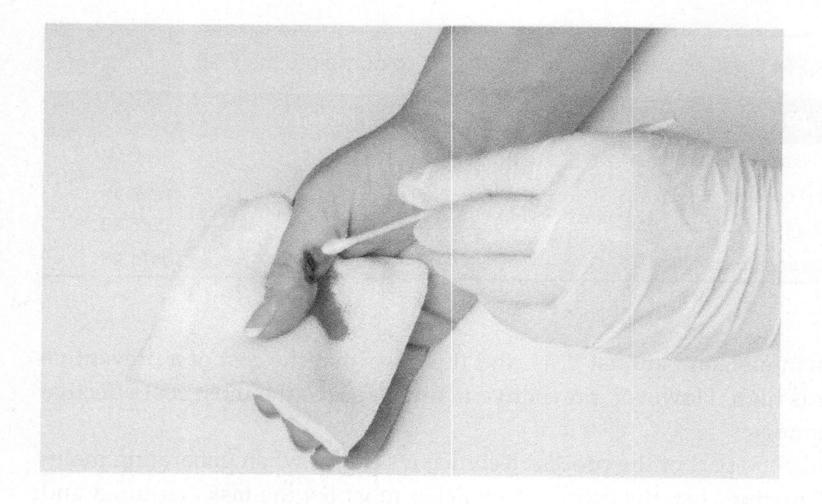

Figure 14.7 Lacerations are the leading hand injury risk in industrial environments and should be prevented.

CREDIT: bmf-foto.de/Shutterstock.

Process technicians must constantly be aware of their surroundings while preparing equipment for maintenance. Many simultaneous operations may continue in areas surrounding the equipment being repaired.

Process technicians are also responsible for placing the equipment back in service or in standby mode. They must ensure that all maintenance isolating devices, such as blinds or locks, have been removed prior to attempting to place the equipment back in service. A procedure normally includes the removal of blinds after maintenance work has been completed. See Figure 14.8.

Figure 14.8 Example checklist for returning equipment to service.

P-101A Pump PM	Time	Tech
17. Once maintenance work has been completed, close and plug the ¾" suction bleeder valve.		
18. Install a nitrogen fitting and hose on the high point bleeder on P-101A discharge line.		
19. Open the ¾" high point bleeder valve and pressure suction and discharge of pump case to fifty pounds (50 PSI) and perform tightness test. **Warning: potential for release of stored energy**		
20. If no leaks are found, Close the ¾" high point bleeder valve; de-pressure and remove the nitrogen hose and fitting.		
21. Open the ¾" case drain valve on P-101A to the closed drain system and vent pump to zero pressure.		
22. Close the ¾" case drain valve on P-101A to the closed drain system.		
23. Open and unlock the ¾" seal oil return valve.		
24. Open and unlock the ¾" seal oil supply valve.		
25. Open and unlock the 4" suction valve.		
26. Open and unlock the 4" discharge valve.		
27. Unlock and energize P-101A 480V breaker.		
28. Hold here until housekeeping around P-101 is completed; then move to Step 29.		
29. Notify control board technician of impending startup of P-101A and shutdown of P-101B.		
30. Start P-101A, allow flow to stabilize, then shutdown P-101B.		
31. Remove all tags, chains and locks and store properly.		
This procedure is completed. Process technician completing this procedure:________________		

14.3 Lubrication

Lubrication is an important part of maintaining equipment in proper operating condition. Lubrication reduces friction and removes heat, which helps to increase longevity and safety of equipment. Lubrication reduces wear of one or both surfaces that are very close to each other and are moving relative to one another. It places a substance between the surfaces to carry, or to help carry, the pressure-generated load between those surfaces. **Lubricants** are vital to maintaining efficient, reliable operation of all types of rotating equipment. Lubricants are necessary because **friction** makes it difficult to keep machine parts in motion. Friction also generates damaging heat and causes wear. The greater the friction, the greater the heat and wear. Lubricants act to reduce friction, heat, and wear, making it easier to keep machines running smoothly.

Process technicians play a key role in ensuring equipment in their area of responsibility is properly lubricated. They may also be responsible for checking oil levels in various types of equipment and adding lubricants as required.

Lubrication the process or technique employed to reduce friction and remove heat for reducing equipment wear and increasing longevity and safety.

Lubricants any substances interposed between two surfaces in motion for the purpose of reducing the friction and/or the wear between them.

Friction force resisting the relative lateral or tangential motion of solid surfaces, fluid layers, or material elements in contact.

Lubricant Storage, Handling, and Disposal

INDOOR AND OUTDOOR STORAGE Lubricants can be delivered in bulk, such as 55-gallon drums, or in smaller containers, depending on the rate of usage. The handling of lubricants between delivery and use is important. After a lubricant has been delivered, it is often stored in its container for an extended period before it is used. The containers can be stored indoors or outdoors, depending on the facility's available storage space. In either case, the lubricant must be protected from weather, contamination, spills, and fire.

Lubricants stored outside should be covered to protect them from the weather. Covering protects metal drums from moisture. An unprotected metal drum will eventually corrode, leak, and waste the lubricant. Any covering, permanent like a shed, or temporary like a canvas or plastic cover, will provide protection. Storing drums on their sides prevents water from collecting on the top of the drum as well. Whatever method is used to cover drums in outside storage, the *drums should be stored on their side* to prevent corrosion and leaks. As the drums heat and cool, their contents expand and contract. Storing them on their side can also prevent air and liquid flow into and from the drum. Container leaks contaminate lubricants, making them unusable. Lubricant spills can also damage the environment. Many states have strict regulations dealing with the storage and handling of lubricants, so extreme care must be taken to avoid drum damage and lubricant spills.

When lubricants are stored indoors, weather is not a concern. However, most lubricants are combustible, and fire is a danger. Lubricants stored outdoors can be kept far enough away from buildings and equipment to minimize damage if a fire should occur. But no matter where lubricants are stored, contamination, spills, and fire danger must be considered and prevented.

HANDLING LUBRICANTS When a lubricant is needed, drums are moved from bulk storage to a central location in the facility where the lubricant can be dispensed. If a lubricant is taken from one extreme condition to another, it should be given plenty of time to become acclimated to the new temperature before use. For example, low temperatures can thicken lubricants and make them unusable until they have been thoroughly warmed.

After a lubricant container has been opened, it should be kept clean. After lubricant is removed from a drum, the drum should be resealed to avoid contamination by dirt and moisture.

When more than one lubricant is dispensed from the same facility location, it is very important to identify each type. The original shipping containers will clearly state what they hold, but any time a lubricant is placed in another container, the new container must be labeled correctly. Incorrect labeling can cause the application of the wrong lubricant to a piece of equipment, leading to extensive damage. It might also lead to accidental mixing of different types of lubricants, which could damage the equipment.

Lubricants should never be returned to the original drum after they have been removed. Returning used lubricants to the original drum may contaminate the entire contents.

In most facilities, opened lubricant containers are stored in a separate room or area to reduce the chance of fire and to make it easier to keep the lubricant clean. Lubricant storage rooms or areas must be clean, well lit, and have some type of fire control equipment.

OIL DISPOSAL Contaminated oil must be disposed of properly. There are three ways to properly dispose of lubricant: returning the contaminated oil to the oil vendor, disposing of the oil in an environmentally safe manner, or purifying and reusing the oil (if all the contaminants can be removed).

REMOVING LUBRICANTS FROM CONTAINERS Lubricant to be used in a facility must be removed from its container for use. There are several ways of handling lubricants. One method uses a barrel pump that connects to the drum. Different types of pumps are used for oil and grease because of their different consistencies. A barrel pump, or drum pump, is an easy way to remove the lubricant without contaminating the rest of the drum. A barrel pump may be hand driven (see Figure 14.9), electric, or driven by nitrogen or air.

In order to install a barrel pump, several steps must be taken. First is to clean off the top of the drum. Second is to remove the larger drum seal to open the drum. The pump is cleaned and inserted into the drum. And finally, the pump is tightened securely to the drum. Most drum pumps screw into the larger threaded opening in the drum.

Figure 14.9 Hand-driven drum pump.

CREDIT: dream02/Shutterstock.

The drum must be vented for the barrel pump to work properly. A drum is usually vented by removing a small cap from the top before the pumping starts. The cap is replaced after the pumping is finished so that the lubricant stays clean.

Another way to remove oil is to attach a spigot to the top of the drum. The drum is laid on its side and vented through the smaller opening, and then oil is drawn from the spigot. The vent is replaced (or closed) when the work is finished.

Drums with spigots are often placed in rocking frames that are stored in an upright position. When oil is required, the drum is rocked over on its side (Figure 14.10), the lubricant is withdrawn, and the drum is returned to its upright position.

Sometimes a funnel is used to transfer oil from the drum, or other container, to a small piece of equipment. Any funnel used must be cleaned after use. Whatever method is used, the following precautions must be observed:

- Add the correct amount of lubricant.
- Use the proper lubricant for that particular job.

Figure 14.10 Steel drum cradle holding orange chemical drum.

CREDIT: Corepics VOF/Shutterstock.

- Make sure all the filler caps are tightly replaced.
- Be careful that the lubricant is not contaminated.

Process Technician's Role in Lubrication

Each process unit has a lubrication schedule that specifies the proper lubricant for each piece of equipment, the intervals for changing and sampling lubricant, and what to look for in lubricant samples. The facility maintenance technicians perform many of these tasks. Each process unit has a lubrication manual with the name of the proper lubricant for each piece of equipment. Using the wrong lubricant can damage the equipment. If unsure about what type of lubricant to add to a piece of equipment, refer to the lubrication manual for clarification. Process technicians are responsible for maintaining this schedule and for making routine, on-shift checks on all lubricated equipment. Bearing failure due to lack of proper lubrication can cause costly equipment repairs as well as process shudown.

The process technician may be responsible for the following checks during their routine rounds:

- Lube oil level
- Oil temperature
- Leak checks
- Oil sampling
- Oil cleanliness (or need for oil change).

LUBE OIL LEVEL The oil level in lubricating systems is critical. The level should be checked before the equipment is started and routinely during operation (Figure 14.11). Reservoir levels in circulating systems, force-feed oilers, and oil-mist systems should also be checked regularly. Sight glasses, oil bottles, and clear reservoirs must be kept clean so levels are clearly visible. Covers should always be closed to prevent contaminants, like dirt and water, from entering the oil. Sight glasses, tattletales, or the bottom of oil-bearing reservoirs must be drained of water because water can be a major cause of bearing failure.

OIL TEMPERATURE One of the first signs of bearing failure is an increase in temperature. Bearing housings can be checked using a documented procedure, visual bearing indicators if installed, or vibration monitoring equipment. If running hotter than normal, report it to a supervisor. Some bearings in very hot service cannot be checked in this manner. In this situation, sample the oil and check its temperature with a thermometer or a temperature gun. Normal temperature should be maintained. It is very important to monitor closely any bearing temperature increase.

LEAK CHECKS Excessive leakage may indicate that the oil is foaming due to contamination or that a plug or oil bottle has loosened. Slinger rings may also cause leakage. Slinger rings are used in bearings using oil sump lubrication. The rings encircle, but are not attached

Figure 14.11 Centrifugal pump with oil bulb for lubrication.

CREDIT: Oil and Gas Photographer/ Shutterstock.

to, the shaft. As the shaft turns, the oil rings rotate with the shaft, extending into the oil sump and "dragging" oil back to the top of the shaft. If the shaft is not perfectly horizontal, the oil ring may ride toward one end of the bearing housing, depositing oil onto the inner surface where the shaft exits the housing. This may result in a leak.

OIL SAMPLING Oil in bearing housings should be sampled as specified in the sample schedule. To sample, drain a small amount of oil into a clear bottle. As with any sampling, be sure to wear the proper PPE for the job. Check the sample for:

- Metal particles, which indicate that the bearing has started to fail
- Dirt, water, or other contaminants; contaminated oil will not provide adequate lubrication and should be changed.

OIL CHANGES At some locations, oil is changed in rotating equipment routinely, or when the oil appears contaminated with water. Dispose of the contaminated oil per guidelines issued by the process facility.

14.4 Turnarounds and Turnaround Maintenance

Turnarounds (TARs) are planned, scheduled shutdowns of a process unit for maintenance and repair. Turnaround maintenance activity is performed on pieces of equipment that cannot be maintained or repaired unless the unit has been shut down and deinventoried. A turnaround is usually scheduled 18 to 24 months in advance. Turnarounds require a great deal of organization, planning, personnel, and data input.

Each facility develops its own maintenance strategy, but all facilities share a goal of having a high-performing maintenance organization and a safely operated plant. Because TAR may involve hazardous operations during shutdown and subsequent startup, many precautions are taken to minimize those hazards.

Turnaround as a blanket term encompasses such specific tasks as debottlenecking (identifying and removing process limitations), revamps (engineered improvements), catalyst regeneration, shutdowns, and outages. Turnarounds are expensive due to labor and equipment costs as well as other materials required to execute them. The cost of turnarounds may approach 30 percent of the facility's annual maintenance budget. As mentioned, there is also the substantial cost of lost production while the unit is offline. If not managed properly, turnarounds can affect a company's bottom line.

Did You Know?

Not every unit is impacted during every turnaround. For example, the industry average is about 4 years between turnarounds for a catalytic cracking unit.

CREDIT: Courtesy of John Zink Company.

There are three phases in a turnaround that must be executed properly to ensure the turnaround is a success. Turnarounds consist of preexecution, execution, and postexecution phases. A turnaround team is appointed specifically for the execution of each turnaround phase. The turnaround team typically consists of the following facility personnel:

- Turnaround manager
- Operations representative (for example, a process technician or leader assigned to the unit)
- Maintenance representative (for example, a craftsperson or maintenance leader assigned to the unit)
- Inspector (multiple inspectors may be required based on the size of the turnaround and the specialty required of the inspectors)
- Turnaround planner (multiple planners may be needed, based on the size of the turnaround or the need for craft-specific planners)
- Engineering representatives (for example, a process engineer, mechanical engineer, and support engineers assigned to the unit)
- Member of the facility leadership team (for example, an operations manager or a delegate)
- Member of the safety department (for example, an industrial hygienist or a safety engineer)
- Member of the facility fire department (for example, fire chief or fire chief's designate).

The turnaround team meets weekly in the early stages of turnaround planning and daily during the execution phase. The team discusses the following topics during the three phases of a turnaround:

- TAR preexecution activities:
  - Defining the work list
  - Planning work activities
  - Purchasing parts and equipment
  - Setting up contracts for contractor services, if required
  - Preparing the worksite and equipment for the turnaround
- TAR execution phase activities:
  - Working the work list
  - Reporting progress on planned work versus actual work
  - Managing the schedule and cost
  - Creating/planning follow-up work

- TAR postexecution phase activities:
  - Demobilizing the work site
  - Materials reconciliation
  - Planned versus actual reconciliation
  - Invoice payments.

Inspection examination of a part or piece of equipment to determine if it conforms to specifications.

One of the key activities during a turnaround is the **inspection** of equipment, piping, and vessels. As towers and other vessels are opened, the inspection department will begin to perform internal and external inspections. Inspections may include ultrasonic thickness tests on pipes or vessels as well as visual inspections by experts from the inspection department.

Relief valves may be recertified or overhauled, depending on the process facility strategy. Many companies send their relief valves to outside vendors for overhaul and certification to prevent unwarranted shutdowns when a relief valve lifts prematurely. Figure 14.12 shows an inspection of equipment.

Figure 14.12 Engineer in a petrochemical maintenance facility inspects pipe components prior to cleaning and refurbishment.

CREDIT: Simon Turner/Alamy Stock Photo.

The goal of a turnaround is to achieve as much work as possible within a narrow window of time to minimize lost production. The turnaround success is gauged not only on whether it came in under budget and on time but also on how safely it was accomplished. A successful turnaround has zero injuries and zero environmental incidents. Machinery starts up properly, and the maintenance performed improves plant performance.

Many facilities conduct postturnaround audits, focused on key participants in the turnaround. They use questionnaires discussing key topics such as planning, safety, engineering, communication, inspection, construction, and operations support. This allows the facility to gauge the effectiveness of the turnaround team. It also provides feedback that may be useful in conducting subsequent turnarounds.

14.5 Process Technician's Role in Turnarounds

The role of the process technician is different for a turnaround versus the role in routine maintenance. In a turnaround, the process technician will be in charge of systems rather than individual pieces of equipment.

In preparation for turnaround, the process technician is responsible for performing preturnaround duties, which include:

- Participating in the planning and scheduling meetings for the turnaround
- Participating in the review of the proposed turnaround work list

- Writing or reviewing shutdown procedures
- Writing or reviewing lockout/tagout (LOTO) procedures
- Acting as a single point of contact for maintenance technicians, participating in job safety analysis or hazard identification, and showing maintenance technicians the blinding locations and equipment locations
- Hanging blind tags associated with blinding locations for turnaround
- Preparing LOTO devices for use (shown in Figure 14.13)
- Ordering miscellaneous supplies such as fittings, hoses, and tags.

Figure 14.13 Example of a locked-out/tagged-out piece of equipment.

CREDIT: King Ropes Access/Shutterstock.

The unit moves toward a controlled shutdown as scheduled on the timeline developed during planning. A controlled shutdown avoids any upset conditions. During the shutown period, process levels may be lowered or raised, material may be pumped out to tanks, pressures may be increased or decreased, unit throughput rates are lowered, and materials are vented or drained as needed. Once the unit is fully shut down, the unit will be completely pumped out, drained, and purged as necessary.

The process technician is responsible for performing the following duties during a turnaround:

- Ensuring that work during the turnaround is executed in a safe manner
- Issuing safe work permits and other permits as needed
- Ensuring that persons performing work wear the proper PPE at all times
- Draining of equipment
- Purging of equipment
- Providing fire watch support for hot work
- Providing entry attendant for confined space entries
- Conducting gas testing for hot work and confined space entry
- Assisting maintenance as required
- Assisting contractor personnel as required
- Assisting inspection personnel as needed
- Witnessing special tests
- Ensuring work is proceeding to plan
- Housekeeping as needed, ensuring the work area and unit remains in a safe condition
- Providing good shift relief.

A turnaround is also an excellent training opportunity for process technicians with limited experience because it provides a greater understanding of process operations, process flow paths, and process equipment operation. A turnaround allows a process technician a chance to learn more about the process unit in preparation for the next turnaround.

Mechanical Integrity

It is important to maintain the mechanical integrity of critical process equipment and ensure it is designed, installed, and operated properly. OSHA PSM mechanical integrity requirements apply to the following equipment:

- Pressure vessels and storage tanks (Figure 14.14)
- Piping systems, including piping components such as valves
- Relief and vent systems and devices
- Emergency shutdown systems
- Controls, including monitoring devices and sensors, alarms, and interlocks
- Pumps and compressors.

Figure 14.14 Tank farms are a vital part of storage within industrial facilities.

CREDIT: Avigator Fortuner/Shutterstock.

The facility must establish and implement written procedures to maintain the ongoing integrity of process equipment. Maintenance technicians involved in maintaining the integrity of process equipment must be trained in an overview of that process and its hazards and in the procedures applicable to a technician's job tasks.

Inspection and testing must be performed on process equipment using procedures that follow recognized good engineering practices. The frequency of inspections and tests must conform to the manufacturer's recommendations, or as determined to be necessary. Each inspection and test on process equipment must be documented, identifying the date of the inspection or test, the name of the person who inspected or tested, the serial number or other equipment identifier, a description of the inspection or test performed, and the results of the inspection or test.

Equipment deficiencies outside the acceptable limits defined by the process safety information must be corrected before further use. In some cases, equipment can continue operating despite deficiencies as long as deficiencies are corrected in a safe and timely manner. At that time, other necessary steps are taken to ensure safe operation.

When constructing new facilities, the employer must ensure that equipment being used is suitable for the application. Appropriate checks and inspections must be performed to ensure that equipment is installed properly and use is consistent with design specifications and the manufacturer's instructions.

Management of Change (MOC)

Management of change (MOC), or change management, is a method of managing and communicating changes to a process, changes in equipment, changes in technology, changes in personnel, or other changes that will impact the safety and health of employees. Written procedures must be established and implemented to manage changes, except for in-kind replacements, to process chemicals, technology, equipment, procedures, and facilities that affect a covered process. These written procedures must ensure that the following considerations are addressed prior to any change:

- The technical basis for the proposed change
- Impact of the change on employee safety and health
- Modifications to operating procedures and all impacted P&IDs
- Time period necessary for the change
- Authorization requirements for the proposed change
- Training requirements due to the change.

Process technicians and maintenance and contract employees whose job tasks are affected by a change in the process must be informed of and trained in the change prior to startup of the process or affected part of the process. If an alteration covered by these procedures results in a change in the required process safety information or requires changes to the required standard operating procedures (SOPs) or practices, documentation must be updated also and all employees notified.

Prestartup Safety Review

Prestartup safety review (PSSR) is conducted to help ensure that certain important considerations have been addressed before hazardous materials are introduced into the process or the modified section(s) of an existing process. Conducting the prestartup safety review (PSSR) is important to ensure that a unit is ready for startup. The review is conducted within days to a week of a unit's expected restart date. It must be completed prior to introducing hazardous materials into the process.

Each process facility has developed a PSSR checklist of tasks that will be completed to ensure compliance with the OSHA regulation. Each process facility's PSSR checklist must confirm the following:

- Construction and equipment are in accordance with design specifications.
- Safety, operating, maintenance, and emergency procedures are adequate and in place.
- A process hazard analysis has been performed on new facilities, and recommendations have been resolved or implemented prior to startup, and modified facilities meet the management of change (MOC) requirements.
- Necessary employee training has been completed.

Shutdown and Startup Safety

The two most critical periods in the operation of a process unit are shutdowns and startups. No two shutdowns are alike, due to the circumstances surrounding each shudown. There are differences between shutting the unit down for a turnaround versus shutting the unit down for inventory control. An inventory shutdown is completed to maintain the unit in such a condition that it can be restarted when required without the additional cost associated with reinventorying.

Depending on the circumstances, circulation is stopped and temperatures and pressures are allowed to decrease in order to maintain the unit in a safe mode while maintaining the chemical composition of the unit. In most turnarounds, all of the material is pumped to tankage or waste, and the unit is purged with an inert gas and prepared for maintenance work. Process technicians should always follow their routine or normal shutdown procedures.

Like shutdowns, startups will differ. If the unit was idled and remained filled, care must be exercised when returning to normal conditions. Material that has been cooled must be heated up to begin a chemical reaction, increasing the likelihood of thermal expansion in flanges and valves. **Thermal expansion** is the tendency of matter to increase in volume in response to an increase in temperature. Figure 14.15 shows a possible reaction to expansion.

Thermal expansion tendency of matter to increase in volume in response to an increase in temperature.

Figure 14.15 Possible reaction of equipment to thermal expansion.

CREDIT: PixMix Images/Alamy Stock Photo.

If the unit has been **deinventoried** for a turnaround, then extreme care must be taken to reinventory the unit. Process technicians must ensure that the equipment has been purged of air prior to introducing any hydrocarbons. During the reinventory process, leaks can occur at flanges and valves, so the process technician should conduct routine unit monitoring during the inventory phase. The unit is then slowly heated or cooled to normal operating conditions. The process technician should follow his or her routine or normal startup operating procedures at all times.

Deinventory reduction or emptying of contents or residuals in a tank to various levels, based on activities and possible exposure of personnel to contents.

Summary

The role of the process technician is critical, providing the first line of defense in maintaining the unit in a safe and reliable state. The process technician must prepare the equipment for maintenance, following standard operating procedures (SOPs) and lockout/tagout (LOTO) procedures, ensuring the equipment is properly de-energized. Frequent communication is necessary with maintenance technicians to ensure that work is proceeding safely. Safety concerns brought up by the maintenance technicians must be addressed promptly to ensure safety.

During shutdowns and turnarounds, process technicians must follow standard operating, deinventory, and maintenance preparation procedures. Willing participation in preturnaround activities will make the turnaround a basis for increased understanding and experience. It will help technicians become familiar with process facility requirements for personal protective equipment (PPE). (Anyone working on equipment must use proper PPE.)

Process technicians must ensure that all isolation devices are removed prior to restarting the unit. For example, neglecting to remove a single TAR installed pipe blank can necessitate returning the unit to cold and beginning the startup all over again. It is important to ensure that all tagging devices are removed and equipment is lined up per procedure.

Technicians should be familiar with process facility requirements for process safety management (PSM). It is advisable to be an active participant in the PSM process and to know the process safety management requirements for a job.

Certain guidelines should be kept in mind. For example, during startups, the process facility's normal startup procedures are used. While reinventorying a unit, everyone must be alert for leaks that may occur. When heating up or cooling down a process, process technicians must remain alert and ready to respond to any situation that may develop.

Checking Your Knowledge

1. Define the following terms:
   a. Friction
   b. Deinventory
   c. Inspection
   d. Isolation
   e. Lubricants
   f. Lubrication
   g. Mechanical integrity
   h. Predictive maintenance (PM)
   i. Preventive maintenance
   j. Reactive maintenance
   k. Routine maintenance
   l. Statistical process control (SPC)
   m. Thermal expansion
   n. Turnaround maintenance
2. Which of the following are strategies that an organization may use as its normal maintenance pattern? (Select all that apply.)
   a. Turnaround (TAR)
   b. Predictive
   c. Reactive
   d. Assertive
3. A reactive maintenance strategy has which of the following disadvantages? (Select all that apply.)
   a. Increased costs associated with repair or replacement of equipment (items purchased under time pressure, overtime labor)
   b. Increased costs due to unplanned downtime of equipment
   c. Increased investment in diagnostic equipment
   d. Increased investment in staff training
   e. Increased risk of accident, injury, or fatality and environmental contamination due to process upsets caused by equipment failure.
4. A preventive maintenance strategy has which of the following advantages? (Select all that apply.)
   a. Increased equipment turnover
   b. Less costly than reactive maintenance
   c. Rigid scheduling for maintenance intervals
   d. Energy savings
   e. Reduced equipment failure
5. The __________ is the overarching document that should incorporate all aspects of work-place hazard controls.
   a. confined space work permit
   b. hot work permit
   c. safe work permit
   d. specialty work permit
6. Match each of the following potential hazards when preparing a piece of equipment for maintenance with the activity where they might occur.

Hazards	Hazardous Activity
I. Chemical exposure	a. Bending or stooping to open or close bleeders, pulling or moving hoses for purging, and closing or opening valves
II. Cuts, scrapes, or bruises	b. Closing or opening valves, tagging out equipment, removing bull plugs, and installing purge hoses
III. Electrical shock	c. Closing or opening, locking, or tagging valves
IV. Slips, trips, and falls	d. De-energizing or re-energizing breakers
V. Strains or sprains	e. Draining and purging

7. Process technicians play key roles in which of the follow aspects of lubrication? (Select all that apply.)
   a. Checking oil levels in equipment
   b. Making sure that any unused lubricant is returned to its original container
   c. Ensuring equipment in their area of responsibility is lubricated
   d. Adding lubricants as required
8. How far in advance is a turnaround usually scheduled?
   a. 3-6 weeks
   b. 2-6 months
   c. 18-24 months
   d. 3-4 years
9. During postturnaround activities, which of the following will occur? (Select all that apply.)
   a. Demobilizing of the TAR work site
   b. Materials reconciliation
   c. Hanging blind tags associated with blinding locations
   d. Invoice payments

10. Which of these statements is false?
    a. Turnaround success is determined both by meeting budget and by having no safety incidents involving personnel or environment.
    b. Turnarounds generally require about two weeks of planning and include personnel from several different departments.
    c. The goal of a turnaround is to safely achieve as much work as possible within a narrow window of time to minimize lost production.
    d. One of the key activities during a turnaround is the inspection of equipment, piping, and vessels.
11. During preturnaround activities, which of the following are a process technician's responsibility? (Select all that apply.)
    a. Acting as a point of contact for maintenance
    b. Delegating a team member to attend meetings about the turnaround
    c. Hazard identification
    d. Writing or reviewing shutdown procedures
    e. Ordering miscellaneous supplies such as fittings, hoses, and tags
12. Identify tasks that are part of the process technician's responsibility during turnaround. (Select all that apply.)
    a. Purging equipment
    b. Issuing safe work permits and other permits as needed
    c. Directing inspection personnel
    d. Authorizing changes in the TAR plan
    e. Providing fire watch support for hot work
    f. Assisting contractor personnel as required
13. Written procedures must ensure that the following considerations are addressed prior to any change: (Select all that apply.)
    a. Technical basis for the proposed change
    b. Roster of training consultants under consideration for the training effort
    c. Upbeat language to assure employees of minimal impact on safety and health
    d. Time period necessary for the change

NOTE: Answers to Checking Your Knowledge questions are in the Appendix.

Student Activities

1. Using the following information, develop a cost estimate for the following work:

   Your unit has just experienced a minor upset due to the loss of a bottoms pump on a distillation tower. You have inspected the pump and have determined that the pump motor is shorted. You have also noticed that the pump would not rotate by hand, and you suspect that the pump has internal damage requiring new bearings and seals. Scaffolding is required for the installation of blinds at both the suction and discharge valves. Use the following labor and equipment costs to develop your cost estimate:

Cost of contract labor:	\$70.00/hr
Cost of company labor:	\$93.65/hr
Cost of new motor:	\$7,500.00
Cost of bearings/seals/etc.:	\$8,500.00
Cost of renting go-devil:	\$43.00/hr

   Develop an estimated timeline for repairs and include in your cost estimate.
2. Perform research on reactive, preventive, and predictive maintenance programs, and write a three-page essay describing your preferred choice of a maintenance program.
3. Perform research on OSHA 1910.119 Process Safety Management of Highly Hazardous Chemicals. Select two of the fourteen elements, and write a two-page essay describing the actions a process technician must take to be in compliance with this regulation.
4. Together with a classmate, develop a preventive maintenance schedule for a process unit with twelve pumps, one compressor, eight control valves (including eight bypass valves around the control valves), and six pump screens. Place your work in a spreadsheet and include the cost per quarter, then annually. Use the cost of labor in Activity 1.

Chapter 15

Abnormal and Emergency Operations

Objectives

After completing this chapter, you will be able to:

15.1 Discuss what types of events could be considered "abnormal operations," and describe their possible causes. (NAPTA Operations, Abnormal Operations: Emergencies 4, 5*) p. 210

15.2 Discuss what types of events could be considered "emergency situations," and describe their possible causes. (NAPTA Operations, Abnormal Operations: Emergencies 1, 5) p. 211

15.3 Discuss the process technician's role in correcting abnormal operations situations. (NAPTA Operations, Abnormal Operations: Emergencies 7, 8, 9) p. 213

15.4 Describe how process personnel prepare for hazardous or emergency situations (for example, drills, exercises). (NAPTA Operations, Abnormal Operations: Emergencies 2) p. 215

15.5 Discuss the process technician's role in emergency situations. (NAPTA Operations, Abnormal Operations: Emergencies 7, 8, 9) p. 217

Key Terms

Abnormal operation—operating a process unit in a mode that is different from normal operations, **p. 210**

Boiling liquid expanding vapor explosion (BLEVE)—explosion resulting from excessive compression of vapor in the container head space and vapor flashing from its release to the atmosphere above its normal boiling point, **p. 220**

*North American Process Technology Alliance (NAPTA) developed curriculum to ensure that Process Technology courses will produce knowledgeable graduates to become entry-level employees in process technology. Objectives from that curriculum are named here in abbreviated form. For example, "(NAPTA Operations, Abnormal Operations: Emergencies 4, 5)" means that this chapter's objective 1 relates to objectives 4 and 5 of NAPTA's curriculum about abnormal operations due to emergencies.

Emergency—sudden, unexpected, or impending situation that may cause injury, loss of life, damage to property, and/or interference with the normal activities of a person or operation, which therefore requires immediate attention and demands remedial action, **p. 210**

Emergency operation—mode of operation or procedure followed when an emergency situation has placed a process unit in an unsafe condition, **p. 210**

Emergency response—effort to mitigate the impact of an incident on the public and the environment, **p. 217**

Explosion—a chemical reaction or change of state which is affected in an exceedingly short space of time with the generation of a high temperature and generally a large quantity of gas, **p. 219**

Fire brigade—local process facility fire department composed of employees who are knowledgeable, trained, and skilled in basic firefighting techniques, **p. 217**

First responders—individuals who likely witness or discover a hazardous substance release and have been trained to initiate an emergency response sequence by notifying the appropriate authorities, **p. 215**

Hazardous Waste Operations and Emergency Response Standard (HAZWOPER)—OSHA's Regulation 29 CFR 1910.120; OSHA standard that applies to personnel who are in a role or position to act as a first responder during an emergency, **p. 215**

Incident response teams—groups of people who prepare for and respond to any emergency incident, such as a fire, spill, explosion, or environmental release that potentially impacts the outlying community, **p. 217**

Mutual aid—agreement among emergency responders to lend assistance across jurisdictional boundaries, **p. 219**

Spill—accidental release of a substance from lines, equipment, or areas, **p. 217**

Train—components of a system; a series of related equipment components, all in an orderly procession or in a line, necessary to accomplish a specific task, for example, distillation or compressor train, **p. 211**

15.1 Introduction

This chapter provides an overview of various types of abnormal and emergency operations that take place in a production facility. Although these two scenarios of operation sound very similar, there are distinct differences between the two. Emergency operations are definitely abnormal, but an abnormal operation should not always be categorized as an **emergency**. After completing this chapter, you will be able to differentiate between abnormal and emergency operations.

- **Abnormal operation** is usually a planned operation that has a specific purpose. It generally is temporary or short term.
- **Emergency operation** is an unplanned situation that can have a severe negative impact on unit personnel and equipment. Depending on the severity of the emergency, site personnel, the environment, and surrounding communities can also be affected.

Emergency sudden, unexpected, or impending situation that may cause injury, loss of life, damage to property, and/or interference with the normal activities of a person or operation, which therefore requires immediate attention and demands remedial action.

Abnormal operation operating a process unit in a mode that is different from normal operations.

Emergency operation mode of operation or procedure followed when an emergency situation has placed a process unit in an unsafe condition.

Abnormal Operations and Causes

Operating a process unit in an abnormal condition can be described in many ways, because process units differ and abnormal conditions are unlimited in scope and purpose. The reasons for abnormal operations are usually equipment related. However, there are times when raw material supply or product demand determines the need to operate a unit in an abnormal mode. Most instances of abnormal operation are planned and temporary. The following are examples of abnormal operation scenarios and causes that a process technician might encounter:

- Depending on process capabilities, and based on the process equipment involved, there may be opportunities to shut down specific pieces of equipment for maintenance or repair and continue to operate the unit. In other words, the production continues even while equipment is removed from service temporarily for repairs. Redundant pieces of equipment (spares) that are performing the same function in a process system can provide these opportunities.
- Some units are constructed using parallel "trains" of operation. A **train** can be described as a parallel system that has been designed and constructed using the exact or similar production equipment, each element of which contributes toward production. If there is a need to operate at reduced rates, or if there is equipment damage within a train, a process design such as this may permit continued production while removing one or more of these parallel trains from service.
- Some units are constructed with an "A side" and a "B side," each with specific design and equipment characteristics that can be operated in different modes or configurations. Normal operation and final product process might take place only with the combination of the two sides and their respective products. Any alternate mode of operation would be considered abnormal.
- Many times, there are process systems from adjacent units that are shared on a permanent or temporary basis. Equipment on Unit A might be in service to process material from Unit B into a viable product. If Unit B is in a shutdown or turnaround mode, then the equipment on Unit A would be temporarily removed from service.
- Rerouting the flare and vent system from one unit to an adjacent unit flare is one of the more common scenarios for sharing systems on a temporary basis.
- Installation and operation of specialty equipment in temporary service is also considered abnormal operation. Temporary equipment must usually be installed per the manufacturer's specifications and can require training and other considerations prior to placing into service. Portable drier and filter systems and portable storage facilities are examples of these.
- Tank farms that contain storage tank, feed tank, effluent tank, and pumping facilities can many times be reconfigured to accommodate numerous temporary conditions that can occur in a production facility Here are some examples of temporary conditions:
  - A feed tank that needs to be removed from service for an annual internal inspection might be bypassed while feed to the unit continues directly from the feed source.
  - A unit system might be shut down and removed from service so that a slop tank can be taken out of service for repair.
  - A spare tank that is not normally used may be placed in service so that another tank can be repaired.
  - Temporary piping might be installed and placed in service so that a tank farm pumping station can be bypassed and removed from service.

Train components of a system; a series of related equipment components, all in an orderly procession or in a line, necessary to accomplish a specific task, for example, distillation or compressor train.

These examples provide a small sample of abnormal operations that can take place on a process unit. Even when events are planned, the hazards associated with abnormal operations are unit specific and are a diversion from normal operations. This diversion carries with it an increased level of risk, which, if not managed properly, can cause injury to personnel, damage to equipment, and damage to the environment and surrounding communities.

15.2 Emergency Situations and Their Causes

There are an unlimited number of variables and scenarios that can cause an emergency situation in a process unit. The presence of large quantities of hazardous, flammable, and explosive materials, coupled with production equipment that can provide an ignition source for these materials, presents continuous danger. Emergency situations are usually caused by the sudden failure of major pieces of process equipment such as compressors, pumps, furnaces, and piping systems (Figure 15.1). Failure of automatic trip or shutdown instrumentation and utilities such as instrument air, steam, or electricity can also cause an emergency situation on a process unit. The simultaneous loss of multiple pieces of major equipment creates one of the most dangerous types of emergency situations. This type of event can have severe health, safety, and environmental consequences. The resulting hazards can include:

- Uncontrolled rapid release of hydrocarbon to flare and vent systems
- Hazardous material spills and environmental releases

Figure 15.1 Gasket failure between flanges can lead to a sudden release of water or hazardous chemicals.

CREDIT: pipicato/Shutterstock.

- Rapid cool down of process equipment in high-temperature service
- Thermal contraction of piping and flanges
- Separation of pipe joints and material release
- Fire or explosion
- Rapid heat up of process equipment in cold or refrigerated service causing additional overpressure as the material heats up
- Equipment and system overpressure leading to a catastrophic event that could impact an entire process facility and the surrounding communities (Figure 15.2).

Figure 15.2 Valves can come apart, releasing dangerous chemicals in the work area.

CREDIT: BasketThought/Shutterstock.

There are many causes of unit specific failures that can result in an emergency situation. Examples of some of the common failures that cause an emergency situation on most process units include:

- Power failure
- Instrument air failure
- Steam failure
- Fuel gas failure
- Compressor failure
- Furnace failure
- Severe weather conditions.

Other causes of emergency situations include the failure of auxiliary systems like hot oil systems, seal oil and dry gas seal systems, water systems, nitrogen, and other utility systems.

Process units are typically designed to handle emergency situations with adequate technology and safety systems that can withstand the effects of an emergency. Flare and vent systems that are adequately sized, piping systems and piping connections that can handle rapid heatup or cooldown without separation, safety instrumented systems, backup power, and redundant refrigeration systems are examples of ways to eliminate or minimize the hazards associated with an emergency situation.

An emergency situation requires immediate attention, as these conditions have the potential to cause serious injury to personnel, damage to equipment, and damage to the environment and surrounding communities if they are not managed correctly. It is critical to the safe operation of every process facility that emergency scenarios are identified and documented, and that mitigation plans are in place to safely manage those scenarios (Figure 15.3).

Figure 15.3 Emergency manuals provide instructions and guidance for responding to emergency conditions.

CREDIT: rchat/Shutterstock.

Engineering controls must be tested on a regular basis to ensure they will work when needed. All of the unit emergency procedures for a given scenario should be reviewed as soon as it is determined that the unit is secure from any immediate safety, health, or environmental hazards.

15.3 Process Technician's Role in Abnormal Operations

The process technician's knowledge of the technology, design, equipment and piping, valves, safety, and control systems, as well as process-specific hazards, plays a key role in the planning and execution of abnormal operations.

At the start of an abnormal condition, an analysis of the abnormal condition should be completed. It must be determined whether it is safe to work or whether the condition is about to escalate into an emergency condition in which workers would be at risk if they stayed in the area. The analysis determines if the mode of operation is safe, and hazards are minimized or eliminated. Standard operating procedures (SOPs) should be written beforehand to address the specific mode of operation and should contain step-by-step instructions, cautions, and hazards for placing the unit in an abnormal condition. The majority of the risks and hazards associated with abnormal operations can be eliminated with the proper development and use of operating procedures that are generated for the specific task.

During abnormal operations, coordination between the process technicians assigned to field duties and the technicians assigned to the control board is critical. The field technician will be engaged in proper lineup of process equipment, piping, and control valves, as well as operation of the rotating equipment. The control board technician will be engaged in establishing and maintaining the operating conditions and process variables for the given mode of operation. Abnormal operations also provide a unique learning experience for the new or inexperienced technician. The diversion from normal operations enables personnel to execute operating procedures and safety, health, and environmental policies and work practices that are seldom encountered during routine or normal operations. These activities give the process technicians an opportunity for hands-on experience that increases their knowledge base. Unit specific as well as site specific knowledge can be gained related to how abnormal unit operations can affect an entire production facility or community.

The process technician's primary responsibilities during abnormal operations include:

- Participation in hazard and operability (HAZOP) studies to review the abnormal operation and complete a hazard evaluation
- Development and execution of unit-operating procedures specific to the abnormal task
- Correct use of personal protective equipment (PPE)
- Proper lineup of process equipment, piping, and control valves
- Monitoring and control of the process during abnormal operation
- Special equipment preparation that may be needed during abnormal operation
- Establishing and maintaining control of process conditions within operating limits
- Coordination of all work activities while the abnormal operation is in progress
- Monitoring all site and contractor craftsmen to ensure safe work practices and that safety, health, and environmental policies are followed during abnormal operation, including those that determine unit PPE
- Ensuring that hazards to personnel, the environment, and equipment are managed correctly and all deviations from the site safety, health, and environmental policies are reported
- Participating in employee health monitoring programs when the potential for unique exposure hazards are present
- Completing control of work (COW) procedures and permitting processes used to manage work activities surrounding process equipment
- Completing lockout/tagout (LOTO) procedures for the purpose of equipment preparation and energy isolation when there is a need to inspect, repair, or replace process equipment
- Maintaining audio visual olfactory (AVO) and equipment monitoring awareness
- Remaining vigilant about first responder roles and responsibilities (for example, hazard identification, radio communications, and emergency evacuation procedures).

15.4 Hazards and Emergency Preparation

Safely managing an emergency situation depends on the knowledge, skills, and abilities of the process technician, emergency responders, and site personnel. Emergency situation management is regulated by the Occupational Safety and Health Administration (OSHA). OSHA is a U.S. government agency created to establish and enforce workplace safety and health standards, conduct workplace inspections and propose penalties for noncompliance, and investigate serious workplace incidents. OSHA's Process Safety Management of Highly Hazardous Materials standard includes guidelines for emergency planning and response (Figure 15.4).

Figure 15.4 Emergency situations must be reported as soon as they are identified. **A.** Alarm box. **B.** First responder calling for medical assistance.

CREDIT: A. Smithsaeng/Shutterstock. **B.** CandyRetriever/Shutterstock.

A.

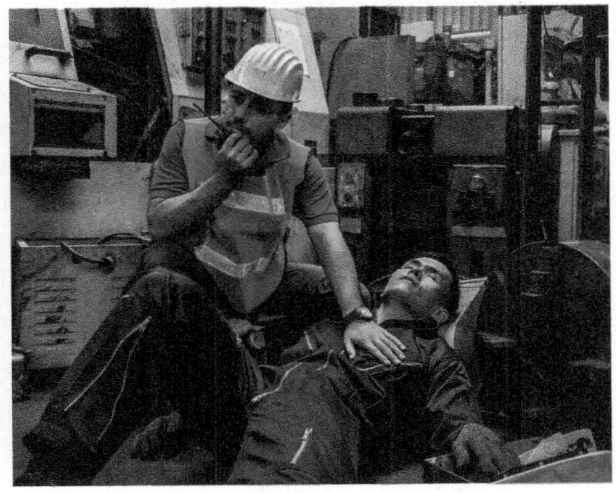

B.

HAZWOPER

Another OSHA standard that applies to the proper management of emergency situations is the **Hazardous Waste Operations and Emergency Response Standard (HAZWOPER)**. **First responders** will take initial steps at the scene. This includes any employees who are exposed or potentially exposed to hazardous substances, including hazardous waste, and those who are engaged in cleanup operations involving hazardous substances. The standard also applies to personnel who are engaged in corrective actions involving cleanup operations at sites covered by the Resource Conservation and Recovery Act (RCRA), which is the U.S. primary law governing the disposal of solid and hazardous waste. It gives the Environmental Protection Agency (EPA) authority to control hazardous waste and emergency response operations for hazardous releases and threats of releases, regardless of location.

Hazardous Waste Operations and Emergency Response Standard (HAZWOPER) OSHA's Regulation 29 CFR 1910.120; OSHA standard that applies to personnel who are in a role or position to act as a first responder during an emergency.

First responders individuals who likely witness or discover a hazardous substance release and have been trained to initiate an emergency response sequence by notifying the appropriate authorities.

HAZWOPER training for emergency responders should include instruction for managing:

- Hazard identification and isolation
- Communication
- Injury to personnel
- Fire
- Spill
- Evacuation.

Emergency Operating Procedures

Planning for an emergency requires that emergency scenarios for each specific process have been identified and that emergency operating procedures have been written and approved. These emergency operating procedures should include step-by-step instructions for securing a process unit for each type of emergency situation and identifying the effect on the process, the environment, and surrounding communities. Emergency operating procedures can also be formatted to identify steps that must be completed by the field technician, the board technician, and the unit personnel.

Process technicians must be trained regularly on these procedures to ensure the appropriate skills and knowledge are exercised when needed. Performing training on simulated emergency scenarios, sometimes called *gun drills,* prepares the process technician for safely managing situations without the immediate need for reference material. Emergency operating procedures are just one type of procedure required by the OSHA PSM standard.

Process Hazard Analysis

OSHA's Process Safety Management of Highly Hazardous Materials standard also includes requirements for process units to conduct periodic process hazard analysis studies. This is a review and study of process systems, equipment, and work processes that determines potential hazards. Unit process hazard analysis and unit specific emergency scenarios are identified during these studies. Operations and engineering team members identify potential emergency scenarios and implement engineering controls, administrative controls, or advanced technology to eliminate or mitigate each hazard. Engineering controls designed to eliminate emergency situations can include:

- Instrumentation controls in the form of alarms, automatic trips, and shutdowns
- Equipment variety, such as using a centrifugal pump rather than a positive displacement pump
- Relief devices where overpressure is a potential hazard
- Properly sized flare and vent systems.

Administrative controls take the form of policies, procedures, and checklists and these include:

- Fire and safety checklists
- Car seal checklists (Car seals are a means of securing devices such as valves, tanks, and so on to ensure they have not been opened or altered.)
- AVO checklists.

Avoiding Potential Hazards

Many of the same hazards present during abnormal and emergency operations are present during unit startups, shutdowns, and even normal operations. The use of an abnormal operating mode is usually infrequent and considered a nonroutine activity. The human element often increases the hazard level when performing unfamiliar activities. Hazardous conditions and emergencies can quickly develop during abnormal operations, and can lead to personnel injury, irreparable damage to equipment, material release, fire, and explosion. Every production facility has the responsibility to:

- Identify scenarios that are considered abnormal operations and perform the appropriate hazard analysis of the operation.
- Develop operating procedures for each abnormal operation.
- Identify potential emergency scenarios and have documented mitigation plans in place to manage such scenarios safely.
- Develop emergency procedures for each emergency scenario.
- Conduct training exercises utilizing emergency procedures.

15.5 Process Technician's Role During Emergencies

A process technician's duties include responding to emergencies of different types. There are many different types of emergencies that may be encountered in the process industry. In situations such as fire, explosion, spills, release of toxic gases, or bomb threats, an **emergency response** by the process technician is needed. **Incident response teams** provide and carry out the emergency response at most process facilities, with the process technician providing support during the emergency (Figure 15.5).

Emergency response effort to mitigate the impact of an incident on the public and the environment.

Incident response teams groups of people who prepare for and respond to any emergency incident, such as a fire, spill, explosion, or environmental release that potentially impacts the outlying community.

Figure 15.5 Incident response team plans an emergency response.

CREDIT: dear2627/Shutterstock.

In many process facilities, the process technician is considered to be the first responder at the *awareness* level. The process technician would take no further action beyond the notification of the incident. (A first responder at the *operations* level has more training and would take further actions to respond to the incident.)

Spills and Releases

While monitoring their assigned area, the process technician might encounter small leaks such as packing leaks, flange leaks, and tubing leaks. If trained and delegated to repair small leaks, the technician must exercise care in order to prevent a more hazardous situation. The process technician must always wear the proper personal protective equipment (PPE) when attempting to repair a leak.

A **spill** involves the uncontrolled discharge of a liquid, and it usually involves more volume than a leak. Each process facility has a determined definition of what volume constitutes a spill. The process technician's primary responsibility when discovering a spill is to report the incident to the proper authorities as soon as possible. The technician must also notify those workers immediately affected by the release to ensure their safety. If the process technician is trained as a first responder at the *awareness level*, they may establish a water spray, from a distance, to prevent harmful vapor leaving the immediate area.

Spill accidental release of a substance from lines, equipment, or areas.

A first responder at the *operations level* is an individual who responds to releases, or potential releases, as part of the initial response team to the incident site for the purpose of protecting nearby persons, property, or the environment. It is a technician trained as an operations level first responder who responds to the spill according to the site emergency response guidelines for spills or releases. Large spills that have the potential to affect the surrounding community or waterways require assistance from the local facility **fire brigade** (Figure 15.6).

Fire brigade local process facility fire department composed of employees who are knowledgeable, trained, and skilled in basic firefighting techniques.

Figure 15.6 Process technicians being trained in firefighting skills.

CREDIT: Soonthorn Wongsaita/Shutterstock.

Fires

Although fires are not common in the process industry, they do happen occasionally. Some fires that have occurred in industry include the following:

- *Dumpster fires*—caused by the improper mixing of chemicals coupled with readily available debris of all types that act as a fuel source.
- *Chemical fires*—caused by the improper mixing of two or more chemicals.
- *Flange fires*—caused primarily by a flange leaking hydrocarbons and then the vapor finding a source of ignition, or by temperatures such that auto-ignition occurs.
- *Furnace fires*—usually occur from a ruptured or leaking furnace tube.
- *Ground fires*—generally occur around the bases of flares and are caused when a large volume of liquid hydrocarbon is dumped into the flare system and it cannot be combusted by the flare.
- *Tank fires*—primarily caused by lightning strikes and generally occur on tanks that are not blanketed with an inert gas.
- *Process unit fires*—caused by leaking hydrocarbons, both liquid and vapor. These include associated piping, pumps, and vessels.

If the process technician encounters a fire, it should be immediately reported according to the process facility's emergency response guidelines. These guidelines typically require that the process technician should:

- Notify coworkers via radio, intercom, or PA system, if available.
- Sound the facility fire alarm to notify the remainder of the facility that an emergency is in progress.
- Call the process facility's emergency number stating the following:
  - Fire location
  - Fire fuel source
  - Fire size
  - If a structure fire, the size of the structure.

It is important to stay on the line with the dispatcher or person taking the call because there will likely be additional questions, such as those given below:

- Do people require medical attention?
- Are hazardous materials present?
- Are nearby properties or storage facilities threatened?

The technician should offer as much detail about the event as possible. After reporting the fire, he or she should stay as calm as possible until the fire brigade responds. Once the fire brigade has responded, the technician should provide whatever assistance the fire chief or incident commander deems necessary. The process technician should show or tell the fire brigade members the location of isolation valves that need to be closed to stop the fuel source feeding the fire. Once the scene is under control, the process technician is responsible for postincident cleanup, as designated by the process facility.

Most fires at large refineries and process facilities are often successfully dealt with by the process facility's firefighting personnel. The role of the local municipal fire department is generally one of reinforcement, helping the process facility firefighters when larger fires occur. If these fires become too large for the local fire brigade, then a mutual aid call may be sounded. A **mutual aid** system consists of many local fire brigades, including the municipal fire department. This type of system may require a regional mutual aid system if conditions warrant.

Mutual aid agreement among emergency responders to lend assistance across jurisdictional boundaries.

The technical knowledge of the process facility's fire officers can be invaluable to a municipal fire officer in the event that the municipal fire officer has command to combat a large refinery or process facility fire.

Explosions

Explosions are a rapid increase in volume followed by a release of energy in an extreme manner, usually with the generation of high temperatures and the release of toxic gases. Less common than fires, explosions can occur in industry if the proper conditions are present. Process explosions are either chemical or physical in nature.

Explosion a chemical reaction or change of state which is affected in an exceedingly short space of time with the generation of a high temperature and generally a large quantity of gas.

CHEMICAL EXPLOSIONS Chemical explosions may be decomposition, combination reaction, vapor cloud, or dust explosions (Figure 15.7). In all these cases, the reaction is exothermic. The explosions release energy, most often in the form of heat. Decomposition reactions occur in material such as TNT and nitroglycerin, both of which contain oxygen molecules. When the molecule decomposes, combustion gases are produced at high temperatures. The volume of the gases is much larger than the volume of the explosive, generating high pressure at the reaction zone. The rapid expansion of the gases forms the shock wave that provides the explosive effect. Even some hydrocarbons without oxygen in their molecules can decompose explosively.

Combination reactions require two or more components that react together exothermically to produce hot gases. Examples include ammonium nitrate and fuel oil, or gunpowder and fireworks.

The damage caused by an explosion depends partly on how fast the explosive reaction occurs. Decomposition reactions generally occur much faster than combination reactions.

Figure 15.7 A large storage tank fire.

CREDIT: Aneese/Shutterstock.

Another type of chemical explosion is the vapor cloud explosion. A vapor cloud explosion can occur when a fuel, such as ordinary propane, is mixed with air in the atmosphere. If the cloud is ignited, the burning rate may be fast enough to form a shock wave. Although the overpressure in the shock wave may not be very high compared to other explosions, it can be strong enough to damage or destroy structures and injure personnel.

A dust explosion, which is very similar to a vapor explosion, takes place when fine combustible particles, such as coal or grain, are distributed in the proper proportion with air and the mixture finds an ignition source.

PHYSICAL EXPLOSIONS Physical explosions are those in which no chemical or nuclear reaction occurs. The most frequent example is the rupture of a vessel whose contents, either gas or liquid, exist under high pressure. If the containment vessel bursts, the contents are free to expand, and a shock wave is formed. A simple example would be the explosion of a common automobile tire when it is overinflated.

Boiling liquid expanding vapor explosion (BLEVE) explosion resulting from excessive compression of vapor in the container head space and vapor flashing from its release to the atmosphere above its normal boiling point.

Liquids that have a normal boiling point well below ambient temperatures are sometimes stored (under their own vapor pressure) at pressures well above atmospheric pressure. If the containment vessel bursts, part of the liquid vaporizes extremely rapidly and expands, forming a shock wave. This process is called a **boiling liquid expanding vapor explosion (BLEVE)**. The resulting explosion can be very destructive (Figure 15.8).

Figure 15.8 Boiling liquid expanding vapor explosion (BLEVE).

CREDIT: konstantin belovtov/Shutterstock.

Explosions can occur during startups and shutdowns when an unplanned event, such as a spill or release occur and lead to an incident such as an explosion. The process technicians should adhere to normal startup and shutdown procedures to prevent the likelihood of any incident. The reporting procedure for explosions is the same as for fire.

Bomb Threats

Most bomb threats are received by phone. However, they can also be generated through handwritten notes or mail. Bomb threats should be taken seriously until proven otherwise. The process facility should have a procedure in place for handling a bomb threat. The process technician must be familiar with bomb threat procedures and know how to access the process facility's bomb threat checklist when needed.

If the process technician receives a bomb threat by phone, she or he should use the following guidelines:

- Remain calm and use the facility bomb threat checklist. The checklist should be located at each phone in the facility as a precaution and for easy access.
- If the phone has a display, note the name or phone number that appears in the display window.

- Obtain the location, appearance, and detonation time of the bomb from the caller.
- Don't hang up; stay on the phone with the caller, get a coworker's attention and have him or her call the emergency response coordinator from another phone. Relay the exact phone number where the threat is being received.

An example of a bomb threat checklist is provided in Figure 15.9.

Bomb Threat Checklist – Any Company, USA

Time of call: ____________

Exact words of threat: ____________

Number where phone call was received: ____________

Try to ask the following questions and get as much information as possible from the caller:

- When will the bomb explode? ____________
- Where is the bomb located? ____________
- What does the bomb look like? ____________
- What kind of bomb is it? ____________
- Did you place the bomb? ____________
- Why? ____________
- What is your name? ____________
- Where are you? ____________

Note any descriptive information from the caller's voice:

Description	*Yes*	*No*	*Description*	*Yes*	*No*
Accent	☐	☐	Lisp	☐	☐
Angry	☐	☐	Loud	☐	☐
Calm	☐	☐	Male	☐	☐
Clearing throat	☐	☐	Nasal	☐	☐
Crackling voice	☐	☐	Normal	☐	☐
Crying	☐	☐	Ragged	☐	☐
Deep	☐	☐	Rapid	☐	☐
Deep breathing	☐	☐	Rasp	☐	☐
Disguised	☐	☐	Slow	☐	☐
Distinct	☐	☐	Slurred	☐	☐
Familiar	☐	☐	Soft	☐	☐
Female	☐	☐	Stutter	☐	☐
Laughter	☐	☐	Other (explain) ____________		

Note any background noises you heard during the conversation:

Description	*Yes*	*No*	*Description*	*Yes*	*No*
Animals	☐	☐	Long distance	☐	☐
Booth	☐	☐	Machinery	☐	☐
Clear	☐	☐	PA system	☐	☐
House	☐	☐	Static	☐	☐
Kitchen	☐	☐	Street	☐	☐
Local	☐	☐	Other (explain) ____________		

Note any information regarding the caller's threat language:

Description	*Yes*	*No*	*Description*	*Yes*	*No*
Incoherent	☐	☐	Taped	☐	☐
Irrational	☐	☐	Well spoken	☐	☐
Reading message	☐	☐	Other (explain) ____________		
Profane	☐	☐			

Additional comments:

____

Figure 15.9 Example of a bomb threat checklist.

Summary

Abnormal and emergency operations are inevitable situations in the operation of a process unit, providing unique learning opportunities for site personnel and the process technician. The risks and hazards may range from none or little for abnormal operations to high or severe for emergency operations. Hazards present during abnormal operations and emergency operations should be sufficiently analyzed to develop procedures that ensure the safety of personnel, unit equipment, the entire facility, and surrounding environment. The procedures followed during abnormal operations and response to emergency situations require all the skills and abilities of a trained process technician.

Abnormal operations and emergency operating procedures should be fully documented in the same form as other operating procedures, so they can be managed safely by site personnel. Regularly practiced process deviation scenarios provide learning experiences for the site staff that only occur during abnormal and emergency operations.

The Process Safety Management of Highly Hazardous Chemicals standard (OSHA 29 CFR 1910.119) requires facility management to develop and document applicable emergency procedures and provide process technician procedure training.

Checking Your Knowledge

1. Define the following terms:
   a. Abnormal operation
   b. Boiling liquid expanding vapor explosion (BLEVE)
   c. Emergency
   d. Emergency operation
   e. Emergency response
   f. Explosion
   g. Fire brigade
   h. First responder
   i. Hazardous Waste Operations and Emergency Response Standard (HAZWOPER)
   j. Incident response teams
   k. Mutual aid
   l. Spill
   m. Train

2. Unlike abnormal operations, emergency operations __________. (Select all that apply.)
   a. have a specific purpose
   b. are unplanned
   c. can sometimes affect the environment and surrounding communities
   d. are temporary or short-term

3. Which of the following can cause an emergency situation? (Select all that apply.)
   a. Failure of process equipment
   b. Overcast weather conditions
   c. Loss of potable water
   d. Loss of instrument air
   e. Noisy equipment
   f. Loss of electricity

4. Some primary responsibilities of the process technicians during startups include: (Select all that apply)
   a. Root cause analysis to identify problems in startup
   b. Proper line-up of process equipment, piping, and control valves
   c. Monitoring and control of rotating equipment
   d. Establishing and maintaining control of process conditions within operating limits

5. Process technicians play a key role in the planning and safe execution of abnormal operations. Their responsibilities include: (Select all that apply.)
   a. Execution of the unit operations procedures
   b. Monitoring the process and equipment while the abnormal operation is in progress
   c. Monitoring all work activities while the abnormal operation is in progress
   d. Special equipment startup and preparation utilizing any normal or special control of work procedures

6. The major focus of OSHA's Process Safety Management of Highly Hazardous Chemicals is:
   a. Fire safety
   b. Response to hazardous spills
   c. Emergency planning and response
   d. Proper use of PPE

7. Which of the following are responsibilities of every production facility? (Select all that apply.)
   a. Develop procedures for abnormal operations.
   b. Remove all flammable chemicals from the facility.
   c. Document mitigation plans for emergency scenarios.
   d. Provide training in emergency procedures.
8. Which of the following need an emergency response by the process technician? (Select all that apply.)
   a. Bomb threats
   b. Explosions
   c. Planned maintenance
   d. Spills
9. Name four types of fires that have occurred in industry.
10. Which of the following is true for process technicians who are first responders at the *awareness* level? (Select all that apply.)
    a. Can respond to spills according to the site emergency guidelines
    b. In a spill, can establish a water spray from a distance to prevent harmful vapor leaving the immediate area
    c. Have more training than a first responder at the *operations* level
    d. Except in a spill, would take no further action beyond the notification of the incident

NOTE: Answers to Checking Your Knowledge questions appear in the Appendix.

Student Activities

1. Select one of the unit emergency shutdown procedures (steam failure, furnace failure, or compressor failure) and, using the scenario below, perform a gun drill together with a classmate to:
   - Determine the responsibilities of individuals in each system or section of the unit.
   - Determine the proper sequence of events necessary to bring the unit to a safe state after the emergency has occurred.
   - Identify any special or critical activities that must take place to mitigate the emergency.
   - Determine if special PPE is required during equipment isolation or shutdown during the emergency.
   - Determine the need for communication and proper contact numbers to adjacent or connecting units.
   - Determine the need for communication and proper contact numbers to site personnel responsible for contacting regulatory authorities.

   Scenario: The unit you are operating consists of two distillation columns, four pumps, one compressor (operating on steam from a third-party vendor), and two small furnaces. During the course of your shift, the third-party vendor supplying your site with steam has to drastically reduce the amount of steam sent to your site; this in turn slows your compressor turbine, causing a unit upset, which trips your furnace offline.

   In addition, your compressor trips offline. You do recover from these upsets—only to be informed later in the shift that the third-party site has lost its steam generation system and, for the remainder of your shift, can no longer send steam to your site.
2. Research at least two (2) of the following disasters that resulted from emergency operations.

   Philadelphia Refinery Explosion June 21, 2019
   Deepwater Horizon oil spill in the Gulf of Mexico, April 20, 2010
   The 2008 Georgia sugar refinery explosion in Port Wentworth, Georgia, February 7, 2008
   Texas City Refinery explosion. March 23, 2005
   Mount Polley mine disaster, near Likely, British Columbia. August 4, 2014
   Three Mile Island accident. March 28, 1979
   Phillips Petroleum Company, Pasadena TX. October 1989

   Prepare a paper explaining what happened, what went wrong, and what could have been done to avoid the event.

Chapter 16

Unit Shutdown

Objectives

After completing this chapter, you will be able to:

16.1 Differentiate between the different types of shutdowns:

- Normal/routine shutdown
- Emergency shutdown
- Shutdown for equipment maintenance
- Shutdown for turnaround. (NAPTA Operations, Normal Shutdown: Overview and Communications 1*) p. 225

16.2 Describe the process technician's role in the execution of unit shutdowns. (NAPTA Operations, Normal Shutdown: Shutdown Process Unit 2–4) p. 230

16.3 Describe the risks and hazards associated with unit shutdowns and the related OSHA standard for Process Safety Management. (NAPTA Operations, Normal Shutdown: Overview and Communications 2, 3) p. 231

Key Terms

Emergency shutdown—sudden failure of major process equipment, such as compressors or furnaces, or failure of utilities such as instrument air, steam, or electricity requiring an immediate shutdown, **p. 227**

Evacuation plan—documentation for the evacuation of a facility, to be used by personnel in the event of an emergency, **p. 227**

Hydrocarbon detector—electronic device that detects, measures, and indicates the concentration of compounds containing only hydrogen and carbon atoms in an area, **p. 232**

*North American Process Technology Alliance (NAPTA) developed curriculum to ensure that Process Technology courses will produce knowledgeable graduates to become entry-level employees in process technology. Objectives from that curriculum are named here in abbreviated form. For example, "(NAPTA Operations, Normal Shutdown: Overview and Communications 1)" means that this chapter's objective 1 relates to objective 1 of NAPTA's curriculum on the overview of and communications during normal shutdown procedures.

Hydrocarbon free—removal of compounds containing only hydrogen and carbon atoms (for example, methane or benzene) from process piping and equipment prior to opening to the atmosphere and introducing air, **p. 232**

Lift plan—documented plan used to evaluate the hazards and define precautions necessary during heavy lifting activities, including lifting over live process equipment or near energized power lines, **p. 231**

Scaffold plan—documented plan for the erecting and dismantling of scaffolding used to access process piping and equipment, **p. 231**

Shutdown—the time period when a unit is not in operation; systematic removal of process equipment from service in order to stop the process, **p. 225**

16.1 Introduction

This chapter provides an overview of various types of unit **shutdowns**. Unit shutdowns are common industry-wide for many reasons, and they are an integral part of process operations. They can include a planned, sequenced event, the details of which are unit specific, or an unexpected shutdown that will require quick thinking, decision making, and action on the part of the process technician. All shutdowns, even those that are planned, are a deviation from normal operations. Those deviations carry an increased level of risk that, if not managed properly, can cause injury to personnel, damage to equipment, and damage to the environment and surrounding communities. The time duration between a unit shutdown when production stops to unit startup when production is resumed is sometimes referred to as *stream-to-stream*.

Shutdown the time period when a unit is not in operation; systematic removal of process equipment from service in order to stop the process.

Many of the activities listed here take place during a unit shutdown and are covered by the OSHA Process Safety Management of Highly Hazardous Materials standard:

- Changes to the process must be managed and documented according to the management of change (MOC) guidelines defined in the OSHA 1910 PSM requirements.
- Process safety information must be updated to reflect changes or modifications to the process, including changes to drawings such as P&IDs, instrument and control loop diagrams, plot plans, and electrical one-line diagrams, as well as operating procedures, training material, and operating manuals.
- Hazard analysis of any changes or additions to the process, or to process safety information, must be completed.
- Subject matter experts (SME) who are associated with the process must be allowed to participate in hazard studies, as well as the development and review of new and revised process safety information.
- Activities related to mechanical integrity and inspection of process equipment are covered, including new and revised maintenance procedures and work practices, updates to equipment files and inspection files, and inspection frequencies and updates to existing equipment health monitoring programs that are intended to prolong or improve equipment integrity.
- Guidelines for emergency planning and response, prestartup safety review, incident investigations, and contractor management are also defined in the OSHA 1910 PSM requirements.

Unit shutdowns can require a vast amount of planning that not only incorporates the skills and abilities of the process technicians but also involves site staffing at almost all levels, as well as various contractor staffing (Figure 16.1). Maintenance planners and staff; process engineers; mechanical engineers; electrical engineers; inspection staff; warehouse and procurement staff; safety, health, and environmental (SHE) staff; project safety management (PSM) staff; and contractor staff all work together with an operations management team to plan and execute most unit shutdowns.

Figure 16.1 Detailed discussions at many levels must occur before unit shutdowns.

CREDIT: Hybrid Images/Cultura Creative (RF)/ Alamy Stock Photo.

A unit shutdown procedure unites personnel from across a facility to work more closely together than during normal operations. A planned shutdown includes an influx of temporary personnel. This provides opportunities to establish working relationships that otherwise might not occur and enables the unique skills of each individual to be shared and better understood. Career paths can even be altered when new opportunities are experienced by personnel related to shutdown planning—such as process hazard analysis; team participation; writing and updating process safety information, procedures, and training materials; and performing maintenance activities not experienced during normal operations.

The steps to shut down any equipment or system should include:

- Clearly defined purpose or task
- References and precautions such as management of change, personal protective equipment (PPE), and chemical properties
- Special equipment such as cranes, specialized blinds, or scaffolds
- Prerequisite tasks or conditions including how shutdown affects upstream and downstream process
- List of all defined steps and associated time needed to complete the job safely
- Steps to verify completion.

This information should be able to be shared to ensure completion and appropriate impact upstream and downstream of the process.

Normal or Routine Shutdowns

Normal or routine shutdowns are those that are planned and have a specific purpose. An example of a normal or routine shutdown might take place after a required amount of a product has been produced. Routine shutdowns of this nature are executed and the unit placed in a secure condition until the next run, or product order, is established. In other cases, normal routine shutdowns can be based on equipment needs and process conditions. In most shutdowns of this type, there are repair, replacement, and inspection opportunities intended to increase the production time of a unit between outages.

Major pieces of process equipment that require shutdown for maintenance and repair may also require the entire unit to be shut down in order to remove the equipment from service. Large centrifugal gas compressors, for example, are sometimes central to a process unit. These compressors may have bearing or seals that require removal of the compressor from service to repair. Monitoring the bearing and seal performance enables operations and maintenance personnel to predict and plan a routine shutdown to repair or replace the worn parts as necessary. Another example of a routine shutdown is a reactor system that needs catalyst replacement or regeneration. Monitoring of the catalyst performance should enable operations and engineering personnel to predict when a routine shutdown is necessary to replace or regenerate the catalyst. Corrosion, design fault, impurities, and equipment lifespan of service are things to mitigate for during shutdown activities.

Planning and executing a routine shutdown for a specific purpose requires a coordinated effort between the operations staff and site staffing at all levels. Maintenance planners and technicians, process engineers, mechanical engineers, electrical engineers, warehouse and procurement staff, and SHE staff all work together with an operations management team to plan and execute routine unit shutdowns and equipment repair or replacement.

Some of the key activities involved in performing a normal or routine shutdown could include the development and communication of a shutdown execution plan that should consider:

- Shutdown purpose and priorities
- Shutdown staffing
- Equipment and system shutdown sequence
- Utility shutdown sequence
- Shutdown timing and hold points
- Notification to regulatory authorities of the potential for flaring, and possible effects on the environment and surrounding communities
- Coordination and planning for the shutdown of various process, auxiliary, and utility systems
- Coordination with connecting units as well as adjacent units
- Controlled deinventory plan for all hazardous chemicals for both pre- and postshutdown
- Execution of operating procedures for shutting down individual pieces of process equipment and systems
- Execution of control of work (COW) procedures on process equipment
- A startup plan.

Emergency Shutdown

Emergency shutdowns may occur at any time but should not occur often. They can result from utility failures and sudden power outages. They can be caused by failure of major pieces of process equipment. Faulty instrumentation can cause emergency shutdowns as well. Safely managing an emergency shutdown depends on the knowledge, skills, and abilities of the process technicians on shift. Although an emergency shutdown is unplanned, emergency scenarios for each specific process can be identified, and emergency operating procedures are written and in place for use by the process technicians and operations support staff. Emergency operating procedures lessen the stress during an emergency operation. These emergency operating procedures should include systematic instructions for securing the process unit during specific types of emergencies and include the expected effect on the process, the environment, and surrounding communities. Emergency operating procedures are only one type of procedure required by the OSHA 1910 PSM standard for managing highly hazardous chemicals. An **evacuation plan** is also an important procedure that should be ready for use. This document details how personnel would evacuate a facility if an emergency event did occur.

Emergency shutdown sudden failure of major process equipment, such as compressors or furnaces, or failure of utilities such as instrument air, steam, or electricity requiring an immediate shutdown.

Evacuation plan documentation for the evacuation of a facility, to be used by personnel in the event of an emergency.

Emergency scenarios are also identified during unit process hazard analysis, also referred to as *HAZOP studies*. The operations and engineering team members are responsible for identifying potential emergency scenarios and implementing engineering or administrative controls to either eliminate or mitigate each hazard. Engineering controls are a preferred method of mitigating emergency scenarios. These can include:

- Instrumentation controls in the form of alarms, trips, and shutdowns
- Utilizing different types of equipment when appropriate, such as using a centrifugal pump rather than a positive displacement pump in a design where there is a closed or blocked discharge.
- Installation of relief devices where overpressure is a potential hazard
- Properly sizing flare and vent systems.

Typical types of administrative controls usually include policy, procedure, and checklists. Some examples of administrative controls are:

- Fire and safety checklists
- Car seal checklists or secure valve checklist
- Audio, visual, olfactory (AVO) checklists (checklists used to check for sounds, sights, and smells that might suggest an abnormal situation).

Utilizing advanced technology can also help to manage or eliminate hazards resulting from emergency scenarios. An example might include the installation of process equipment that would yield a specific product without the use of chemicals that have characteristics that are difficult to manage safely.

Shutdown for Equipment Maintenance

A number of situations may require an entire process unit shutdown. Examples of the need for equipment maintenance that could require an entire process unit shutdown include centrifugal compressor seal repair and reactor catalyst replacement or regeneration. Another type of shutdown could entail removing individual pieces of auxiliary equipment from a process or system without the need for an entire unit shutdown. Circumstances occur where a compressor, a furnace, a process pump, a centrifuge, and a fractionation tower or reactor system can be isolated and removed from service and have only a minimum effect on the rest of the process. This enables the process to continue to operate and produce product, but possibly at reduced rates. The associated hazards will be present but are much more limited than shutting down an entire process unit.

The complexity of the equipment removed from service determines the level of planning required, possible hazards, level of communication, and necessary execution steps (Figure 16.2). Safely managing such a shutdown can be an extensive undertaking and involve

Figure 16.2 Technician working on a heat exchanger. Equipment will be emptied, cleaned, and flushed. The system will be inspected closely for cracks and other anomalies.

CREDIT: Patrick Pleul/dpa-Zentralbild/ZB/dpa picture alliance/Alamy Stock Photo.

many personnel, or it can be a simple task that requires only the on-shift process technician and maintenance staff.

Examples of auxiliary equipment within process units that can be removed from service with little effect on the unit as a whole include the following:

- Spare pumps and compressors
- Storage tanks
- Multiple pieces of process equipment that perform the same function, such as filters, centrifuges, crystallizers, or lube oil systems

Most operations activities require precautions in order to mitigate the hazards associated with the shutdown and repair of auxiliary equipment. Proper communication and coordination between operations and maintenance personnel is critical to ensure the safe removal and efficient repair of auxiliary equipment.

The hazards associated with even minor repairs to equipment in hydrocarbon and process service can be severe. A process technician should make sure of the following:

- Standard operating procedures (SOPs) are used to remove equipment safely from service and minimize the impact on the rest of the unit.
- Control of work procedures are used to identify, isolate, energy free, and prepare the equipment.
- Proper PPE must be used to protect individuals from exposure and associated hazards.
- Maintenance procedures are used to ensure a quality repair and maximize equipment integrity.

Entire Unit Shutdown for Turnaround

An entire unit shutdown for turnaround (TAR) can be the most complex type of shutdown. TAR planning requires maximum coordination and communication among process facility personnel. It also requires prolonged communication between the operations staff and site staff at almost all levels, and may include these groups, among others:

- Maintenance planners
- Maintenance technicians
- Inspection staff
- Process engineers
- Control engineers
- Mechanical engineers
- Electrical engineers
- Warehouse and procurement staff
- Safety, health, and environmental staff
- PSM staff
- Contractor managers
- Contractor engineers
- Contractor staff.

All the staff works together with an operations management team to plan and execute a whole unit shutdown and turnaround. Planning can begin many months before the actual shutdown date. The duration of a TAR can be many months, depending on the type and amount of work scheduled.

16.2 Process Technician's Role in the Planning and Execution of Shutdowns

Process technicians play a key role in the planning and safe execution of unit shutdowns. Their knowledge of the process technology and design; process equipment; and interconnecting piping, valves, safety and control systems; as well as process-specific hazards makes process technicians one of the most important groups on the unit during shutdown. The primary responsibilities of a process technician during shutdown include the following:

- Execute the unit shutdown and deinventory as needed to facilitate a safe, efficient, and controlled shutdown.
- Ensure that potential hazards to personnel, the environment, and equipment are managed correctly.
- Report all deviations from the site safety, health, and environmental policies.
- Shut down special equipment, and prepare to use any normal or special control of work (COW) procedures.
- Monitor the process and equipment while the shutdown is in progress.
- Maintain all the unit safety equipment in good order so that it is available in case of an emergency situation or an environmental hazard requiring personnel protection.
- Monitor all work activities while the shutdown is in progress. Issue safe work permits as necessary for work taking place in the area.
- Monitor all site and contractor staff to ensure safe work practices and SHE policy compliance, including appropriate use of PPE.
- Participate in employee health monitoring programs when the potential for unique exposure hazards is present.
- Maintain the facility in a clean, orderly, and safe condition.

Shutdowns provide a unique learning experience for new or inexperienced technicians. The deviation from normal operations enables personnel to execute operating procedures, SHE policies, and work practices that are seldom encountered during normal operations. These activities provide process technicians an opportunity to increase unit and site specific knowledge about how a unit shutdown can affect an entire facility or community.

During shutdowns, the process technician may also find opportunities to revise operating procedures where corrections or deviations to previously established work practices are necessary. Each facility should have guidelines in its SHE policies for operating procedures that define the steps required for corrections to operating procedures. Certain corrections fall within the OSHA PSM 1910.119 guidelines for management of change (MOC) and may require hazard analysis of the change prior to execution.

Employee participation is invaluable to improve operating procedures for the benefit and future use of all personnel. Many of the risks and hazards associated with the process industry can be eliminated with the proper development and use of operating procedures, and it is critical that the process safety information contained in the procedures be correct and without omissions. Revision of process safety information by a process technician provides the opportunity to make work practice safer for all site personnel.

A process technician should be familiar with the site safety, health, and environmental policies for safe execution of unit shutdowns. These policies should be readily available in electronic form or hard copy. Process technicians should use these policies to understand and implement the established safe work practices for a given activity.

The following list provides an example of several typical SHE policies that are utilized during unit shutdowns:

- *Blinding*—policy that defines the process and procedure to isolate equipment for hot work or specific activities that require equipment removal

- *Confined space entry*—policy that defines the process and procedure for entering confined spaces such as equipment, storage tanks, and excavations below grade
- *Employee health monitoring*—policy that defines the need for employee health monitoring while activities are conducted in hazardous areas, during hazardous chemical sampling, or where the extended exposure to hazardous chemicals can occur
- *Environmental reporting*—policy that defines the requirements and reportable quantities for chemicals that, when released to the atmosphere, require reporting to the proper regulatory authorities
- *Hot work*—policy that defines process and procedure for conducting hot work, such as welding or grinding in, on, or around process equipment
- *Housekeeping*—policy that defines activities that must be completed in order to maintain the facility in a clean, orderly, and safe condition
- *Lockout/tagout (LOTO)*—procedure used in industry to isolate energy sources from a piece of equipment
- **Lift plan**—documented plan to be used during heavy lifting activities above active process equipment or near live power lines
- *Management of change (MOC)*—method of managing and communicating changes to a process, changes in equipment, changes in technology, changes in personnel, or other changes that will impact the safety and health of employees
- *Material release reporting*—policy that defines reporting requirements of regulatory authorities when venting, purging, or draining equipment or in the event of a material release
- *Standard operating procedures (SOPs)*—unit specific procedures used for the purpose of equipment and system startup or shutdown in normal operations, as well as emergency operations
- *Personal protective equipment (PPE)*—specialized gear that provides a barrier between hazards and the worker
- *Process hazard analysis*—systematic assessment of the potential hazards associated with an industrial process, taking into account specific hazards and locations of highest potential for exposure
- *Process safety information*—policy that defines the type of documentation that is considered process safety information in support of the OSHA PSM regulation, including operating procedures, inspection and maintenance procedures, operating manuals and training material, process drawings (P&IDs), electrical one-line diagrams, instrument loop drawings, and electrical classification drawings
- *Process safety management (PSM)*—OSHA standard 1910.119 that contains the requirements for management of hazards associated with processes using highly hazardous materials
- **Scaffold plan**—plan for providing scaffolding to access process piping and equipment
- *Vehicle entry*—policy that defines process and procedure for vehicle entry into process areas
- *Working at heights*— policy that defines requirements for working at elevated heights.

Lift plan documented plan used to evaluate the hazards and define precautions necessary during heavy lifting activities, including lifting over live process equipment or near energized power lines.

Scaffold plan documented plan for the erecting and dismantling of scaffolding used to access process piping and equipment.

16.3 Potential Hazards

There are many hazards associated with unit shutdowns, which occur infrequently and are a nonroutine activity. Performing unfamiliar activities increases hazards. Different types of shutdowns can cause a variety of hazards.

An emergency shutdown in which the loss of multiple pieces of major equipment occurs is one of the most dangerous types of shutdown. Recycle gas compressors, refrigeration compressors, fractionation towers, boilers, furnaces, hot oil systems, reboilers, and reactor systems are some of the systems and equipment that, when shutdown simultaneously in an emergency, can have severe health, safety, and environmental consequences. Typical hazards include these:

- Uncontrolled rapid release of chemicals to flare and vent systems can overload the vent system and result in an environmental release.
- Rapid cooldown of process equipment in high-temperature service can result in thermal contraction of piping and flanges, separation of pipe joints, and material release, which can lead to a fire or an explosion.
- Rapid heating of process equipment that is in cold or refrigerated service can lead to overpressure conditions. In cases involving specialty chemicals, such as hydrofluoric acid, rapid heating can result in overpressure and lead to a catastrophic event that could affect an entire process facility and the surrounding communities.

Emergency shutdowns can also lead to personnel injury, irreparable damage to equipment, material release, fire, and explosion. A *hydrocarbon atmospheric release*, otherwise known as a *material release*, can affect the air quality of the entire facility and surrounding communities, as well as local ground water and waterways.

Process units are typically designed to handle emergency shutdowns by the installation of adequate technology and safety systems. Some ways to eliminate or minimize hazards associated with emergency shutdown include the following:

- Adequately sized flare and vent systems, piping systems, and piping connections that can handle rapid heating or cooldown without separation
- Safety instrumented systems
- Backup power
- Redundant refrigeration systems.

An emergency shutdown, if it is not managed correctly, has the potential to cause serious injury to personnel, damage to equipment, and damage to the environment and the surrounding community. It is critical to the safe operation of every process facility that potential emergency scenarios be identified and documented and that mitigation plans be in place to manage them safely. Engineering controls must be tested regularly to ensure they will work when needed. Emergency procedures must be in place, and process technicians should be trained in their use. Continued practice of emergency procedures provides the skills for dealing with an emergency shutdown. All unit emergency procedures for a given scenario should be reviewed periodically and following an emergency shutdown.

Normal or routine shutdowns, however infrequent, are typically a planned activity and do not pose the same threat of uncontrolled hazards as emergency shutdowns do. A normal or routine shutdown for a large process unit with many systems, and sometimes hundreds of pieces of equipment, can take days or weeks to plan. Hazards are similar to those of an emergency shutdown, but planning, use of SOPs, and safe work practices will minimize or eliminate the chance of a hazardous accident. Process units can be systematically shut down while carefully managing the hazards. Hydrocarbon removal and deinventorying can minimize flaring and material release. A **hydrocarbon detector** (Figure 16.3) can be useful in a shutdown to make sure that piping is **hydrocarbon free**. Controlled cooldown of process furnaces eliminates thermal stress on the furnace tubes and equipment in high-temperature service. Controlled shutdown and isolation of large frame centrifugal compressors and associated seal and lube oil systems should eliminate the possibility of damage to the equipment and help ensure equipment integrity for a later startup.

Hydrocarbon detector electronic device that detects, measures, and indicates the concentration of compounds containing only hydrogen and carbon atoms in an area.

Hydrocarbon free removal of compounds containing only hydrogen and carbon atoms (for example, methane or benzene) from process piping and equipment prior to opening to the atmosphere and introducing air.

Normal or routine shutdowns can also lead to personnel injury due to nonroutine tasks performed. Back sprains, tripping hazards, and elevated work are potential hazards. When

Figure 16.3 A hydrocarbon detector ("gas sniffer").

CREDIT: Oil and Gas Photographer/ Shutterstock.

reviewing and implementing shutdown activities, technicians should be alert to such dangers as damage to equipment, material release due to deinventory methods, engineered obstacles, or confined versus ventilated areas.

Auxiliary systems, such as hot oil systems, seal oil and dry gas seal systems, water systems, steam and condensate systems, and air, nitrogen and other utility systems, can all be shut down safely with adequate planning and procedures. These are typically sequenced events based on the needs of the process. For example, a steam and condensate system is not shut down until after the process systems that require them are shut down and secure.

Shutdown for a turnaround (TAR) is managed like a normal, routine shutdown and includes the same potential hazards and planning effort. The scope of work surrounding TARs includes:

- Equipment preparation for confined space entry and internal inspection activities
- Piping and equipment external inspections
- Hot work
- Vehicle entry
- Inspection and x-ray
- Excavations
- Heavy lifting
- Demolition of piping, equipment, and support structures
- Installation of new piping, equipment, and support structures
- Installation of new technology.

The difference between a TAR shutdown and a normal, routine shutdown is the level of activity and work scope involved. The scope of work and TAR duration require many site and contractor employees working on the unit simultaneously. Mitigating the hazards associated with such a high level of work and so many workers on the unit at the same time requires a well-orchestrated plan for work activities. The TAR work scope and demand for product dictates the shutdown timing, TAR duration, work schedule, and startup timing. Isolation, de-energizing, draining, and purging of process equipment for maintenance, inspection, and repair are some of the primary activities during a major unit TAR. No matter the type of shutdown, planned or emergency, the process technicians must communicate important information to upstream and downstream units that will be affected.

Summary

Unit shutdowns are an important part of process operations and provide unique learning opportunities for process facilities, site personnel, and process technicians. Activities that take place during shutdowns require a high level of knowledge and focus from process technicians and site personnel.

The deviation from the routine duties of normal operation provides learning experiences for all staff that only occur during shutdowns. Maintenance, repair, and replacement of process equipment should improve equipment health, longevity, and integrity. New technology, when implemented, can result in safer operating processes, with decreased risks to personnel, the facility, and the surrounding community. In many cases, the installation of new technology can reduce operating costs, leading to higher profitability for a process facility.

There are many hazards associated with unit shutdowns. Performing unfamiliar activities increases hazards. Emergency shutdowns carry the greatest risk to personnel, equipment, and the environment.

Shutdown planning and execution, when managed and completed safely, constitutes some of the most gratifying work opportunities in the field of process technology. New experiences gained from the efforts of many, working as a team, and working with other site and contract personnel for the common benefit of a facility, are very rewarding for individual employees.

Checking Your Knowledge

1. Define the following terms:
   a. Emergency shutdown
   b. Evacuation plan
   c. Hydrocarbon detector
   d. Hydrocarbon free
   e. Lift plan
   f. Scaffold plan
   g. Shutdown

2. The sudden failure of major pieces of equipment such as compressors or furnaces, or the failure of utilities such as instrument air, steam, or electricity is called ______.
   a. a turnaround
   b. unscheduled maintenance
   c. an emergency shutdown
   d. steam-to-stream response

3. Many activities during shutdown or unit turnaround are covered by:
   a. FDA regulation 22
   b. OSHA Process Safety Management of Highly Hazardous Chemicals standard
   c. EPA 1993.220
   d. Hazardous Waste Operations and Emergency Response Standard (HAZWOPER).

4. Elements of the OSHA 1910 Process Safety Management regulation include: (Select all that apply.)
   a. Management of change (MOC)
   b. Community open houses
   c. Subject matter experts (SME)
   d. Emergency planning and response

5. AVO checklists check for _______, __________, and ________ that might suggest an abnormal situation.

6. Safe removal and efficient repair of auxiliary equipment require proper communication and coordination between operations and _______ personnel.

7. Welding and grinding are considered to be ______.
   a. material release hazards
   b. hot work
   c. routine maintenance work
   d. scaffold work

8. Process technicians play a key role in the planning and safe execution of unit shutdowns. Their responsibilities include: (Select all that apply.)
   a. Execution of the unit shutdown and de-inventory procedures
   b. Starting up any special equipment using normal standard operating procedures (SOP).

c. Monitoring the process and equipment while the shutdown is in progress
d. Monitoring all work activities while the shutdown is in progress

9. Match each of the following five typical SHE policies to the appropriate description.

SHE Policy		Description
I. Blinding	_______	a. defines activities required to maintain the facility in a clean, orderly, and safe condition
II. Employee health monitoring	_______	b. employee health monitoring requirements for working, sampling, or being exposed to hazardous chemicals
III. Hot work	_______	c. procedure required to isolate energy sources from a piece of equipment
IV. Housekeeping	_______	d. process and procedure for conducting such work as welding or grinding in, on, or around process equipment
V. Lockout/tagout (LOTO)	_______	e. process and procedure for isolating equipment for hot work or specific activities that require equipment removal

10. Match each of the following five descriptions to the appropriate SHE policy:

Description		SHE Policy
I. Procedure to be used during heavy lifting activities above active process equipment or near live power lines	_______	a. confined space entry
II. Process and procedure for entering tight spaces such as equipment, storage tanks, and excavations below grade	_______	b. environmental reporting
III. Reporting requirements of regulatory authorities when venting, purging, or draining equipment	_______	c. lift plan
IV. Requirements and reportable quantities for chemicals that, when released to the atmosphere, require reporting to the proper regulatory authorities	_______	d. material release reporting
V. Unit-specific procedures used for the purpose of equipment and system startup or shutdown in normal operations, as well as emergency operations	_______	e. standard operating procedures (SOPs)

11. Name three ways to eliminate or minimize hazards associated with emergency shutdowns.

12. What must process technicians and operations support do prior to an emergency shutdown? (Select all that apply.)
a. Nothing. They are unplanned occurrences.
b. Identify emergency scenarios.
c. Put emergency operating procedures in place.
d. Plan to use their personal cell phones for communication during the crisis.

Note: Answers to Checking Your Knowledge questions appear in the Appendix.

Student Activities

1. Select a unit emergency shutdown procedure, and perform a practice drill together with shift personnel. Be able to meet the following objectives as outcomes of this exercise:
    - Explain the responsibilities of individuals in each system or section of the unit.
    - Define the proper sequence of events necessary to bring the unit to a safe state after the emergency has occurred.
    - Identify any special or critical activities that must take place to mitigate the emergency.
    - Determine whether or not any special PPE is required during equipment shutdown and isolation during the emergency.
    - Specify the need for communication and proper contact numbers to adjacent or connecting units.
    - Determine the need for communication and proper contact numbers to site personnel responsible for contacting regulatory authorities.

2. Form small shutdown planning teams:
    - Identify the procedures and documentation needed to execute the desired work shown. Include conditions that may affect upstream and/or downstream systems or sections of the plant and any shutdown-specific safety concerns.
    - Determine a method to deinventory equipment for shutdown and isolation:
        - Large hydrocarbon storage tank in the tank farm
        - Steam turbine-driven compressor
        - Binary distillation column
    - Next, identify any special or critical activities that might take place in preparation for reinventory and deisolation.
    - Share your planning with the class.

Appendix

Answers to Checking Your Knowledge Questions

Chapter 1	Answer	LO #
1.	See Key Terms list.	1.1, 1.2, 1.3
2.	Process technicians must understand the following about the equipment in their area: • its function • potential problems • safety, health, and environ-mental (SHE) concerns • potential quality issues • related operating and emergency procedures.	1.1
3.	I. Distillation system — b. process that separates feed stream components by repeated vaporization and condensation with separate recovery of vapor and liquids. II. Flare system — c. device to burn unwanted process gases before they are released into the atmosphere. III. Reactor system — f. process that chemically alters materials by the application of heat and pressure. IV. Refrigeration system — a. system for the removal of heat. V. Relief valve system — d. system designed to open if the pressure of a liquid in a closed space exceeds a preset level. VI. Steam generation system — e. process that converts high-purity water to high-pressure, high-temperature steam for heating process streams. VII. Utility systems — h. critical systems such as wastewater disposal, process sewers, and systems that safely dispose of liquid and gaseous wastes in an environmentally sound manner. VIII. Water systems — g. systems including fire water, process water, potable (drinkable) water, cooling water, demineralized water, and boiler feed water systems, among others.	1.1
4.	c. shift supervisors	1.2
5.	a. SHE, b. electrical/instrumentation, e. maintenance superintendant.	1.2
6.	c. perform safety verification checks as required.	1.3
7.	b. interpret lab results and adjust process parameters to maintain product specifications.	1.3
8.	a. participating in PSSR, e.	1.3
9.	c.	1.3
10.	Shift change communication topics include any of the following: • Safety and environmental issues that exist or were corrected • Process and equipment problems, including corrective actions taken • Material transfers in progress • Special operating instructions • Items being coordinated with other process areas • Ongoing or upcoming unit maintenance or contract work • Technical support personnel working on the unit	1.3
11.	c. process technician	1.3
12.	b. distributed control system (DCS)	1.3
13.	a. specific process function	1.1
14.	b. logbook	1.3

Chapter 2		
1.	See Key Terms list.	2.1, 2.2, 2.3
2.	On-the-job training	2.1
3.	communication	2.1
4.	mentor	2.2
5.	a. operating manual, b. P&ID, e. safety, health and environment (SHE) policies.	2.3
6.	d. piping and instrumentation design pressure.	2.3
7.	b. unit layout, c. piping and equipment material of construction.	2.3
8.	b. process simulators	2.3
9.	a. benzene, b. nitrogen, c. asbestos.	2.3
10.	hazards	2.3

Chapter 3	Answer	LO #
1.	See Key Terms list.	3.1, 3.2
2.	a. pump capacities, b. equipment symbols.	3.1
3.	False. P&IDs have greater detail than process flow diagrams.	3.1
4.	b. left to right.	3.1
5.	c. shows the entire electrical system of interconnecting generators, transformers, transmission and distribution lines, and so on.	3.1
6.	I. d. P&ID needle valve. II. a. P&ID hydraulic actuator. III. c. P&ID butterfly valve. IV. b. P&ID pneumatic actuator.	3.2
7.	I. b. heat exchanger. II. d. trayed tower. III. a. pump. IV. c. compressor.	3.2
8.	hand valve.	3.2
9.	b. flow.	3.2
10.	c. ISA.	3.2

Chapter 4		
1.	See Key Terms list.	4.1, 4.2, 4.3, 4.5
2.	b. SHE.	4.1
3.	c. hot work.	4.1
4.	risks, hazards.	4.2
5.	b. area cleanup after maintenance and repair activities.	4.3
6.	a. 30-pound dry powder fire extinguisher.	4.4
7.	d. foam addition system for applying AFFF (aqueous film forming foam).	4.4
8.	b. overpressure scenarios, d. planned unit startups.	4.5
9.	a. unit layout, b. piping and equipment material of construction, c. piping and equipment design temperature, d. piping and equipment design pressure.	4.5
10.	I. b. confined space entry. II. a. Lockout/tagout. III. d. Sampling. IV. c. Vehicle entry.	4.6
11.	True. Process technicians must be able to access and use reference material in order to respond quickly and effectively in a potentially hazardous situation.	4.6

Chapter 5		
1.	See Key Terms list.	5.1, 5.2, 5.3
2.	d. combination for each lock.	5.1
3.	b. control of hazardous energy.	5.1
4.	chains / cables.	5.2
5.	a. DO NOT CLOSE, c. DO NOT START, d. DO NOT OPERATE.	5.2
6.	zero-energy.	5.2
7.	c. the authorized employee.	5.3
8.	b. line leaks, c. unusual noises, e. high vibration.	5.3
9.	b. perform pressure tests of equipment if necessary, c. notify affected employees of impending startup, f. remove locks and tags from the circuit breakers.	5.3

Chapter 6		
1.	See Key Terms list.	6.1, 6.2, 6.3, 6.4, 6.5, 6.6
2.	d. sender interrupts the receiver's questions.	6.1
3.	Tips for good verbal communication include any of the following: • Use an animated voice. • Avoid ambiguity. • Do not send mixed messages. • Listen. • Make eye contact. • Pronounce words correctly. • Slow down. • Speak clearly. • Use appropriate volume. • Use gestures. • Use the right words.	6.1
4.	Principles of technical and business writing include any of the following: • Convey a single idea within a given sentence. • Create sentences that are short and specific. • Create the message with the most inexperienced person in the unit in mind. • Define any technical terms that may be unfamiliar to the receiver. • If possible, identify members of the target audience to review the message to enable revisions before it is sent. • Provide a background for context in a short paragraph before writing the core message or requesting action or information.	6.2
5.	a. avoid clichés, d. avoid inappropriate or slang words, f. use correct spelling, h. use technical and business writing principles.	6.2
6.	nonverbal.	6.3
7.	b. hand signals.	6.3
8.	The four work groups with whom the process technician is expected to communicate are: • Operations • Maintenance • Engineering • Safety.	6.3
9.	safety.	6.3
10.	Intercoms.	6.4
11.	b. paging personnel/emergency notification.	6.4

Chapter 6	Answer	LO #
12.	I. Engineering — c. process related issues such as temperatures, pressures, or catalyst activation. II. Maintenance — d. status of any ongoing repairs and other housekeeping work. III. Operations — b. current equipment status, including lockout/tagout status of equipment. IV. Safety — a. any startup related issues that may be industrial hygiene related, such as leaks or spills.	6.6
13.	b. Issue a confined space entry (CSE) permit, c. Follow an isolation and lockout tagout procedure to prepare the confined space for entry.	6.6
14.	c. process technician.	6.6

Chapter 7		
1.	See Key Terms list.	7.1, 7.2, 7.3, 7.4, 7.5
2.	b. executable, error-preventing, c. audience specific, accurate, e. well organized, clear	7.1
3.	True. Badly written procedures can sometimes be the cause of industrial accidents.	7.2
4.	a. Open the block valve 5 full turns, c. Open the block valve gradually until you observe condensate coming out of the downstream drain valve, d. Close the vent when the pressure is below 20 PSI.	7.2
5.	b. It helps ensure that each step is completed in the correct order.	7.2
6.	b. encourages creativity.	7.3
7.	Words to avoid when writing procedure instructions include the following: • about • approximately • could • may • might • ought • should	7.4
8.	b. colorful, descriptive language.	7.4
9.	Safety.	7.5
10.	c. Disconnect the power cable from the electrical source.	7.5
11.	c. PPE for the procedure.	7.2

Chapter 8		
1.	See Key Terms list.	8.1, 8.2, 8.3, 8.4, 8.5
2.	a. alarms and their current status, b. procedures in progress, d. process status, f. status of permits in force.	8.1
3.	a. shift handover, c. shift passdown, d. making relief.	8.1
4.	b. be identical.	8.2
5.	a. process safety management limits exceeded, b. contractors remaining on the unit after shift change, c. equipment on order through purchasing, d. operational requirements exceeded, f. abnormal situations on unit.	8.2
6.	d. verbal communication.	8.3
7.	c. used when eLogs are not in use.	8.3
8.	b. technician-to-technician.	8.4
9.	b. paraphrasing, d. reflection.	8.4
10.	a. be attentive, c. be a team player, d. be willing.	8.5

Chapter 9		
1.	See Key Terms list.	9.1 through 9.10
2.	a. operating pumps, c. providing process heating.	9.1
3.	b. water hammer can occur when steam condenses and the accumulated condensate is carried along with the steam flowing within the piping, d. can occur when condensate is forced to stop or change direction.	9.1
4.	b. drinking water.	9.2
5.	explosion.	9.2
6.	c. The statement "Eductors are used for processing firewater" does not describe an eductor.	9.3
7.	b. Sanitary sewer systems should be reseeded periodically with a bioaugmentation product.	9.3
8.	a. refrigerant, b. evaporator, d. compressor, e. expansion valve.	9.4
9.	a. evacuating the system, c. purging the system with nitrogen or dry gas, d. injecting methanol.	9.4
10.	c. Counterflow towers have process water and air flowing in opposite parallel directions from one another.	9.5
11.	Problems that must be addressed to maintain a cooling water system are any of the following: • leaks • loss of power • contamination • freezing concerns	9.5
12.	d. withholding chlorination.	9.5
13.	a. transformers, c. feed-circuit switches.	9.6
14.	b. Safety release valves are not a component of uninterruptible power systems.	9.6
15.	b. purging equipment containing inert gas to allow entry for maintenance, d. operating pneumatic tools and pumps.	9.7

Chapter 9	Answer	LO #
16.	Instrument air.	9.7
17.	I. Loss of cooling water to condensers and coolers — d. pressure increase in process units. II. Loss of reflux — b. increased pressure in distillation towers and disruption of the volume of vapor leaving the distillation tower. III. Sudden vapor and pressure increases — a. potential equipment internal explosion, uncontrolled chemical reactions, thermal acceleration, or amassed gases in vessels. IV. Excessive steam pressure — c. heater upset, damaged equipment, fire, upset automated controls or heat exchangers.	9.8
18.	b. failure of threshold pressure to open due to corrosion or plugged valve inlets or outlets, c. failure to reseat after opening, d. chattering and early opening due to operating pressure too near the valve threshold pressure.	9.8
19.	b. inerting.	9.9
20.	c. H_2S.	9.10

Chapter 10		
1.	See Key Terms list.	10.1, 10.2
2.	c. planning.	10.1
3.	d. took more than 14 days.	10.1
4.	b. line blows of process piping, c. pressure testing of vessels, piping, and other equipment.	10.2
5.	d. learning the procedures for each instrument.	10.2
6.	a. are post-startup items, d. must be addressed by the contracted construction firm.	10.2
7.	b. taking reading as required, d. monitoring vessels and equipment, e. starting pumps, agitators, mixers, and compressors as required.	10.2
8.	b. mechanical completion.	10.2
9.	a. electrical instrumentation loop checks, b. flushing and cleaning of vessels and equipment, d. issuing work permits, f. performing or providing fire watch duty as requested.	10.2
10.	c. Several process technicians are included.	10.2

Chapter 11		
1.	See Key Terms list.	11.1, 11.2, 11.3
2.	feedstock.	11.1
3.	c. Unit startups entail systematically putting process equipment into service in order to start the process, d. Unit startups are a diversion from normal operations and carry increased levels of risk.	11.1

Chapter 11		
4.	a. abilities of process technicians, c. knowledge of process technicians.	11.1
5.	major process equipment	11.1
6.	b. PSM, e. trade secrets, f. PSSR, g. compliance audits, h. training.	11.2
7.	a. execution of unit startup and inventory procedures to facilitate a safe, efficient, and controlled startup, d. establishing and maintaining control of process conditions within operating limits.	11.2
8.	a. know the equipment, b. know operating limits and design criteria, c. understand potential hazards.	11.2
9.	I. Atmospheric release or a material release — c. poor testing of equipment, valve, and flange tightness. II. Pipe and equipment damage — d. thermal expansion or contraction. III. Hazards from slips, trips, or falls — b. poor housekeeping. IV. Personal injury — a. failure to wear proper PPE.	11.3
10.	a. rotating equipment is properly installed and ready, c. electrical equipment and controls are properly installed and ready, e. piping and structural equipment are properly installed and ready for startup.	11.4

Chapter 12		
1.	See Key Terms list.	12.1, 12.3, 12.4
2.	a. overseeing and assisting maintenance personnel, contractors, and technical personnel, b. notifying control board technician of any process or equipment problems and suggesting corrective actions, e. making a thorough inspection of the unit at the beginning of the shift and at regular intervals during the shift.	12.1
3.	c. inspecting the unit utilizing the control system at the beginning and throughout the shift at regular intervals, e. recording shift activities in the unit logbook or eLog, including all personnel, f. recording all lab data.	12.1
4.	c. cutting pliers.	12.2
5.	a. add correct viscosity oil, b. check seal pots for level and pressure, f. check for proper lubrication.	12.3
6.	a. verify pressure and temperature, b. check motor amperage, c. check oil flow, e. verify control valve positions.	12.3
7.	c. check temperature and/or pressure differential across exchangers, f. perform periodic surveys to verify exchangers to not have fouling.	12.3

Chapter 12	Answer	LO #
8.	c. increase production time, decrease cost, and monitor equipment after repair or startup.	12.4
9.	1. d. Verify that the spare pump has proper oil levels 2. a. Ensure that the spare pump suction and discharge valve are in the open position. 3. c. Ensure that any seal oil, external or internal, is lined up properly. 4. f. Notify the control board technician and ensure that the unit is prepared for pump swap. 5. e. Start the spare pump. 6. b. Allow time for flow conditions to stabilize, then shut down and secure the main pump to be prepared for maintenance or service.	12.5
10.	personal protective equipment (PPE).	12.6
11.	c. PPE costs are not something that OSHA requires employers to provide to employees.	12.6
12.	Routine work activities that process technicians are required to document in the unit logbook include the following: • Process alarms that activated during the shift • Vibration and temperature alarms experienced on equipment • Equipment oil levels and the amount of oil added • Chemicals used in the process • Maintenance activities performed and planned • Unusual events, such as process upsets, and equipment malfunctions.	12.7

Chapter 13		
1.	See Key Terms list.	13.1, 13.2, 13.3, 13.4, 13.5, 13.6
2.	a. to ensure reliability of continuous stream analyzers, b. to maintain correct process parameters, d. the ensure specifications are met.	13.1
3.	d. all of the above.	13.2, 13.3, 13.4
4.	b. using a nonmetallic sample container.	13.2
5.	septum.	13.3
6.	c. bombs.	13.3
7.	loops.	13.3
8.	b. protective eyewear (goggles or face shield).	13.4
9.	d. steel-toed shoes.	13.4
10.	a. process unit name, b. date and time, d. vessel of origin or sample point ID, f. quantity of sample, h. current analyzer result, i. variables to be tested.	13.5

Chapter 13		
11.	I. Basic sediment and water (BS&W) — g. if crude oil meets the technical specification limits of the buyer. II. Color — c. determine product color purity. III. Gas chromatography (GC) — a. determine the exact makeup of the gas sample. IV. Gas detector tubes — e. determine if the targeted chemical is present. V. Karl Fischer (KF) titration method — d. quantifying water content. VI. Lead acetate test — b. identifying the presence of sulfur or sulfur-based compounds. VII. pH — f. measurement of acid or base level of a solution.	13.6

Chapter 14		
1.	See Key Terms list.	14.1, 14.2, 14.3, 14.4, 14.5
2.	b. predictive, c. reactive.	14.1
3.	a. increased costs of repair or replacement, b. costs of unplanned downtime, e. increased risk of accident, injury, fatality, or environmental contramination.	14.1
4.	b. less costly maintenance budget, d. energy savings, e. reduced equipment failure.	14.1
5.	c. safe work permit.	14.2
6.	I. Chemical exposure — e. draining and purging. II. Cuts, scrapes, or bruises — b. closing/opening valves, tagging out equipment, removing bull plugs, installing purge hoses. III. Electrical shock — d. de-energizing or re-energizing breakers. IV. Slips, trips, and falls — c. closing or opening, locking, or tagging valves. V. Strains or sprains — a. bending or stooping, pulling or moving hoses, closing or opening valves.	14.2
7.	a. checking oil levels, c. ensuring equipment is lubricated, d. adding lubricants as required.	14.3
8.	c. 18 to 24 months.	14.4
9.	a. demobilizing the TAR work site, b. materials reconciliation, d. invoice payments.	14.4
10.	b. It is false that turnarounds generally require about two weeks of planning.	14.4
11.	a. acting as point of contact for maintenance, c. hazard identification, d. writing or reviewing shutdown procedures, e. ordering miscellaneous supplies.	14.5

Chapter 14	Answer	LO #
12.	a. purging equipment, b. issuing safe work permits and other permits, e. providing fire watch support, f. assisting contractor personnel as required.	14.5
13.	a. technical basis for the proposed change, d. time period necessary for the change.	14.5

Chapter 15		
1.	See Key Terms list.	15.1, 15.2, 15.3, 15.4, 15.5
2.	b. are unplanned, c. can sometimes affect the environment and surrounding communities.	15.1
3.	a. failure of process equipment, c. loss of potable water, d. loss of instrument air, f. loss of electricity.	15.2
4.	b. proper line-up of equipment, piping, and valves, c. monitoring and control of rotating equipment, d. controlling process conditions within operating limits.	15.3
5.	a. execution of unit procedures, b. monitoring the process and equipment while the abnormal operation is in progress, c. monitoring all work activities, d. special equipment startup and preparation.	15.3
6.	c. emergency planning and response.	15.4
7.	a. develop procedures for abnormal operations, c. document mitigation plans for emergency scenarios, d. provide training in emergency procedures.	15.4
8.	a. bomb threats, b. explosions, d. spills.	15.5
9.	Types of fires that have occurred in industry are: • Chemical fires • Dumpster fires • Flange fires • Furnace fires • Ground fires • Process unit fires • Tank fires	15.5
10.	b. in a spill, can establish a water spray from a distance, d. except in a spill, would take no further action beyond notification.	15.5

Chapter 16		
1.	See Key Terms list.	16.1, 16.2, 16.3
2.	c. an emergency shutdown.	16.1
3.	b. OSHA Process Safety Management of Highly Hazardous Chemicals standard.	16.1
4.	a. management of change (MOC), c. subject matter experts (SMEs), d. emergency planning and response.	16.1
5.	sounds, sights, smells.	16.1
6.	maintenance.	16.1
7.	b. hot work.	16.2
8.	a. execution of the unit shutdown and de-inventory procedures, b. startup of any special equipment using normal SOP, d. monitoring work activities while shutdown is in progress.	16.2
9.	I. Blinding — e. process and procedure for isolating equipment for hot work or specific activities that require equipment removal. II. Employee health monitoring — b. employee health monitoring requirements for working, sampling, or being exposed to hazardous chemicals. III. Hot work — d. process and procedure for conducting such work as welding or grinding. IV. Housekeeping — a. activities required to maintain the facility in a clean, orderly, and safe condition. V. Lockout/tagout (LOTO) — c. procedure required to isolate energy sources from a piece of equipment.	16.2
10.	I. Procedure to be used during heavy lifting activities above active process equipment or near live power lines — c. lift plan. II. Process and procedure for entering tight spaces such as equipment, storage tanks, and excavations below grade — a. confined space entry. III. Reporting requirements of regulatory authorities when venting, purging, or draining equipment — d. material release reporting. IV. Requirements and reportable quantities for chemicals that, when released to the atmosphere, require reporting to the proper regulatory authorities — b. environmental reporting. V. Unit-specific procedures used for the purpose of equipment and system startup or shutdown in normal operations, as well as emergency operations — e. standard operating procedures (SOP)	16.2
11.	Ways to eliminate or minimize hazards associated with emergency shutdowns include the following: • Adequately sized flare and vent systems, piping systems, and piping connections • Safety instrumented systems • Backup power • Redundant refrigeration systems	16.3
12.	b. identify emergency scenarios, c. put emergency operating procedures in place.	16.3

Glossary

Abnormal operation operating a process unit in a mode that is different from normal operations.

Acceptance documentation the formal written validation that the unit has achieved its design capacity and specifications, and the facility agrees that the unit will function as engineered.

Affected employee process technician or other employee whose job requirement is to operate or use a machine or piece of equipment that is being serviced or maintained under lockout or tagout conditions, or whose job requires them to work in an area in which servicing or maintenance is being performed.

Air free purged of any oxygen from process piping and equipment prior to the introduction of process chemicals.

American National Standards Institute (ANSI) organization that oversees and coordinates the voluntary standards in the United States. ANSI develops and approves norms and guidelines that impact many business sectors. The coordination of U.S. standards with international standards allows American products to be used worldwide.

American Petroleum Institute (API) trade association that represents the oil and natural gas industry in the areas of advocacy, research, standards, certification, and education.

American Society of Mechanical Engineers (ASME) organization that specifies requirements and standards for pressure vessels, piping, and their fabrication.

Analyzer a device used to measure physical and/or chemical compositions of materials.

Application block main part of a drawing that contains symbols and defines elements such as relative position, types of materials, descriptions, and functions.

Audio, visual, olfactory (AVO) method used by process technicians to monitor the sounds, sights, and smells of a process unit or area during unit walkthrough inspections.

Authorized employee process technician or other employee who locks out or tags out a piece of equipment for required service or maintenance on that particular piece of equipment.

Blinding the process and procedure to isolate equipment for hot work, cold work (like changing out a seal or bearings), or specific activities that require equipment removal.

Blinding/unblinding permit work permit that allows equipment isolation via the installation of blinds and blind flanges.

Block flow diagrams (BFDs) simple drawings that show a general overview of a process, indicating the parts of a process and their relationships.

Body harness fall protection device worn while working at heights.

Boilers vessels in which water is boiled and converted into steam under controlled conditions.

Boiling liquid expanding vapor explosion (BLEVE) explosion resulting from excessive compression of vapor in the container head space and vapor flashing from its release to the atmosphere above its normal boiling point.

Bunker gear protective clothing worn for firefighting.

Capable of being locked out a piece of equipment that has a multiple padlock attachment or other means of attachment to which, or through which, multiple locks can be affixed.

Cell phone long-range electronic device used for mobile communication, text messaging, or data transmission across a cellular network of specialized base stations known as cell sites.

Checklist procedure written in a list format that requires the user to initial or check the completion of each step.

Chromatography an analytical technique used by the laboratory and unit analyzers for determining the individual components in a sample of liquid or gas.

Color visual comparison scale used in the process industry to determine product color purity; generally refers to the color of liquid chemicals (for example, the clarity and hue of yellow represents a poor quality; a high quality is water-white or clear); also known as the *ASTM International* or the *Saybolt color scale*.

Commissioning systematic course of action by which process units are placed into active service, whether it is the initial startup of newly built unit or the recommissioning of a revised process unit.

Commissioning team group of individuals who play a key role in the planning and implementing of the commissioning or decommissioning of a process unit or facility.

Communication verbal, nonverbal, or written transfer of information between people.

Computer-based training (CBT) delivers training material through a facility computing system.

Confined space entry (CSE) policy that defines the process and procedure for entering confined spaces such as equipment, storage tanks, and excavations below grade.

Confined space entry (CSE) permit permit that allows human entry and work within an OSHA-defined confined space, the issuance of which indicates all regulated and pertinent safety measures have been taken and/or are active.

Construction phase building phase of an initial process unit or facility.

Control board technician process technician whose primary job function is to remotely monitor and control the process unit within normal operating parameters; also called *console operator*, *board operator*, or *inside operator*.

Control of work (COW) work practice that identifies the means of safely controlling maintenance, demolition, remediation, construction, operating tasks, and similar work.

Control room room from which operators and technical personnel control the unit; it houses the facility's distributed control system (DCS), which may incorporate all of the facility's operating control boards.

Deinventory reduction or emptying of contents or residuals in a tank to various levels, based on activities and possible exposure of personnel to contents.

Distributed control system (DCS) a subsystem of a supervisory control system used to control a process unit: consists of field instruments and field controllers connected by wiring that carries a signal from the controller transmitter to a central control monitoring screen.

Effective communication communication skills that help convey the intended meaning efficiently.

Electric heat tracing series of self-regulating heating cables designed to provide freeze protection and temperature maintenance to metallic and nonmetallic pipes, tanks, and equipment.

Electrical diagrams diagrams that help process technicians understand power distribution, and how it relates to the process.

Electronic logbook (eLog) computer based event logging program developed to assist the process technician record and report significant shift activities.

Emergency sudden, unexpected, or impending situation that may cause injury, loss of life, damage to property, and/or interference with the normal activities of a person or operation, which therefore requires immediate attention and demands remedial action.

Emergency operation mode of operation or procedure followed when an emergency situation has placed a process unit in an unsafe condition.

Emergency response effort to mitigate the impact of an incident on the public and the environment.

Emergency shutdown sudden failure of major process equipment, such as compressors or furnaces, or failure of utilities such as instrument air, steam, or electricity requiring an immediate shutdown.

Energized connected to an energy source; containing residual or stored energy.

Energy-isolating device mechanical device that physically prevents the transmission or release of energy.

Energy sources any source of electrical, mechanical, hydraulic, pneumatic, chemical, thermal, or other energy.

Environmental Protection Agency (EPA) a federal agency charged with authority to make and enforce the national environmental policy.

Equipment health monitoring (EHM) efficient system for protecting rotating equipment and facility operations from unscheduled downtime that provides personnel with the equipment knowledge they need to schedule maintenance, manage inventories, and support efficient workflow scheduling.

Equipment symbols set of symbols located on one sheet of a set of process flow diagrams (PFDs) for every piece of equipment found in industry for the user to review.

Evacuation plan documentation for the evacuation of a facility, to be used by personnel in the event of an emergency.

Explosion a chemical reaction or change of state which is affected in an exceedingly short space of time with the generation of a high temperature and generally a large quantity of gas.

Feed forward the process of feeding a process stream to the next processing area.

Field technician process technician whose primary job is to monitor the fixed and rotating field equipment, perform sampling, and ensure that the unit operates within normal operating parameters; also called *field operator* or *outside operator*.

Fire brigade local process facility fire department composed of employees who are knowledgeable, trained, and skilled in basic firefighting techniques.

Fire retardant clothing (FRC) wearing apparel for use in situations where there is a risk of arc, flash, or thermal burns that is regulated by NFPA-70E, ASTM, and OSHA.

Firewater water from plant firewater lines used for emergencies.

First responders individuals who likely witness or discover a hazardous substance release and have been trained to initiate an emergency response sequence by notifying the appropriate authorities.

Flame resistant characteristic of a fabric to resist ignition and to self-extinguish if ignited.

Flame retardant characteristic of a fabric that has had a chemical substance to impart flame resistance.

Flashback situation in which gas vapors ignite and return to the source of the vapors.

Friction force resisting the relative lateral or tangential motion of solid surfaces, fluid layers, or material elements in contact.

Gas chromatography (GC) a system of identifying and quantifying a gas sample's compounds by their boiling points. It consists of an oven, column, autosampler, and detector. The autosampler injects the sample into the column, which is packed with special material to help separate the components of the sample. A carrier gas sweeps the sample through the column while the oven adds temperature to speed the sample along. The mixture separates in the column and elutes as individual compounds into the detector where each is individually quantified. As a result, each component leaves the column separately and in a predictable sequence and rate.

General work permit permit that allows work activity other than blinding/unblinding, hot work, lockout/tagout, and confined space.

Hazard and operability (HAZOP) formal and structured review and study method used to determine potential hazards associated with process systems, equipment, process materials, and work processes.

Hazardous Waste Operations and Emergency Response Standard (HAZWOPER) —OSHA's Regulation 29 CFR 1910.120; OSHA standard that applies to personnel who are in a role or position to act as a first responder during an emergency.

Hot work permit permit that allows hot work, such as welding, grinding, or vehicle entry in or around process equipment.

Housekeeping act of keeping a work area and equipment in a safe, clean, usable condition.

Hydro test strength and integrity test, using water, for process piping and equipment.

Hydrocarbon detector electronic device that detects, measures, and indicates the concentration of compounds containing only hydrogen and carbon atoms in an area.

Hydrocarbon free removal of compounds containing only hydrogen and carbon atoms (for example, methane or benzene) from process piping and equipment prior to opening to the atmosphere and introducing air.

Hydrogen sulfide (H_2S) highly toxic, highly flammable, colorless gas with a very distinctive, rotten egg-like odor.

Immediately dangerous to life and health (IDLH) condition from which serious injury or death to personnel can occur in a short amount of time.

Incident response teams groups of people who prepare for and respond to any emergency incident, such as a fire, spill, explosion, or environmental release that potentially impacts the outlying community.

Initial startup the first commissioning of the unit that involves the introduction of feedstock to produce a defined product at a given purity.

Initial startup procedures set of guidelines or instructions used to perform the initial startup of a new process unit or facility.

Inspection examination of a part or piece of equipment to determine if it conforms to specifications.

Instrumentation any device used to measure or control flow, temperature, level pressure, analytical data, and so on.

Intercom stand-alone electronic communication system intended for limited or private conversation.

Interim test test of equipment requiring removal of lockout/tagout devices prior to completion of maintenance or repair of equipment.

Internal procedure company specific procedure.

International Organization for Standardization (ISO) regulates safety and health standards internationally.

Intrinsically safe devices electronic devices certified safe for use in explosive atmospheres.

ISA a global, nonprofit technical society that develops standards for automation, instrumentation, control, and measurement.

Isolation separation; requirements for preventing stored energy (electrical, compressed air, kinetic, pressurized liquid, and so on) from entering or leaving a piece of equipment; may also be referred to as *lockout/tagout (LOTO)* or *lock/tag/try*.

Isometric drawings (Isoms) perspective drawings that depict objects, such as equipment and piping, as a 3D image, as they would appear to the viewer.

Karl Fischer (KF) water method a process that uses a chemical solution to determine the quantity of water in refined chemicals; also known as *Karl Fischer titration*.

Legend section of a drawing that defines the information or symbols contained within the drawing.

Lift plan documented plan used to evaluate the hazards and define precautions necessary during heavy lifting activities, including lifting over live process equipment or near energized power lines.

Lockbox safety device ensuring no lockout/tagout (LOTO) devices are removed while work is performed. Lockboxes have multiple locks into which all keys and/or tags from the LOTO devices securing the equipment are inserted, and a single authorized employee using a LOTO device and a job lock during multishift operations then secures the box.

Lockout a safety term used to describe the isolation of equipment for maintenance; a federally mandated safety precaution.

Lockout device a device that utilizes a positive means such as keyed or combination type lock to hold equipment in a zero-energy state.

Lockout/tagout (LOTO) a safety term used to describe the isolation of equipment for maintenance; this is a federally mandated safety precaution; procedure of tagging valves, breakers, etc. in preparation of equipment for maintenance.

Logbook typically, hardbound ledgers used to handwrite significant activities that have occurred during the shift.

Lubricants any substances interposed between two surfaces in motion for the purpose of reducing the friction and/or the wear between them.

Lubrication the process or technique employed to reduce friction and remove heat for reducing equipment wear and increasing longevity and safety.

Maintenance activity events performed by the maintenance personnel in a process facility, such as pump or compressor repair, pipe work, and routine general maintenance.

Management of change (MOC) method of managing and communicating changes to a process, changes in equipment, changes in technology, changes in personnel, or other changes that will impact the safety and health of employees.

Mechanical completion documented checking and testing of the construction to confirm the installation is in accordance with construction drawings and specifications, and is ready for commissioning in a safe manner in compliance with project requirements.

Mechanical integrity the state of being whole, sound, and undamaged; capable of functioning at design specification.

Mentor influential senior sponsor or trainer, usually in the form of a training coordinator or a chief or lead operator, who delivers the training material, tracks material completion, provides feedback, and conducts written and performance evaluations to verify knowledge.

Monitor to observe or watch; observing and listening to the equipment routinely to prevent process upsets.

Multiple padlock attachment clamp-like device used to install multiple locks on a lockout device.

Mutual aid agreement among emergency responders to lend assistance across jurisdictional boundaries.

Nameplate capacity designed capacity of the unit.

National Electric Code (NEC) a standard that specifies electrical cable sizing requirements and installation practices.

National Emissions Standards for Hazardous Air Pollutants (NESHAP) emissions standards set by the Environmental Protection Agency (EPA) for air pollutants that may cause fatalities or serious, irreversible, or incapacitating illness if not regulated.

National Fire Protection Association (NFPA) international organization that specifies fire codes including building construction codes, fire suppression systems, and firefighting capabilities required at facilities.

Natural gas a flammable gas associated with gas and oil fields and consisting principally of methane and the lower saturated paraffin hydrocarbons. It may also include impurities such as water vapor, hydrogen sulfide, and carbon dioxide.

Nitrogen an odorless, invisible, inert gas, forming approximately 80% of the atmosphere. An important purging and blanketing medium.

Nonoperating personnel personnel other than process technicians who are performing work on the unit; these may include people visiting the unit, such as engineers, members of the management team, maintenance staff, and contractors.

Nonverbal communication (NVC) unspoken communication, such as gesture, expression, or body language.

Normal operations actions performed or procedures followed when a process unit is operating within design parameters.

Occupational Safety and Health Administration (OSHA) U.S. government agency created to establish and enforce workplace safety and health standards, conduct workplace inspections and propose penalties for noncompliance, and investigate serious workplace incidents.

On-the-job training (OJT) programs objective-oriented training and qualification programs for process technicians to master process equipment, control systems, safety, and hazard management.

One-line diagram a single page document that represents a facility's electrical distribution infrastructure; also known as the single-line diagram.

Operations procedures unit specific procedures used for the purpose of equipment and system startup, shutdown, normal operation, as well as emergency situations.

Paraphrasing summarizes the information received to clarify understanding.

Performance testing step test of the unit to determine if the process unit is able to achieve its maximum design intent.

Personal protective equipment (PPE) specialized gear that provides a barrier between hazards and the worker.

Piping and instrumentation diagrams (P&IDs) detailed drawings that graphically represent the equipment, piping, and instrumentation contained within a process facility.

Planning phase phase of the project where justification and plans are developed for the construction of a new process unit.

Plot plans diagrams that show the layout and dimensions of equipment, units, and buildings, drawn to scale, so that everything is of the correct relative size.

Postcommissioning last phase of the commissioning process, which begins after initial startup is completed.

Potable water water that is safe to drink and use for cooking.

Precommissioning activities that must be completed prior to moving into the startup phase of a new process unit.

Predictive maintenance (PM) maintenance strategy that helps determine the condition of in-service equipment to predict when maintenance should be performed.

Pre-startup safety review (PSSR) an element of the process safety management program; the PSSR helps to ensure that the new or modified process or facility is safe and operable before startup.

Preventive maintenance equipment maintenance strategy based on replacing, overhauling, or remanufacturing an item at a fixed interval, regardless of its condition at the time.

Procedures series of actions that must be done in the specified manner and sequence to obtain the desired result under the same circumstances each time the work is performed.

Procedure owner individual who is accountable for the accurate development and maintenance of a procedure.

Procedure template form or guide that accurately and effectively shapes procedure presentation and content.

Procedure user process technician trained and qualified on the subject matter of the procedure prior to use.

Process drawings diagrams that provide a visual description and explanation of the processes, equipment, and other important items in a facility.

Process flow diagrams (PFDs) basic drawings that use symbols and directional arrows to show primary product flow through a process, including such information as operating conditions, the location of main instruments, and major pieces of equipment.

Process hazard analysis (PHA) systematic assessment of the potential hazards associated with an industrial process, taking into consideration specific hazards and locations of highest potential for exposure.

Process safety management (PSM) OSHA standard that contains the requirements for management of hazards associated with processes using highly hazardous materials.

Process simulator stand-alone, computer-generated simulation of a process unit or process system that emulates process equipment, piping systems, control mechanisms, and behaviors that control the process.

Process technician worker in a process facility who monitors and controls mechanical, physical, and/or chemical changes throughout a process in order to create a product from raw materials; term reflects increased competence and skills required; also called *process operator* or *operator*.

Public address system (PA system) system that reinforces and distributes a given sound throughout a venue.

Punchlist list of uncompleted construction items from contracted design that are not safety critical but must be addressed by the contracted construction firm.

Reactive maintenance equipment maintenance strategy in which equipment and facilities are repaired only in response to a breakdown or a fault.

Recommissioning returning existing process units or equipment to active service after an extended idled period.

Reflection act of repeating what was communicated to confirm the information was understood correctly.

Repeatable able to be done in exactly the same manner each time so that results can be scientifically compared.

Resource Conservation and Recovery Act (RCRA) primary federal law whose purpose to protect human health and the environment and to conserve natural resources. It completes this goal by regulating all aspects of hazardous waste management; generation, storage, treatment, and disposal.

Rounds routine walkthrough of the unit, monitoring the fixed and rotating equipment, and performing other routine tasks.

Route sequential path followed in order to perform equipment health monitoring (EHM).

Routine duties duties performed that are rigidly prescribed by control over the work or by written or verbal procedures, or well-defined, constant, and repetitively performed duties that preclude the need for procedures or substantial controls.

Routine maintenance work routinely performed to maintain equipment in its original manufactured condition and maintain operability.

Safety data sheet (SDS) a document that contains information related to the safety, hazards, and handling of a specific material.

Safety, health, and environmental (SHE) policies policies implemented by process facilities to minimize or prevent risks and/or hazards associated with the process industry and to ensure that the facility is in compliance with applicable regulatory agencies.

Safety, health, environmental diagrams a visual layout of the emergency access, personnel safety equipment, fire protection systems, and environmental systems.

Sample loop a continuous circulation of process liquid or gas from a higher-pressure source to a lower-pressure return, to ensure capture of a representative sample.

Sample point section of small diameter-valved tubing that extends from the main process piping system for collecting samples.

Sampling process of collecting a representative portion of a material for analysis.

Scaffold plan documented plan for the erecting and dismantling of scaffolding used to access process piping and equipment.

Self-contained breathing apparatus (SCBA) independent breathing device worn by rescue workers, firefighters, process technicians, and others to provide breathable air in a hostile environment.

Shift change/relief handing off the responsibility for operation and maintenance of a facility from one crew to another at a designated time; also known as *shift handover, shift pass-down, shift turnover, making relief,* and by other terms.

Shutdown the time period when a unit is not in operation; systematic removal of process equipment from service in order to stop the process.

Sound-powered phones phones containing electromechanical transducers that convert voice directly into electrical energy.

Specifications the quality limitations for a product; product purity parameters that have been agreed on by the company and the customers or regulated by governmental agencies; commonly called *specs*.

Spill accidental release of a substance from lines, equipment, or areas.

Standard operating procedures (SOPs) a set of directions or instructions that has a recognized and permanent value and that defines the particular steps to take when a certain situation or condition occurs.

Standards guidelines established by authority as a rule to measure quantity, weight, extent, value, or quality; OSHA standards are rules that describe the methods employers must use to protect their employees from hazards.

Startup bringing into operation a piece or pieces of equipment or a process facility.

Statistical process control (SPC) statistical procedures that keep track of a process in order to reduce variation and improve quality.

Steam water vapor, or water in its gaseous state. The gas or vapor into which water is converted when heated to the boiling point.

Steam clouds tiny drops of water that have condensed from steam and are carried along by the invisible vapor.

Steam generators any shell and tube exchanger or kettle type exchangers using boiler feedwater (BFW) to remove process heat, convert BFW to steam, and then pressure control that steam to a supply header.

Steam jets essentially a steam nozzle that discharges a high velocity jet and used to create a particular pattern of spray steam.

Steam tracing tubing that is installed adjacent to a pipeline and is enclosed with the pipeline by insulation. Steam is then passed through the tubing providing heat.

Steam turbines (1) prime movers for the conversion of heat energy of steam into work on a revolving shaft, utilizing fluid acceleration principles in jet and vane machinery. (2) Turbines driven by the pressure of steam discharged at high velocity against the turbine vanes.

Subject matter experts (SMEs) individuals within an organization possessing a very high level of expertise regarding a particular job, task, or process.

Symbology various graphical representations used to identify equipment, lines, instrumentation, or process configurations.

Symbols letters used to designate chemical elements or equipment classes; figures used to designate types of equipment.

System set of interacting or interdependent equipment and process elements that work together to deliver a specific process function.

Tagout procedure of tagging valves, breakers, and so on, in preparation of equipment for maintenance; placement of a tag to indicate that the energy-isolating device and the equipment being controlled may not be operated until the tagout device is removed.

Tagout device prominent warning device, such as a tag and a means of attachment, which can be securely fastened to an energy-isolating device in accordance with an established procedure, to indicate that the energy-isolating device and the equipment being controlled may not be operated until the tagout device is removed.

Thermal expansion tendency of matter to increase in volume in response to an increase in temperature.

Tightness test pressurization test, typically using nitrogen or other inert gas, for process piping and equipment to ensure that equipment is leak free prior to the introduction of hydrocarbons; also known as the *leak test*.

Title block section of a drawing that contains information such as drawing title, drawing number, revision number, sheet number, originator signature, and approval signatures.

Train components of a system; a series of related equipment components, all in an orderly procession or in a line, necessary to accomplish a specific task, for example, distillation or compressor train.

Trunked radio system complex type of computer-controlled radio system.

Turnaround (TAR) the shutdown period for an operation unit, usually for mechanical reconditioning. The period from the end of one run to the beginning of the next, that is, the offstream to onstream period.

Turnaround maintenance work done during the shutdown period for an operation unit, usually for mechanical reconditioning.

Two-way radio radio that can transmit and receive voice communication.

Unit status report information gathered by the current operating shift for reporting to the oncoming shift during shift change.

Utility flow diagrams (UFD) drawings that provide process technicians a P&ID-type view of the utilities used for a process.

Verbal communication dialogue or conversation between two or more people for transferring information.

Vibration readings measurement and documentation of rotating equipment to ensure vibrations are within an acceptable range.

Water hammer hydraulic action associated with a noncompressible fluid in a pipe, so named because it sounds like a pipe being hit with a hammer; the energy developed by the sudden stoppage of fluid in motion.

Work permits documents that allow individuals or groups to perform work on a process unit.

Written communication communication by means of written or printed symbols or letters.

Zero-energy state the state of equipment following specific process isolation and clearing procedures, followed by isolating all hazardous energy sources using lockout/tagout devices.

Index

R

S

T

U

V

W

Z